PROGRESS IN CLINICAL AND BIOLOGICAL RESEARCH

RECENT TITLES

Vol 227: **Advances in Chronobiology**, John E. Pauly, Lawrence E. Scheving, *Editors*. Published in two volumes.

Vol 228: **Environmental Toxicity and the Aging Processes,** Scott R. Baker, Marvin Rogul, *Editors*

Vol 229: **Animal Models: Assessing the Scope of Their Use in Biomedical Research,** Junichi Kawamata, Edward C. Melby, Jr., *Editors*

Vol 230: **Cardiac Electrophysiology and Pharmacology of Adenosine and ATP: Basic and Clinical Aspects,** Amir Pelleg, Eric L. Michelson, Leonard S. Dreifus, *Editors*

Vol 231: **Detection of Bacterial Endotoxins With the Limulus Amebocyte Lysate Test,** Stanley W. Watson, Jack Levin, Thomas J. Novitsky, *Editors*

Vol 232: **Enzymology and Molecular Biology of Carbonyl Metabolism: Aldehyde Dehydrogenase, Aldo-Keto Reductase, and Alcohol Dehydrogenase**, Henry Weiner, T. Geoffrey Flynn, *Editors*

Vol 233: **Developmental and Comparative Immunology,** Edwin L. Cooper, Claude Langlet, Jacques Bierne, *Editors*

Vol 234: **The Hepatitis Delta Virus and Its Infection**, Mario Rizzetto, John L. Gerin, Robert H. Purcell, *Editors*

Vol 235: **Preclinical Safety of Biotechnology Products Intended for Human Use,** Charles E. Graham, *Editor*

Vol 236: **First Vienna Shock Forum,** Günther Schlag, Heinz Redl, *Editors*. Published in two volumes: Part A: *Pathophysiological Role of Mediators and Mediator Inhibitors in Shock.* Part B: *Monitoring and Treatment of Shock.*

Vol 237: **The Use of Transrectal Ultrasound in the Diagnosis and Management of Prostate Cancer,** Fred Lee, Richard McLeary, *Editors*

Vol 238: **Avian Immunology,** W.T. Weber, D.L. Ewert, *Editors*

Vol 239: **Current Concepts and Approaches to the Study of Prostate Cancer,** Donald S. Coffey, Nicholas Bruchovsky, William A. Gardner, Jr., Martin I. Resnick, James P. Karr, *Editors*

Vol 240: **Pathophysiological Aspects of Sickle Cell Vaso-Occlusion,** Ronald L. Nagel, *Editor*

Vol 241: **Genetics and Alcoholism,** H. Werner Goedde, Dharam P. Agarwal, *Editors*

Vol 242: **Prostaglandins in Clinical Research,** Helmut Sinzinger, Karsten Schrör, *Editors*

Vol 243: **Prostate Cancer,** Gerald P. Murphy, Saad Khoury, Réne Küss, Christian Chatelain, Louis Denis, *Editors*. Published in two volumes: Part A: *Research, Endocrine Treatment, and Histopathology.* Part B: *Imaging Techniques, Radiotherapy, Chemotherapy, and Management Issues*

Vol 244: **Cellular Immunotherapy of Cancer,** Robert L. Truitt, Robert P. Gale, Mortimer M. Bortin, *Editors*

Vol 245: **Regulation and Contraction of Smooth Muscle,** Marion J. Siegman, Andrew P. Somlyo, Newman L. Stephens, *Editors*

Vol 246: **Oncology and Immunology of Down Syndrome,** Ernest E. McCoy, Charles J. Epstein, *Editors*

Vol 247: **Degenerative Retinal Disorders: Clinical and Laboratory Investigations,** Joe G. Hollyfield, Robert E. Anderson, Matthew M.LaVail, *Editors*

Vol 248: **Advances in Cancer Control: The War on Cancer—15 Years of Progress,** Paul F. Engstrom, Lee E. Mortenson, Paul N. Anderson, *Editors*

Vol 249: **Mechanisms of Signal Transduction by Hormones and Growth Factors,** Myles C. Cabot, Wallace L. McKeehan, *Editors*

Vol 250: **Kawasaki Disease,** Stanford T. Shulman, *Editor*

Vol 251: **Developmental Control of Globin Gene Expression,** George Stamatoyannopoulos, Arthur W. Nienhuis, *Editors*

Vol 252: **Cellular Calcium and Phosphate Transport in Health and Disease,** Felix Bronner, Meinrad Peterlik, *Editors*

Vol 253: **Model Systems in Neurotoxicology: Alternative Approaches to Animal Testing,** Abraham Shahar, Alan M. Goldberg, *Editors*

Vol 254: **Genetics and Epithelial Cell Dysfunction in Cystic Fibrosis,** John R. Riordan, Manuel Buchwald, *Editors*

Vol 255: **Recent Aspects of Diagnosis and Treatment of Lipoprotein Disorders: Impact on Prevention of Atherosclerotic Diseases,** Kurt Widhalm, Herbert K. Naito, *Editors*

Vol 256: **Advances in Pigment Cell Research,** Joseph T. Bagnara, *Editor*

Vol 257: **Electromagnetic Fields and Neurobehavioral Function,** Mary Ellen O'Connor, Richard H. Lovely, *Editors*

Vol 258: **Membrane Biophysics III: Biological Transport,** Mumtaz A. Dinno, William McD. Armstrong, *Editors*

Vol 259: **Nutrition, Growth, and Cancer,** George P. Tryfiates, Kedar N. Prasad, *Editors*

Vol 260: **EORTC Genitourinary Group Monograph 4: Management of Advanced Cancer of Prostate and Bladder,** Philip H. Smith, Michele Pavone-Macaluso, *Editors*

Vol 261: **Nicotine Replacement: A Critical Evaluation,** Ovide F. Pomerleau, Cynthia S. Pomerleau, *Editors*

Vol 262: **Hormones, Cell Biology, and Cancer: Perspectives and Potentials,** W. David Hankins, David Puett, *Editors*

Vol 263: **Mechanisms in Asthma: Pharmacology, Physiology, and Management,** Carol L. Armour, Judith L. Black, *Editors*

Vol 264: **Perspectives in Shock Research,** Robert F. Bond, *Editor*

Vol 265: **Pathogenesis and New Approaches to the Study of Noninsulin-Dependent Diabetes Mellitus,** Albert Y. Chang, Arthur R. Diani, *Editors*

Vol 266: **Growth Factors and Other Aspects of Wound Healing: Biological and Clinical Implications,** Adrian Barbul, Eli Pines, Michael Caldwell, Thomas K. Hunt, *Editors*

Vol 267: **Meiotic Inhibition: Molecular Control of Meiosis,** Florence P. Haseltine, *Editor*

Vol 268: **The Na, K-Pump,** J.C. Skou, J.G. Nørby, A.B. Maunsbach, M. Esmann, *Editors*. Published in two volumes: Part A: *Molecular Aspects*. Part B: *Cellular Aspects*.

Vol 269: **EORTC Genitourinary Group Monograph 5: Progress and Controversies in Oncological Urology II,** Fritz H. Schröder, Jan G.M. Klijn, Karl H. Kurth, Herbert M. Pinedo, Ted A.W. Splinter, Herman J. de Voogt, *Editors*

Vol 270: **Cell-Free Analysis of Membrane Traffic,** D. James Morré, Kathryn E. Howell, Geoffrey M.W. Cook, W. Howard Evans, *Editors*

Vol 271: **Advances in Neuroblastoma Research 2,** Audrey E. Evans, *Editor*

Vol 272: **Bacterial Endotoxins: Pathophysiological Effects, Clinical Significance, and Pharmacological Control,** Jack Levin, Harry R. Büller, Jan W. ten Cate, Sander J.H. van Deventer, Augueste Sturk, *Editors*

Vol 273: **Ion Pumps: Structure, Function, and Regulation,** Wilfred D. Stein, *Editor*

Vol 274: **Oxidases and Related Redox Systems,** Tsoo E. King, Howard S. Mason, Martin Morrison, *Editors*

Please contact the publisher for information about previous titles in this series.

BACTERIAL ENDOTOXINS
Pathophysiological Effects, Clinical Significance, and Pharmacological Control

BACTERIAL ENDOTOXINS
Pathophysiological Effects, Clinical Significance, and Pharmacological Control

Proceedings of an International Conference held in Amsterdam, The Netherlands, May 21–23, 1987

Editors

Jack Levin
Departments of Laboratory Medicine
and Medicine
University of California School of Medicine
San Francisco, California
and
Veterans Administration Medical Center
San Francisco, California

Jan W. ten Cate
Division of Hemostasis and Thrombosis
Academic Medical Centre
Amsterdam, The Netherlands

Harry R. Büller
Division of Hemostasis
and Thrombosis
Academic Medical Centre
Amsterdam, The Netherlands

Sander J.H. van Deventer
Division of Hemostasis
and Thrombosis
Academic Medical Centre
Amsterdam, The Netherlands

Augueste Sturk
Division of Hemostasis
and Thrombosis
Academic Medical Centre
Amsterdam, The Netherlands

ALAN R. LISS, INC. • NEW YORK

Address all Inquiries to the Publisher
Alan R. Liss, Inc., 41 East 11th Street, New York, NY 10003

Printed in the United States of America

Library of Congress Cataloging-in-Publication Data

Bacterial endotoxins.

(Progress in clinical and biological research ; v. 272)
Includes bibliographies and index.
1. Endotoxins—Physiological effect—Cogresses.
2. Endotoxins—Toxicology—Congresses. I. Levin, Jack, 1932- . II. Series. [DNLM: 1. Bacterial Toxins—congresses. 2. Endotoxins—congresses.
W1 PR668E v.272 / QW 630 B1306 1987]
QP632.E4B29 1988 616'.0145 88-6759
ISBN 0-8451-5122-3

Contents

Contributors

Ben J. Appelmelk, Department of Medical Microbiology, Vrije Universiteit, 1081 BT Amsterdam, The Netherlands **[373]**

Françoise Bacle, INSERM U28, Hôpital Broussais, 75014 Paris, France **[89]**

Soheyl Bahrami, Ludwig Boltzmann Institute for Experimental Traumatology, A-1200 Vienna, Austria **[283]**

Hans G. Beger, Department of General Surgery, University of Ulm, 7900 Ulm, Federal Republic of Germany **[115]**

Dieter Berger, Department of General Surgery, University of Ulm, 7900 Ulm, Federal Republic of Germany **[115]**

Helmut Brade, Forschungsinstitut Borstel, Institut für Experimentelle Biologie und Medizin, D-2061 Borstel, Federal Republic of Germany **[17,373]**

Lore Brade, Forschungsinstitut Borstel, Institut für Experimentelle Biologie und Medizin, D-2061 Borstel, Federal Republic of Germany **[17]**

Petter Brandtzaeg, Department of Infectious Diseases, Ullevål University Hospital, Oslo, Norway **[263]**

P. Braquet, Department of Research, IHB Research Labs, 92350 Le Plessis-Robinson, France **[135]**

M. Kennedy Browne, Surgical Unit, Monklands District General Hospital, Airdrie ML6 0JS, Scotland **[361]**

Harry R. Büller, Division of Hemostasis and Thrombosis, Academic Medical Centre, Amsterdam, The Netherlands **[213]**

Ingrid Burmeister, Forschungsinstitut Borstel, Institut für Experimentelle Biologie und Medizin, D-2061 Borstel, Federal Republic of Germany **[125]**

Martine Caroff, Institut de Biochimie, Université Paris XI, 91405 Orsay, France **[89]**

Lisa Carter, School of Microbiology, University of New South Wales, Kensington 2033, N.S.W., Australia **[173]**

Jean-Marc Cavaillon, Unité d'Immuno-Allergie, Institut Pasteur, 75015 Paris, France **[89]**

Anthony Cerami, Laboratory of Medical Biochemistry, Rockefeller University, New York, NY 10021 **[77]**

The numbers in brackets are the opening page numbers of the contributors' articles.

John K.S. Chia, Department of Medicine, F. Edward Hebert School of Medicine, Uniformed Services University of the Health Sciences, Bethesda, MD 20814 **[327]**

K.T. Chong, Department of Pharmacology-Toxicology, Cetus Corporation, Emeryville, CA 94608 **[383]**

J. Cohen, Department of Medicine and Bacteriology, Royal Postgraduate Medical School, Hammersmith Hospital, London W12 0HS, England **[195,351]**

A. Coumbos, Max-Planck-Institut für Immunbiologie, 7800 Freiburg, Federal Republic of Germany **[295]**

Tracy Deinhart, Department of Immunology, Cetus Corporation, Palo Alto, CA 94303 **[383]**

I. de Vries, Department of Thrombosis and Hemostasis, Academic Medical Center, Amsterdam, and Department of Bacteriology, Slotervaart Hospital, Amsterdam, The Netherlands **[213]**

Jan Dobloug, Department of Infectious Diseases, Ullevål University Hospital, Oslo, Norway **[263]**

S. Edlin, Intensive Therapy Unit, Western Infirmary, Glasgow G11 6NT, Scotland **[225]**

Regina Engel, Forschungsinstitut Borstel, Institut für Experimentelle Biologie und Medizin, D-2061 Borstel, Federal Republic of Germany **[125]**

A. Etienne, Department of Pharmacology, IHB Research Labs, 91940 Les Ulis, France **[135]**

L. Faulkner, Department of Medicine and Bacteriology, Royal Postgraduate Medical School, Hammersmith Hospital, London W12 0HS, England **[195]**

Jeffrey Flatgaard, Department of Process Development, Cetus Corporation, Palo Alto, CA 94303 **[383]**

Marina A. Freudenberg, Max-Planck-Institut für Immunbiologie, 7800 Freiburg, Federal Republic of Germany **[63,295]**

Ichiro Fukuda, Department of Cell Chemistry, Institute of Medical Science, University of Tokyo, Tokyo 108, Japan; Present Address: 3rd Department of Internal Medicine, Tohoku University School of Medicine, Sendai 980, Japan **[185]**

Chris Galanos, Max-Planck-Institut für Immunbiologie, 7800 Freiburg, Federal Republic of Germany **[63,295]**

Peter Gaustad, Department of Microbiology, Ullevål University Hospital, Oslo, Norway **[263]**

K.E. Giercksky, Department of Surgery, University Hospital of Tromsø, 9012 Tromsø, Norway **[247]**

Oanh Gray, Department of Immunology, Cetus Corporation, Palo Alto, CA 94303 **[383]**

Richard Greenberg, XOMA Corporation, Berkeley, CA, and St. Louis University, St. Louis, MO **[395]**

Reinhart Grundmann, Chirurgische Universitätsklinik Köln-Lindenthal, D-5000 Köln 41, Federal Republic of Germany **[339]**

Gretchen Guelde, Department of Medicine, F. Edward Hebert School of Medicine, Uniformed Services University of the Health Sciences, Bethesda, MD 20814 **[327]**

Nicole Haeffner-Cavaillon, INSERM U28, Hôpital Broussais, 75014 Paris, France **[89]**

Scott Harkonen, XOMA Corporation, Berkeley, CA, and St. Louis University, St. Louis, MO **[395]**

F. Hecquet, Department of Pharmacology, IHB Research Labs, 91940 Les Ulis, France **[135]**

Otto Holst, Forschungsinstitut Borstel, Institut für Experimentelle Biologie und Medizin, D-2061 Borstel, Federal Republic of Germany **[17]**

Michael Hornung, Chirurgische Universitätsklinik Köln-Lindenthal, D-5000 Köln 41, Federal Republic of Germany **[339]**

R. Huby, Department of Pathology, Royal Postgraduate Medical School, Hammersmith Hospital, London W12 0HS, England **[351]**

Graham D.F. Jackson, School of Microbiology, University of New South Wales, Kensington 2033, N.S.W., Australia **[173]**

Margaret A. Johns, Department of Infectious Diseases, Maxwell Finland Laboratory, Boston University School of Medicine, Boston, MA 02118 **[103]**

Shiro Kanegaseki, Department of Cell Chemistry, Institute of Medical Science, University of Tokyo, Tokyo 108, Japan **[185]**

Donald Kennedy, XOMA Corporation, Berkeley, CA, and St. Louis University, St. Louis, MO **[395]**

Peter Kierulf, Department of Clinical Chemistry, Ullevål University Hospital, Oslo, Norway **[263]**

Nancy L. Koles, Department of Medicine, F. Edward Hebert School of Medicine, Uniformed Services University of the Health Sciences, Bethesda, MD 20814 **[327]**

Hella-Monika Kuhn, Forschungsinstitut Borstel, Institut für Experimentelle Biologie und Medizin, D-2061 Borstel, Federal Republic of Germany **[17]**

I.A. Lampert, Department of Pathology, Royal Postgraduate Medical School, Hammersmith Hospital, London W12 0HS, England **[351]**

James W. Larrick, Department of Immunology, Cetus Corporation, Palo Alto, CA 94303 **[383]**

I.McA. Ledingham, Intensive Therapy Unit and University Department of Surgery, Western Infirmary, Glasgow G11 6NT, Scotland **[225, 237]**

Jack Levin, Departments of Laboratory Medicine and Medicine, University of California School of Medicine, San Francisco, CA 94143; Veterans Administration Medical Center, San Francisco, CA 94121; Marine Biological Laboratory, Woods Hole, MA 02543 **[3]**

Stephen F. Lowry, Laboratory of Surgical Metabolism, New York Hospital–Cornell University Medical Center, New York, NY 10021 **[77]**

Janneke Maaskant, Department of Medical Microbiology, Vrije Universiteit, 1081 BT Amsterdam, The Netherlands **[373]**

David MacLaren, Department of Medical Microbiology, Vrije Universiteit, 1081 HV Amsterdam, The Netherlands **[373]**

M. Matsuura, Max-Planck-Institut für Immunbiologie, 7800 Freiburg, Federal Republic of Germany **[295]**

William R. McCabe, Department of Infectious Diseases, Maxwell Finland Laboratory, Boston University School of Medicine, Boston, MA 02118 **[103]**

A. Christine McCartney, Department of Bacteriology, Royal Infirmary, University of Glasgow, Glasgow G4 0SF, Scotland **[225, 237,361]**

Fred Mihm, Stanford University Hospital, Stanford University Medical Center, Stanford, CA 94305 **[283]**

Maryalice Miller, Department of Medicine, F. Edward Hebert School of Medicine, Uniformed Services University of the Health Sciences, Bethesda, MD 20814 **[327]**

Tom Eirik Mollnes, Bloodbank and Department of Immunology, Ullevål University Hospital, Oslo, Norway **[263]**

R.H. Moore, Department of Medicine, Royal Postgraduate Medical School, Hammersmith Hospital, London W12 0HS, England **[195,351]**

David C. Morrison, Department of Microbiology, Kansas University Medical Center, Kansas City, KS 66103 **[47]**

P.M. Newman, University Department of Surgery, Western Infirmary, Glasgow G11 6NT, Scotland **[237]**

James P. Nolan, Department of Medicine, State University of New York at Buffalo, Erie County Medical Center, Buffalo, NY 14215 **[147]**

Bjarne Østerud, Department of Biochemistry, University Hospital of Tromsø, 9000 Tromsø, Norway **[247]**

Eva Paul, Ludwig Boltzmann Institute for Experimental Traumatology, A-1200 Vienna, Austria **[283]**

Sharon A. Paull, School of Microbiology, University of New South Wales, Kensington 2033, N.S.W., Australia **[173]**

W. Pauw, Department of Thrombosis and Hemostasis, Academic Medical Center, Amsterdam, and Department of Bacteriology, Slotervaart Hospital, Amsterdam, The Netherlands **[213]**

Boguslawa I. Piotrowicz, Department of Bacteriology, Royal Infirmary, University of Glasgow, Glasgow G4 0SF, Scotland **[225]**

Matthew Pollack, Department of Medicine, F. Edward Hebert School of Medicine, Uniformed Services University of the Health Sciences, Bethesda, MD 20814 **[327]**

Reuven Porat, Department of Internal Medicine, Division of Infectious Diseases, Ichilov Hospital, Tel-Aviv Medical Center, Tel-Aviv 64239, Israel **[103]**

G. Ramsay, University Department of Surgery, Western Infirmary, Glasgow G11 6NT, Scotland **[237]**

N. Rapson, Department of Experimental Immunobiology, Wellcome Research Laboratories, Kent, England **[195]**

Heinz Redl, Ludwig Boltzmann Institute for Experimental Traumatology, A-1200 Vienna, Austria **[283]**

A. Revhaug, Department of Surgery, University Hospital of Tromsø, 9012 Tromsø, Norway **[247]**

Ernst Th. Rietschel, Forschungsinstitut Borstel, Institut für Experimentelle Biologie und Medizin, D-2061 Borstel, Federal Republic of Germany **[17]**

Eva Röhrscheidt, Forschungsinstitut Borstel, Institut für Experimentelle Biologie und Medizin, D-2061 Borstel, Federal Republic of Germany **[17]**

O. Røkke, Department of Surgery, University Hospital of Tromsø, 9012 Tromsø, Norway **[247]**

Antoni Rozalski, Forschungsinstitut Borstel, Institut für Experimentelle Biologie und Medizin, D-2061 Borstel, Federal Republic of Germany **[17]**

Ulrich F. Schade, Forschungsinstitut Borstel, Institut für Experimentelle Biologie und Medizin, D-2061 Borstel, Federal Republic of Germany **[17,125]**

Günther Schlag, Ludwig Boltzmann Institute for Experimental Traumatology, A-1200 Vienna, Austria **[283]**

Wim F. Schouten, Department of Microbiology, Vrije Universiteit, 1081 BT Amsterdam, The Netherlands **[373]**

Jon W. Scopes, Department of Pediatrics, St. Thomas' Hospital, London SE1, England **[273]**

Kari Sirnes, Bloodbank and Department of Immunology, Ullevål University Hospital, Oslo, Norway **[263]**

Colin M.M. Stern, Department of Pediatrics, St. Thomas' Hospital, London SE1, England **[273]**

Augueste Sturk, Division of Hemostasis and Thrombosis, Academic Medical Centre, Amsterdam, The Netherlands **[213]**

Ken-ichi Tanamoto, Institute of Medical Science, University of Tokyo, Tokyo 108, Japan; Present Address: Department of Biological Chemistry, National Institute of Hygienic Sciences, Tokyo 158, Japan **[185]**

Jan W. ten Cate, Division of Hemostasis and Thrombosis, Academic Medical Centre, Amsterdam, The Netherlands **[213]**

Vernon L. Tesh, Department of Microbiology and Immunology, Emory University School of Medicine, Atlanta, GA 30322 **[47]**

Bert Thijs, Medical Intensive Care Unit, Vrije Universiteit, 1081 HV Amsterdam, The Netherlands **[373]**

Alistair P.J. Thomson, Department of Pediatrics, St. Thomas' Hospital, London SE1, England **[273]**

Peter S. Tobias, Department of Immunology, Scripps Clinic and Research Foundation, La Jolla, CA 92037 **[309]**

Kevin J. Tracey, Laboratory of Surgical Metabolism, New York Hospital–Cornell University Medical Center, New York, NY 10021 **[77]**

Richard J. Ulevitch, Department of Immunology, Scripps Clinic and Research Foundation, La Jolla, CA 92037 **[309]**

Hans Van Bossuyt, Laboratory for Cell Biology and Histology, Free University of Brussels (VUB), 1090 Brussels-Jette, Belgium **[161]**

Sander J.H. van Deventer, Division of Hemostasis and Thrombosis, Academic Medical Centre, Amsterdam, The Netherlands **[213]**

L.W. Statius van Eps, Department of Internal Medicine, Slotervaart Hospital, Amsterdam, The Netherlands **[213]**

Marian Verweij-van Vught, Department of Medical Microbiology, Vrije Universiteit, 1081 BT Amsterdam, The Netherlands **[373]**

Stanley W. Vukajlovich, Department of Microbiology, Kansas University Medical Center, Kansas City, KS 66103 **[47]**

Jeff Wang, Department of Immunology, Cetus Corporation, Palo Alto, CA 94303 **[383]**

Chris West, Department of Community Health, University of Liverpool Medical School, Liverpool, England **[273]**

E. Wisse, Laboratory for Cell Biology and Histology, Free University of Brussels (VUB), 1090 Brussels-Jette, Belgium **[161]**

Lowell S. Young, Kuzell Institute for Arthritis and Infectious Diseases, San Francisco, CA 94115 **[209]**

Ulrich Zähringer, Forschungsinstitut Borstel, Institut für Experimentelle Biologie und Medizin, D-2061 Borstel, Federal Republic of Germany **[17]**

Elizabeth J. Ziegler, Department of Medicine, Division of Infectious Diseases, University of California at San Diego, San Diego, CA 92103 **[319]**

Introduction

A century ago William H. Howell published his observations on the coagulation of the blood of *Limulus polyphemus* [1]. Almost three quarters of a century later, Frederik B. Bang discovered the profound effects of bacteria on the coagulation system of *Limulus* [2]. Twelve years later, Jack Levin and Frederik Bang [3] published their landmark observations on the role of bacterial endotoxins in this coagulation process. These investigations can be considered harbingers of current basic and clinical endotoxin research. Since 1968 more than 300 key publications on the structure, detection, and biomedical significance of endotoxins have appeared.

Our understanding of endotoxin and its structure–function relationships has increased dramatically during the past ten years. Likewise, knowledge of interactions of endotoxin with cellular components of various host-defense systems and of the role of endotoxin in Gram-negative septicemia has also grown. Moreover, it is now possible to detect endotoxin in different body fluids, and effective strategies for the control of endotoxemia are on the horizon.

This volume contains the Proceedings of the Second International Conference on Endotoxins Amsterdam, entitled "Pathophysiological Effects, Clinical Significance and Pharmacological control of Bacterial Endotoxins," held in May, 1987 at the Academic Medical Centre, Amsterdam, The Netherlands. The aim of this conference was to bring together clinicians and basic scientists and to provide them with a comprehensive review of current clinical and basic research on endotoxins. The conference was divided into four sessions, each of which was introduced by a "state-of-the-art" lecture followed by several original presentations, selected from more than two hundred submitted abstracts. In addition, poster discussions, not published here, provided participants the opportunity to establish valuable contacts and to discuss the latest developments.

In his opening address, Dr. Jack Levin provided an extensive review of the intriguing similarity of the effects of bacterial endotoxins and bacteria upon the coagulation systems of the horseshoe crab (Limulus), other marine organisms, and humans. The ability of endotoxin to activate blood coagulation mechanisms has persisted during evolution. In this regard, Limulus amebocytes and mammalian platelets share a wide range of characteristics and functions.

The first session of the conference focussed on the biological effects of endotoxins. Endotoxins (lipopolysaccharides) of various Gram-negative bacteria share a common architecture: a polysaccharide and a lipid portion (lipid A). The use of chemically synthesized lipid A has provided evidence that the endotoxic principle—i.e., the biological activity that produces fever, hypotension and skin necrosis—is embedded in the lipid A portion of lipopolysaccharides (LPS). The minimal structural requirements for these effects have been clearly defined. Interestingly, not all of the biological effects of LPS are limited to lipid A, as previously assumed. For instance, the inner core oligosaccharide triggers the production of immunostimulatory mediators. One such mediator, cachectin, which is identical to tumor necrosis factor (TNF), is produced by macrophages in response to endotoxin and causes fever and shock. Cachectin shares some of these properties with interleukin-1 (IL-1), which is released by monocytes after activation by LPS.

Important potential mediators of inflammation, such as the lipoxygenase product 13-hydroxylinoleic acid and the phospholipid platelet-activating factor (PAF), which are produced by LPS-activated cell systems, also may contribute to the pathogenesis of septic shock. Animal experiments that have demonstrated a protective effect of PAF-antagonists when they are given before or soon after LPS injection are of interest in this regard. The intriguing process of endotoxin—cell interactions, the ensuing mediator synthesis/release, and the subsequent mediator—cell interactions, with their deleterious effects, will undoubtedly soon be elucidated. This information could facilitate development of new approaches to the prevention of septic shock.

The release of endotoxins from the outer cell membrane of Gram-negative bacteria and the subsequent distribution, degradation (or inactivation secondary to binding with soluble factors present in plasma), and elimination are becoming better understood. Recent experimental data suggest that endotoxin release can occur without bacterial killing. This release of LPS requires terminal complement components. LPS present in the plasma compartment is partially bound to high density lipoproteins. Endotoxins are mainly cleared by the liver macrophages (Kupffer cells), granulocytes, and hepatocytes, in which they persist for long periods. After partial degradation, the endotoxin remnants are secreted in the bile and eliminated in the stool.

The second session of the conference delineated the role of endotoxins in gastrointestinal and liver disease. Although the gastrointestinal tract contains a large reservoir of endotoxins, under physiological circumstances, almost no endotoxin passes through the intestinal mucosa. However, intestinal disease and, more convincingly, severe hepatic disorders such as alcohol-induced liver cirrhosis or massive hepatic necrosis are associated with gut flora-derived endotoxemia. This may result in further deterioration of liver function, as profound effects of endotoxins on the ultrastructure of Kupffer cells have been demonstrated: swelling, disappearance of primary lysosomes, disintegration of mitochondrial matrix and cristae, and the presence of cytoplasmic fat droplets.

In contrast, LPS stimulates and increases the number of Kupffer cells and anti-LPS IgM synthesis within the liver. Circulating endotoxins that have passed through the damaged intestinal mucosa may also damage other organ systems and further aggravate liver failure as they cause renal insufficiency and respiratory distress.

Whether endotoxins are present in portal blood of healthy individuals remains controversial. Recent data also have focussed attention upon the influence of endotoxin on graft-versus-host disease. Investigations in animals and in humans have identified areas for future anti-endotoxin (or anti-mediator) intervention studies.

The clinical significance and the detection of endotoxemia were the subject of the third conference session. Early diagnosis of endotoxemia, which is a prerequisite for successful intervention, now seems feasible utilizing the Limulus amebocyte lysate test. One of the main problems in the clinical application of a diagnostic endotoxin test is the lack of a gold standard. It has now become clear and accepted that endotoxemia can occur without concurrent bacteremia, and refinements in the Limulus assay have provided evidence for this concept. The clinical utility of this endotoxin assay for the prediction of septicemia in febrile patients is derived from new data that reveal a high predictive value of endotoxemia for the development of sepsis. Moreover, endotoxemia precedes the septic period. These findings imply that detection of biologically active endotoxins in blood is a more important indicator of impending septicemia than a positive blood culture. Circulating endotoxins can now be detected in patients during the acute phase of Gram-negative sepsis and septic shock; and animal experiments suggest that the LPS plasma concentrations further increase after gentamicin treatment. However, no increase was observed after antibiotic treatment in patients with systemic meningococcal disease. Further clinical research is required to solve this intriguing and relevant controversy.

The fourth major conference session considered anti-endotoxin mechanisms and treatment. The immunochemistry of the core and lipid A region is now better understood on the molecular level. This enables research on immunoprophylactic and immunotherapeutic approaches for the control of endotoxemia. A novel acute phase protein designated lipopolysaccharide binding protein (LBP) has been isolated and characterized. It directly binds to a wide variety of rough and smooth forms of LPS; only the lipid A domain is required for binding. The pathophysiological role of this interesting protein will be delineated in the near future.

For many years, clinicians have wanted to protect patients from the severe complications of sepsis by passive immunization. The studies performed so far suggest that human polyclonal *Escherichia coli* J5 antiserum is antitoxic rather than antibacterial and that it may be effective in reducing the death rate of patients with Gram-negative bacteremia. Recent advances have been made in the production of monoclonal antibodies, with anticipation of increased specificity, ease of production, and less risk of sensitization. However, it must be stressed that these theoretical advantages and the efficacy of monoclonal antibodies must be established by rigorously performed clinical trials. Several studies are now in progress.

In conclusion, major steps have been made in our understanding of the molecular structure required for the production of the pathophysiologic effects of endotoxins, the fate of endotoxins in the host, substances mediating their effects, and the relevance of endotoxins in certain clinical settings. These new insights have led to potential prophylactic strategies that may be useful in the near future (and that may be an appropriate topic for the next conference on bacterial endotoxins).

The Editors

REFERENCES

1. Howell, W.H. (1885) Observations upon the chemical composition and coagulation of the blood of *Limulus polyphemus, Callinectes hastatus* and *Cucumaria* sp. Johns Hopkins Univ. Circ. 5:4–5.
2. Bang, F.B. (1956) A bacterial disease of *Limulus polyphemus*. Bull. Johns Hopkins Hosp. 98:325–351.
3. Levin, J., Bang, F.B. (1968) Clottable protein in *Limulus:* Its localization and kinetics of its coagulation by endotoxin. Thromb. Diath. Haemorrh. 19:186–197.

Acknowledgments

We acknowledge the generous support of the following corporate sponsors:

KabiVitrum AB, Stockholm, Sweden
Sandoz Pharmaceutical Company, Basle, Switzerland
Armour Pharmaceutical Company, Kankakee, Illinois
Wako Chemicals GMbH, Neuss, Federal Republic of Germany
Whittaker M.A. Bioproducts Inc., Walkersville, Maryland
Byk Nederland BV, Zwanenburg, The Netherlands
Labsystems Benelux BV, Waddinxveen, The Netherlands
The Upjohn Company, Kalamazoo, Michigan
The Perkin Elmer Corporation, Norwalk, Connecticut
American Diagnostica Inc., Greenwich, Connecticut
Marine Biologicals Inc., Marmora, New Jersey

We wish to thank the following session chairmen for their excellent work: D.C. Morrison, E.Th. Rietschel, J.P. Nolan, E. Wisse, J. Levin, L.S. Young, H. Brade, and H.R. Büller.

We sincerely acknowledge the scientific support of the advisory board, which consisted of the following members: H. Brade, D.C. Morrison, E.Th. Rietschel, J. Verhoef, and E. Wisse.

Members of the Editorial Board were: H. Brade, H.R. Büller, J.W. ten Cate, S. van Deventer, J. Levin, D.C. Morrison, E.Th. Rietschel, and E. Wisse. Their efforts are highly appreciated.

The staff, technicians, and medical students of the Division of Hemostasis and Thrombosis provided superior assistance in the planning and organization of this second International Conference on Endotoxins in Amsterdam (ICEA).

We are grateful for the help of the staff of Alan R. Liss, Inc., with this publication.

Biological Effects of Endotoxins

Bacterial Endotoxins: Pathophysiological Effects, Clinical Significance, and Pharmacological Control, pages 3–15

THE HORSESHOE CRAB: A MODEL FOR GRAM-NEGATIVE SEPSIS IN MARINE ORGANISMS AND HUMANS

Jack Levin

Departments of Laboratory Medicine and Medicine, University of California School of Medicine, San Francisco, CA 94143; the Veterans Administration Medical Center, San Francisco, CA 94121; and the Marine Biological Laboratory, Woods Hole, MA 02543

In 1956, Frederik Bang observed that the injection of a marine vibrio species, a gram-negative organism, into *Limulus* caused intravascular coagulation and death (Bang, 1956). The same effect was produced by a heat stable extract of this organism. He also noted that the level of circulating amebocytes, the only type of circulating blood cell in *Limulus*, was markedly reduced. Furthermore, the amebocytes that remained in the circulation were degranulated. Dr. Bang correctly concluded that intravascular coagulation had occurred, and that this could be caused by either the living gram-negative rod or a toxin prepared by heating and disrupting the organisms. The latter technique is the classical method for the production of crude bacterial endotoxin.

The effects of gram-negative infection in *Limulus* are summarized in Table 1. There is marked similarity to some of the effects of gram-negative infection in humans.

Table 1. The Effects of Gram-Negative Infection in *Limulus*

Amebocytopenia

Intravascular coagulation

Clots contain Gram-negative organisms

Blood becomes incoagulable

Fatal

Endotoxin can produce these effects

Since the entire blood coagulation mechanism in *Limulus* is contained within the circulating amebocytes, activation of the coagulation mechanism in *Limulus* by gram-negative organisms or their products requires that amebocytes be sensitive to low concentrations of bacterial endotoxin. The cytoplasm of normal amebocytes is packed with granules, but after exposure to bacterial endotoxin, amebocytes become flattened and degranulated (Figure 1). During this process, amebocytes release the components of the coagulation mechanism into the surrounding medium (Levin and Bang, 1964a, 1964b, 1968). The clots that form can immobilize bacteria. Figure 2 compares endotoxin-free *Limulus* blood with a sample after addition of a minute concentration of endotoxin. Coagulation has occurred. Furthermore, it has been shown that lysates of *Limulus* amebocytes are bactericidal for a variety of bacteria (Stagner and Redmond, 1975; Furman and Pistole, 1976; Nachum et al., 1979; Pistole and Graf, 1984).

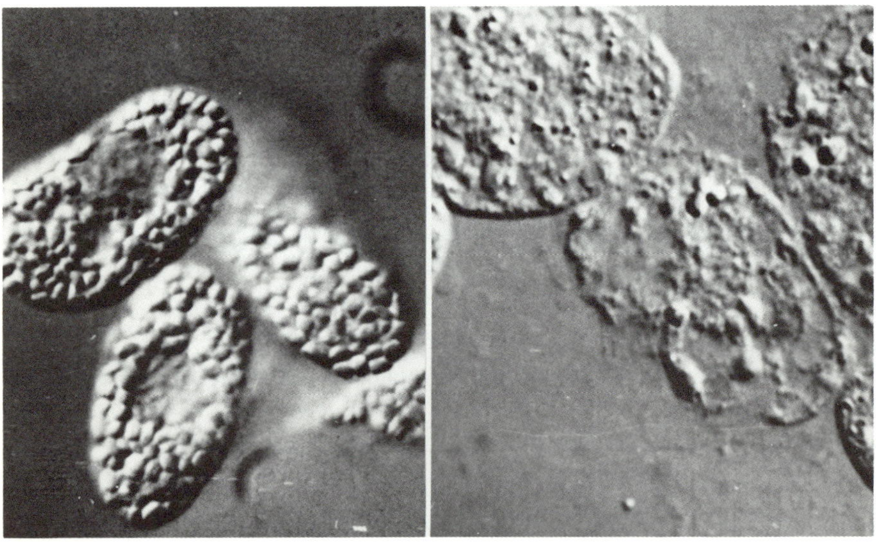

Figure 1. *Limulus* amebocytes. Three intact normal cells are shown on the left. Flattened, spread, degranulated cells, after exposure to bacterial endotoxin, are shown on the right. (Differential interference phase microscopy; original magnif. x 1,000.)

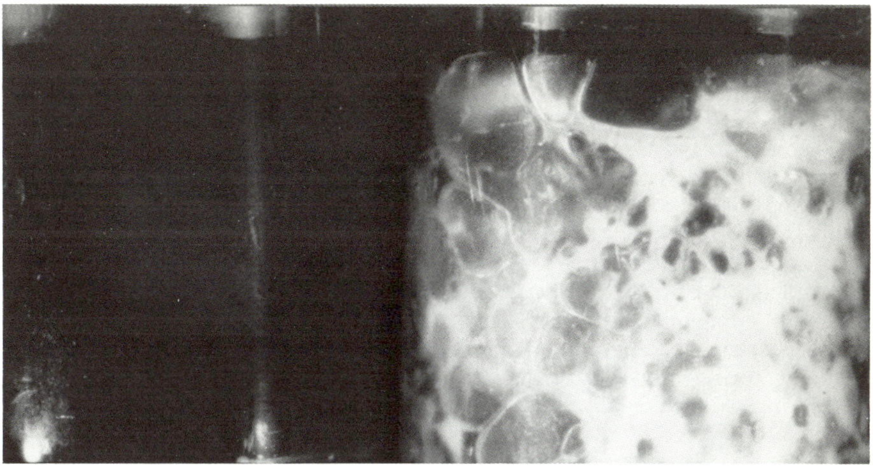

Figure 2. *Limulus* blood. Endotoxin-free (left). After addition of bacterial endotoxin (final concentration, 0.1 μg/ml) (right).

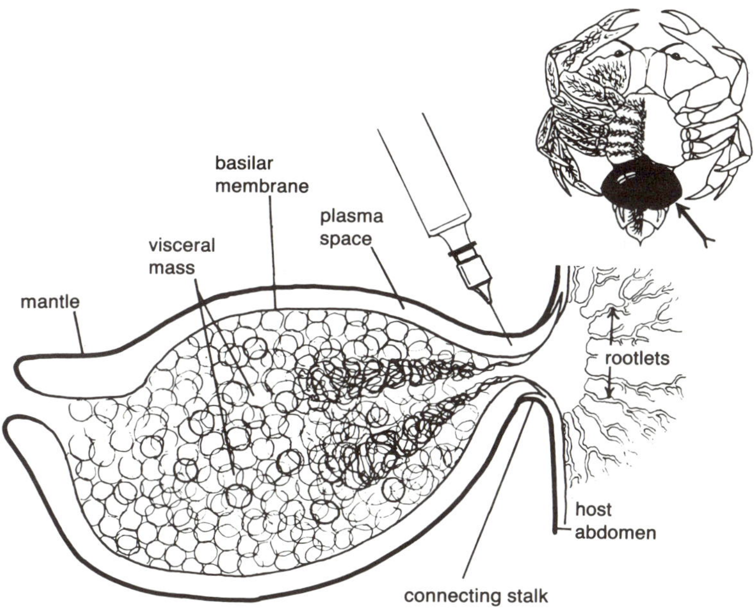

Figure 3. Schematic sagittal view of external *Sacculina*. Inset: *Sacculina* attached to ventral surface of host crab. (From Barker and Bang, 1966; with permission.)

Horseshoe crabs are not unique among marine organisms in this response, since similar alterations in blood coagulation have been shown in other marine invertebrates. A particularly interesting example is provided by *Sacculina carcini*, a parasitic crustacean. This parasite, shown in Figure 3, has a clear plasma in the space between its basilar membrane and mantle. When marine ciliates are incubated with this plasma, the plasma gels and immobilizes the ciliates (Figure 4). Barker and Bang have reported that introduction of a marine vibrio species into the plasma space caused the plasma to become incoagulable (Barker and Bang, 1966). The animals subsequently died. Examination revealed that there were gross masses of gelled material within the plasma space, which contained immobilized bacteria. Exhaustion of the coagulation mechanism resulted in bacteremia, death of the parasite, and strikingly, also death of the host crab.

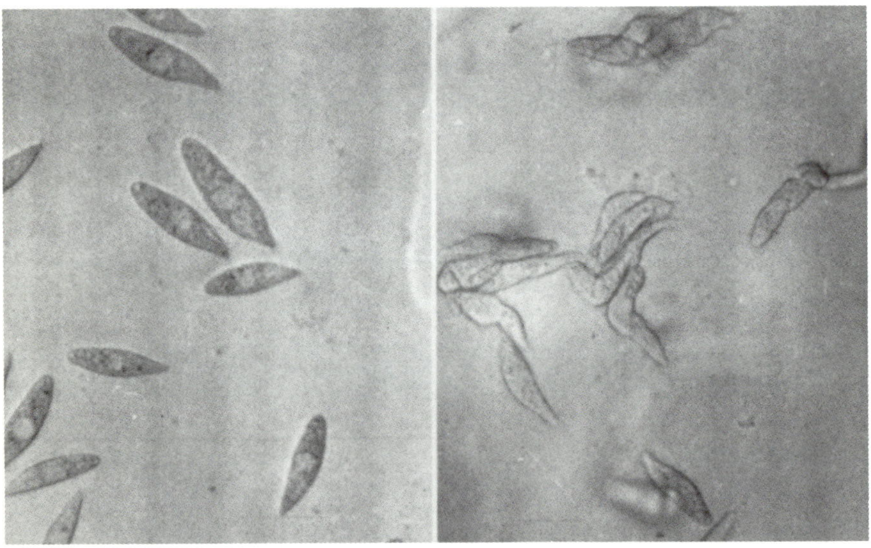

Figure 4. Immobilization of marine ciliates by gelled *Sacculina* plasma. Left: Control ciliates swimming freely in unclotted *Sacculina* plasma. Right: Immobilized ciliates in gelled plasma. (Phase contrast microscopy; original magnif. x 100.) (From Barker and Bang, 1966; with permission.)

The administration of endotoxin to *Limulus* causes a reduction in the numbers of circulating amebocytes. A similar response occurs in the lobster, *Homarus americanus*. Injection of gram-negative organisms produces a reduction in numbers of circulating hemocytes (Figure 5) (Paterson et al., 1976). Similar responses are observed in other marine organisms, as summarized in Table 2. Infection of the crab, *Carcinus maenus*, with a gram-negative bacillus, is associated with amebocytopenia, loss of coagulability of the blood, and disappearance of the blue color of the plasma which is normally present (Cantacuzene, 1925). Recovery is associated with return of the property of coagulability. *Maia squinado*, the spider crab, develops amebocytopenia and abnormalities of coagulation after infection with an enteric gram-positive bacillus (Cantacuzene, 1923). In the animals that survive, coagulation is only slightly altered, and returns to normal as the amebocyte count rises. In animals that develop a fatal infection, there is a marked

Table 2. Initiation of Blood Coagulation in Invertebrates by Endotoxin or Bacteria*

Animal	Agent
Limulus polyphemus	*Vibrio sp.*
	E. coli
	Shigella
Sacculina carcini	*Vibrio sp.*
	E. coli
	Gaffkya (?)
	Gram-negative bacillus
Homarus americanus	*Shigella*
	Gaffkya
Homarus vulgaris	*Aerococcus viridans*
Carcinus maenus	Gram-negative bacillus
Callinectes sapidus	*Vibrio sp.*
	Gram-negative bacillus
Maia squinado	Gram-positive bacillus
Cancer irroratus	*Vibrio sp.*
Penaeid shrimp	*Vibrio alginolyticus*
(P. aztecus, P. setiferus,	*Vibrio anguillarum*
P. duorarum)	*Aeromonas sp.*

*The original citations from which most of these data were collected are listed in Levin (1967). See also Cantacuzene (1921), Stewart (1975), Lightner and Lewis (1975), Johnson (1976), Newman and Feng (1982), and Durliat and Vranckx (1983a, 1983b).

decrease in amebocytes; and just before the terminal septicemic episode, the blood becomes incoagulable. Similar observations have been made in penaeid shrimp (Lightner and Lewis, 1975), and blue crabs (Johnson, 1976). Hemocyte levels decrease in *Cancer irroratus*, the rock crab, and in *Homarus vulgaris*, the European lobster, following infection with a *Vibrio species* or *Aerococcus viridans*, respectively; and blood coagulation becomes impaired in *Homarus* (Newman and Feng, 1982; Durliat and Vranckx (1983a, 1983b; Durliat, 1985).

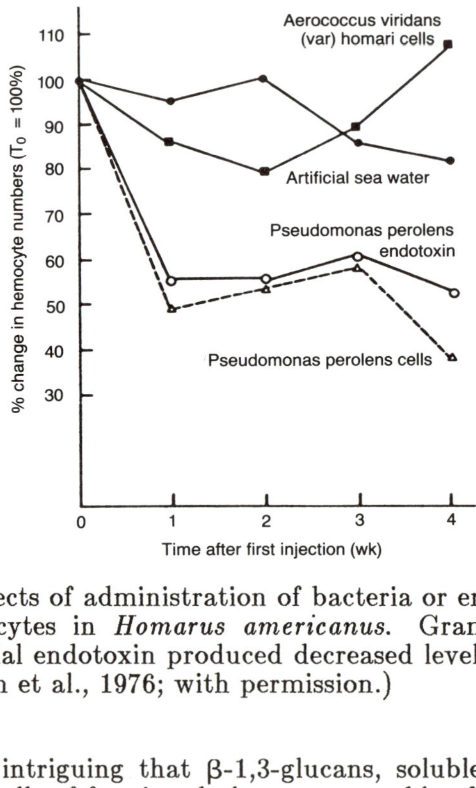

Figure 5. Effects of administration of bacteria or endotoxin on circulating hemocytes in *Homarus americanus*. Gram-negative bacteria or bacterial endotoxin produced decreased levels of hemocytes. (From Paterson et al., 1976; with permission.)

It is also intriguing that β-1,3-glucans, soluble carbohydrates from the cell walls of fungi and algae, are capable of reducing circulating hemocyte levels in *Carcinus maenus*, the shore crab and *Astacus astacus*, the freshwater crayfish, for an approximately three-hour period following injection (Smith and Soderhall, 1983). Pertinently, samples of hemolymph obtained from *Carcinus* or *Astacus*, after administration of β-1,3-glucans, were described as clotting more rapidly. Furthermore, in the crayfish, β-1,3-glucans and bacterial endotoxins activate the prophenoloxidase activating system and the blood coagulation mechanism detectable in lysates of hemocytes (Soderhall, 1981; Smith and Soderhall, 1983; Soderhall and Hall, 1984).

The descriptions of gram-negative infection in marine invertebrates are all essentially identical: severe gram-negative infection produces incoagulability of the blood in association with disappearance of the circulating blood cells. From this I have inferred that there are many other organisms that, like *Limulus*, have either all or key components of their blood coagulation system present in blood cells, rather than free in the plasma. Furthermore, it is likely that the response of the blood coagulation mechanism, following activation of the circulating blood cells by bacteria and/or their products, represents a very ancient defense mechanism.

Table 3. Comparison of *Limulus* Amebocytes and Mammalian Platelets

Characteristics or functions	*Limulus* amebocyte	Mammalian platelet
Hemostasis	Essential	Essential
Blood coagulation	Essential	Ancillary
Clottable protein	Yes	Yes
Nucleus	Yes	No
Viscous metamorphosis	Yes (?)	Yes
Cellular processes	Yes	Yes
Granules	Yes	Yes
Response to endotoxin	Yes	Yes
Phagocytosis	Yes	Yes
Antibacterial function	Yes	Yes
Motility	Yes	Yes

The similarities between the described responses of invertebrates and the production of disseminated intravascular coagulopathy during gram-negative sepsis in humans are obvious. A comparison of the functions of *Limulus* amebocytes and mammalian platelets suggests at least one aspect of the biological mechanism by which humans have retained this response despite approximately 400,000,000 years of evolution (Table 3). Amebocytes and mammalian platelets share many characteristics. Platelets possess rudimentary bactericidal and some phagocytic activity (Glynn et al., 1965; Levin, 1976; Lewis et al., 1976). It is noteworthy that platelets contain endotoxin-binding substances (Springer and Adye,

1975) and have been shown to interact with bacteria, endotoxin, and viruses (Clawson, 1973; MacIntyre et al., 1977; Levin, 1984). An interesting recent paper has reported that human platelets are cytotoxic for parasites, based on an IgE receptor on the platelet surface (Capron et al., 1985). Under certain circumstances, platelets demonstrate a chemotactic response and migrate (Lowenhaupt et al., 1973, 1977). None of these functions would seem to be necessary for the current role of platelets in mammalian hemostasis. Accordingly, it is likely that some of the capabilities of mammalian platelets represent the residua of functions originally present in the more primitive and multicompetent cells from which mammalian blood cells have evolved.

Table 4. Activation of Mammalian Blood Coagulation Mechanism(s) by Bacterial Endotoxin

Platelets
White blood cells
Factor XII
Vascular endothelium

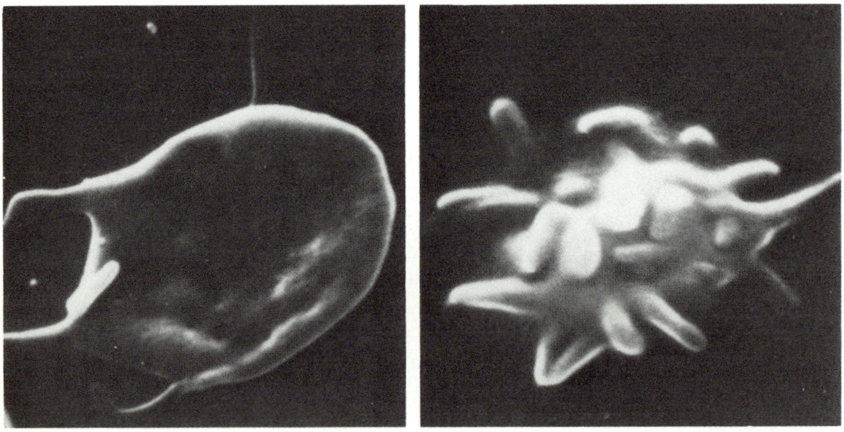

Figure 6. Effects of bacterial endotoxin on the surface characteristics of rabbit platelets. Left: Control platelet. Right: Platelet incubated with *S. minnesota* R595 endotoxin (final concentration, 400 μg/ml). (Scanning electron microscopy.) (From Morrison et al., 1981.)

Bacterial endotoxins can initiate mammalian coagulation in a variety of ways (Table 4). Platelets can be activated; white blood cells can be stimulated to produce tissue factor, an important procoagulant; and Factor XII can be activated by endotoxin. Subsequent to incubation with endotoxin, platelets become aggregated and show marked shape changes (Ausprunk and Das, 1978; Morrison et al., 1981). Figure 6 demonstrates the effect of endotoxin on the surface characteristics of platelets. In addition, vascular damage can be produced by bacterial endotoxins, providing another mechanism by which platelets and the blood coagulation mechanism can become activated.

Table 5 summarizes changes in components of hemostasis in humans during the syndrome designated disseminated intravascular coagulopathy (DIC). The pertinent circulating blood cell (i.e., platelets) decreases in number, in association with evidence of activation of blood coagulation. If the process is sufficiently severe, the blood becomes incoagulable, as has been described in marine invertebrates (Tables 1 and 2). Just as gram-negative organisms and their endotoxins activate the blood coagulation mechanisms of invertebrates, it is noteworthy that a major cause of DIC in adults is infection. An example of the effects of endotoxemia upon blood coagulation in a mammal is provided by the generalized Shwartzman reaction. This is a classical experimental model in which after appropriate preparation of the rabbit, the intravenous injection of bacterial endotoxin activates blood coagulation and eventually produces renal thrombi and death (McKay and Shapiro, 1958; Levin and Cluff, 1965). Development of the generalized Shwartzman reaction is associated with thrombocytopenia and hypofibrinogenemia, and can be prevented by anticoagulation (Levin and Beck, 1966; Gurewich et al., 1976).

Table 5. Disseminated Intravascular Coagulopathy (DIC)

- Thrombocytopenia
- Hypofibrinogenemia
- Increased prothrombin time
- Increased partial thromboplastin time
- Fibrinogen-fibrin degradation products
- Microangiopathic hemolytic anemia
- Cryoprotein(s)

Summary. The roles of the amebocyte in providing hemostasis and controlling infection, and its reaction to endotoxin, suggest that the response of platelets and the blood coagulation system in various mammals to gram-negative infection or endotoxin is an evolutionary remnant of this ancient mechanism. In humans, this mechanism occasionally subverts its presumed protective function by overresponding in a manner that results in pathophysiologic thrombosis or hemorrhage. (In this regard, it is interesting that human platelets are much more resistant to the effects of bacterial endotoxins than are other species.) Similarly, the rudimentary ability of mammalian platelets to phagocytose particles and kill bacteria may be another remnant of functions that are more important in amebocytes (or the thrombocytes of other invertebrates). Thus, these two cells, one from an ancient invertebrate and the other from mammals, have remarkably similar characteristics, although the relative importance of their various functions has changed as evolution has taken place. Nevertheless, after at least 400,000,000 years of evolution, coagulation and anti-bacterial mechanisms remain at least partially linked.

REFERENCES

Ausprunk DH, Das J (1978). Endotoxin-induced changes in human platelet membranes: Morphologic evidence. Blood 51:487-495.

Bang FB (1956). A bacterial disease of *Limulus polyphemus*. Bull Johns Hopkins Hosp 98:325-351.

Barker WH Jr, Bang FB (1966). The effect of infection by gram-negative bacteria, and their endotoxins, on the blood-clotting mechanism of the crustacean *Sacculina carcini*, a parasite of the crab *Carcinus maenas*. J Invertebr Pathol 8:88-97.

Cantacuzene J (1921). Quelques remarques au sujet d'une infection experimentale chez *Maia squinado*. Compt Rend Soc Biol 84:1007-1010.

Cantacuzene J (1923). Le probleme de l'immunite chez les invertebres. Compt Rend Soc Biol, 75th Anniv 88:(Suppl):48-119.

Cantacuzene A (1925). Reaction du Crabe Sacculine vis-a-vis d'une infection experimentale de la Sacculine. Compt Rend Soc Biol 93:1417-1419.

Capron A, Ameisen JC, Joseph M, Auriault C, Tonnel AB, Caen J (1985). New functions for platelets and their pathological implications. Int Archs Allergy Appl Immunol 77:107-114.

Clawson CC (1973). Platelet interaction with bacteria. III. Ultrastructure. Am J Pathol 70:449-472.

Durliat M (1985). Clotting processes in Crustacea Decapoda. Biol Rev 60:473-498.

Durliat M, Vranckx R (1983a). Analysis of clotting defects in diseased lobsters. I. Alterations in blood parameters. Comp Biochem Physiol 76A:95-101.

Durliat M, Vranckx R (1983b). Analysis of clotting defects in diseased lobsters. II. Immunochemical study of blood proteins. Comp Biochem Physiol 76A:103-108.

Furman RM, Pistole TG (1976). Bactericidal activity of hemolymph from the horseshoe crab, *Limulus polyphemus*. J Invertebr Pathol 28:239-244.

Glynn MF, Movat HZ, Murphy EA, Mustard JF (1965). Study of platelet adhesiveness and aggregation, with latex particles. J Lab Clin Med 65:179-201.

Gurewich V, Lipinski B, Wetmore R (1976). A study of the large heparin requirements in the generalized Shwartzman reaction. Blood 47:297-304.

Johnson PT (1976). Bacterial infection in the blue crab, *Callinectes sapidus*: Course of infection and histopathology. J Invertebr Pathol 28:25-36.

Levin J (1976). Blood coagulation in the horseshoe crab (*Limulus polyphemus*): A model for mammalian coagulation and hemostasis. In "Animal Models of Thrombosis and Hemorrhagic Diseases." (Proceedings of a Symposium of the National Academy of Sciences.) Washington, DC: U.S. Department of Health, Education and Welfare. DHEW Publication No. (NIH) 76-982, pp 87-96.

Levin J (1984). Bleeding with infectious diseases. In Ratnoff OD, Forbes CD (eds): "Disorders of Hemostasis." Orlando, FL: Grune & Stratton, pp 367-378.

Levin J, Bang FB (1964a). The role of endotoxin in the extracellular coagulation of Limulus blood. Bull Johns Hopkins Hosp 115:265-274.

Levin J, Bang FB (1964b). A description of cellular coagulation in the Limulus. Bull Johns Hopkins Hosp 115:337-345.

Levin J, Bang FB (1968). Clottable protein in Limulus: Its localization and kinetics of its coagulation by endotoxin. Thromb Diath Haemorrh 19:186-197.

Levin J, Beck E (1966). A re-appraisal of the role of blood coagulation and platelets in the generalized Shwartzman phenomenon. Thromb Diath Haemorrh 15:519-538.

Levin J, Cluff LE (1965). Platelets and the Shwartzman phenomenon. J Exp Med 121:235-246.

Lewis JC, Maldonado JE, Mann KG (1976). Phagocytosis in human platelets: Localization of acid phosphatase-positive phagosomes following latex uptake. Blood 47:833-840.

Lightner DV, Lewis DH (1975). A septicemic bacterial disease syndrome of penaeid shrimp. Marine Fisheries Rev 37:25-28.

Lowenhaupt RW, Miller MA, Glueck HI (1973). Platelet migration and chemotaxis demonstrated *in vitro*. Thromb Res 3:477-487.

Lowenhaupt RW, Glueck HI, Miller MA, Kline DL (1977). Factors which influence blood platelet migration. J Lab Clin Med 90:37-45.

MacIntyre DE, Allen AP, Thorne KJI, Glauert AM, Gordon JL (1977). Endotoxin-induced platelet aggregation and secretion I. Morphological changes and pharmacological effects. J Cell Sci 28:211-223.

McKay DG, Shapiro SS (1958). Alterations in the blood coagulation system induced by bacterial endotoxin. I *In vivo* (generalized Shwartzman reaction). J Exp Med 107:353-376.

Morrison DC, Oades ZG, Di Pietro D (1981). Endotoxin-initiated membrane changes in rabbit platelets. In Majde JA, Person RJ (eds): "Pathophysiological Effects of Endotoxins at the Cellular Level," New York: Alan R. Liss, pp 47-64.

Nachum R, Watson SW, Sullivan JD Jr, Siegel SE (1979). Antimicrobial defense mechanisms in the horseshoe crab, *Limulus polyphemus*: Preliminary observations with heat-derived extracts of *Limulus* amoebocyte lysate. J Invertebr Pathol 33:290-299.

Newman MC, Feng SY (1982). Susceptibility and resistance of the rock crab, *Cancer irroratus*, to natural and experimental bacterial infection. J Invertebr Pathol 40:75-88.

Paterson WD, Stewart JE, Zwicker BM (1976). Phagocytosis as a cellular immune response mechanism in the American lobster, *Homarus americanus*. J Invertebr Pathol 27:95-104.

Pistole TG, Graf SA (1984). Bactericidal activity of Limulus lectins and amebocytes. In Cohen E (ed): "Recognition Proteins, Receptors, and Probes: Invertebrates," New York: Alan R. Liss, pp 71-81.

Smith VJ, Soderhall K (1983). β-1,3 glucan activation of crustacean hemocytes *in vitro* and *in vivo*. Biol Bull 164:299-314.

Soderhall K (1981). Fungal cell wall β-1,3-glucans induce clotting and phenoloxidase attachment to foreign surfaces of crayfish hemocyte lysate. Dev Comp Immunol 5:565-573.

Soderhall K, Hall L (1984). Lipopolysaccharide-induced activation of prophenoloxidase activating system in crayfish haemocyte lysate. Biochim Biophys Acta 797:99-104.

Springer GF, Adye JC (1975). Endotoxin-binding substances from human leukocytes and platelets. Infect Immun 12:978-986.

Stagner JI, Redmond JR (1975). The immunological mechanisms of the horseshoe crab, *Limulus polyphemus*. Mar Fish Rev 37:11-19.

Stewart JE (1975). Gaffkemia, the fatal infection of lobsters (Genus *Homarus*) caused by *Aerococcus viridans* (var.) *homari*: A review. Mar Fish Rev 37:20-24.

Wardle EN (1985). Endotoxaemia, septicaemia, shock lung and acute renal failure (chapter 13). In "Cell Surface Science in Medicine and Pathology." New York: Elsevier, pp 381-411.

Bacterial Endotoxins: Pathophysiological Effects, Clinical Significance, and Pharmacological Control, pages 17–45

STRUCTURE, ENDOTOXICITY, IMMUNOGENICITY AND ANTIGENICITY OF BACTERIAL LIPOPOLYSACCHARIDES (ENDOTOXINS, O-ANTIGENS)

Helmut Brade, Lore Brade, Ulrich Schade, Ulrich Zähringer, Otto Holst, Hella-Monika Kuhn, Antoni Rozalski, Eva Röhrscheidt, and Ernst Th. Rietschel

Forschungsinstitut Borstel, Institut für Experimentelle Biologie und Medizin, D-2061 Borstel, FRG

INTRODUCTION

In his classical studies, performed at the end of the 19th century, Richard Pfeiffer found that lysates of heat-inactivated Vibrio cholerae bacteria induced in experimental animals shock and death. Later it was recognized that also lysates of other gram-negative bacteria (Enterobacteriacae, Pseudomonadaceae, Neisseriaceae) exhibited toxicity and, furthermore, caused a great number of other pathophysiological effects such as fever, hypotension and disseminated intravascular coagulation. Pfeiffer postulated that the toxic principle was present within the bacterial cell and, therefore, termed it endotoxin (for literature see Westphal et al., 1986). In subsequent studies, however, endotoxins were shown to be localized on the surface of bacterial cells and to form, together with phospholipids and proteins, the outer membrane of gram-negative bacteria. As surface structures, endotoxins play an important role in the interaction of gram-negative bacteria with higher organisms. Thus, the host's defense system recognizes invading bacteria by their endotoxin structures, and it reacts with the formation of antibacterial antibodies directed against endotoxin structures. Endotoxins, therefore, represent immunoreactive surface antigens, and if this property is to be emphasized they are also called O-antigens (Lüderitz et al., 1982).

This paper is dedicated to Prof. Dr. Dr. hc. Otto Westphal on the occasion of his 75th birthday (1.2.1988) as an expression of our gratitude.

On the other hand endotoxins, released from bacteria, significantly contribute to the symptoms of bacteremia and sepsis and, thus, represent important pathogenetic factors in gram-negative infections.

CHEMICAL NATURE AND STRUCTURE OF ENDOTOXIN

Their exposed position on the cell surface allows the gentle isolation of endotoxins. Various extraction methods have been described, among which the phenol-water and the phenol-chloroform-petroleum ether procedures have received wide application (Galanos et al., 1977). These methods, combined with purification steps, yielded endotoxin preparations which were free of nucleic acids, proteins, phospholipids and other bacterial cell components. Studies on highly purified endotoxins then revealed that they consist of a polysaccharide and a lipid component, termed lipid A, and hence chemically represent lipopolysaccharides. Today, the general structure of many lipopolysaccharides is known, and in the following some of the structural principles are briefly summarized.

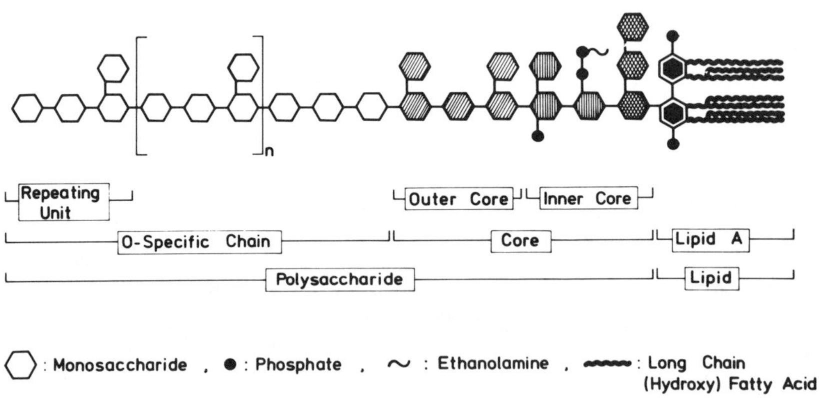

Figure 1. Schematic representation of an enterobacterial lipopolysaccharide (Rietschel et al., 1984)

Polysaccharide component

The polysaccharide component of enterobacterial lipopolysaccharides consists of two regions which differ in their genetic determination, biosynthesis and architecture. These regions are the O-specific chain and the core oligosaccharide (Fig. 1).

O-specific chain

The O-specific chain is made up of repeating oligosaccharide units which contain up to five sugar residues. A large diversity of the constituent components of repeating units has been revealed within different gram-negative bacteria. These include neutral pyranosidic and furanosidic sugars (hexoses, pentoses, and deoxy- and O-methyl derivatives) and charged sugars (aminohexoses and aminopentoses, hexuronic acids, and hexosaminuronic acids), which may carry a variety of substituents such as amino acyl, phosphoryl, glyceryl, lactyl, and acetyl groups (for literature see Lüderitz et al., 1982; Rietschel et al., 1982). The nature, sequence, type of linkage, and type of substitution of the individual monosaccharide residues within a repeating unit is characteristic and unique for a given lipopolysaccharide and the parental bacterial strain. Thus, the O-specific chain is species-specific. Because of the diversity of constituents and their linkages an enormous number of structures of O-specific chains is conceivable and also verified in nature. Therefore, an immense structural variability is revealed if the O-specific chains of distinct bacterial origin are compared.

Core oligosaccharide

The core region consists of a heterooligosaccharide which is subdivided into the lipid A-proximal inner core and the lipid A-distal outer core (Fig. 1). The outer core contains the frequently occurring sugars D-glucose, D-galactose, and N-acetyl-D-glucosamine, whereas the inner core region is composed of the unusual sugars heptose, mainly in the L-glycero-D-manno and the D-glycero-D-manno configuration and 2-keto-3-deoxyoctonic acid (KDO, also termed 3-deoxy-D-mannooctulosonic acid). These are often substituted by charged groups such as phosphoryl, pyrophosphoryl, phosphorylethanolamin and pyrophosphorylethanolamin. In *Bordetella pertussis* (Chaby and Szabo, 1975) and *Vibrio cholerae* (Brade,

1985) lipopolysaccharide, KDO-phosphate has been detected, and in *Acinetobacter calcoaceticus* the unusual sugars heptulosonaric acid (Brade and Rietschel, 1985) and octulosonic acid (Kawahara et al., 1987) are present. Thus, the inner core region exhibits a considerable accumulation of charged residues. The structural variation of the core within different bacterial species is low as compared to that of the O-specific chain. Thus, in *Salmonella* only one core type exists for all serotypes which is termed the Ra-core, and in *E. coli* so far six core types have been described for more than hundred different serotypes (Jansson et al., 1981; Jann and Jann, 1984).

In these core types some structural variation is seen in the outer region while the KDO-containing inner core appears to be more conserved. The structural elucidation of the KDO-containing inner core was hampered for a long time because of the difficulties encountered with the analytical chemistry of KDO. KDO is a polyfunctional sugar acid with eight carbon atoms which carries carboxyl, keto, deoxy and hydroxyl groups rendering it extremely sensitive for the action of acid and other chemical reagents. Thus, under the experimental conditions of its isolation, purification or derivatization, KDO undergoes numerous side reactions yielding many artefacts. In addition, the extreme sensitivity of the ketosidic bond towards acid-catalyzed hydrolysis seemed to prevent the isolation of KDO oligosaccharides. Therefore, classical approaches such as methylation analysis which were successfully applied in the structural elucidation of other complex carbohydrates were not applicable to the KDO region. Likewise, NMR spectroscopy of lipopolysaccharide did not lead to unequivocal results (Strain et al., 1983). These techniques, however, became amenable after a new methodology to handle KDO had been developed. The first isolation of a KDO oligosaccharide was achieved when we found appropriate hydrolysis conditions, mild enough not to degrade KDO. After hydrolysis of lipopolysaccharide from a *S. godesberg* Re mutant in acetate buffer (pH 4.4 at 70°C) an oligosaccharide was obtained which, by use of high-voltage paper electrophoresis and the thiobarbiturate assay was characterized as a disaccharide (Brade et al., 1983). Combined gas-liquid chromatography/mass spectrometry and ^{1}H- and ^{13}C-NMR spectroscopy of the KDO disaccharide showed that it was α(2–4)-interlinked (Brade et al., 1984). Using this compound and synthetic standards as reference substances, a gas-liquid chromatography/mass spectrometry method was established which allowed to ascertain this structure as a common constituent of many lipopolysaccharides (Brade and Rietschel, 1984). Subsequently, a modified methylation analysis was developed which allowed the linkages between the KDO residues and those of adjacent neutral sugars to be determined (Tacken et al., 1986). All structural data obtained during these studies are condensed in Fig. 2 which shows the chemical structure of the inner core region of phosphate-less Rc mutant (strain R5) of S. minnesota.

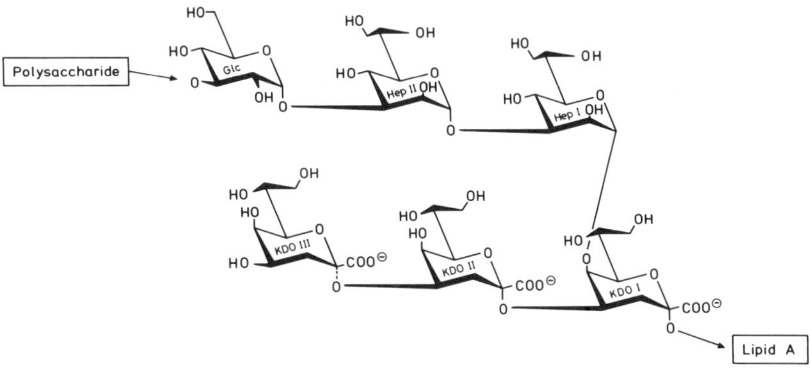

Figure 2. Chemical structure of the core oligosaccharide in the lipopolysaccharide of S. minnesota chemotype RcP⁻ (strain R5) (Tacken et al., 1986). KDO I and KDO II are the only core constituents in Re-type LPS. KDO III is only found in mutants other than Re where it is not always present in stoichiometric amounts (as indicated by the dotted line).

In Re mutants, only two KDO molecules (KDO I and KDO II) are found which are both present in the α-pyranoid ring form and which are (2-4)-interlinked. In mutants other than Re, the lipid A proximal KDO residue (KDO I) is substituted in position 5 by a heptosyl residue while KDO II carries in position 4 in α-pyranosidic linkage, a third KDO group (KDO III). The substitution with this third KDO is - in most lipopolysaccharides studied - not quantitative; it is, however, present in stoichiometric amounts in an Rb_2 mutant (strain R345) of S. minnesota.

Lipid A component

The chemical structure of, notably, enterobacterial lipid A's is known in great detail (Rietschel et al., 1984). As an example Fig. 3 shows the structure of lipid A of Escherichia coli and S. minnesota which will be discussed below (together with lipid A of Proteus mirabilis).

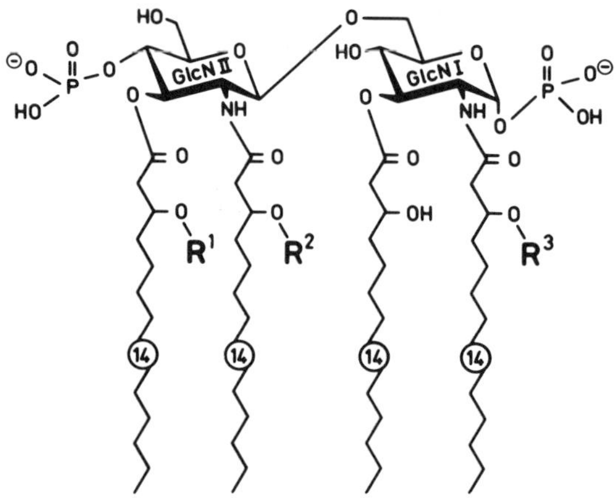

Source of Lipid A	Designation of Synthetic Lipid A	Chemical Nature of		
		R1	R2	R3
E. coli	506	14:0	12:0	H
S. minnesota	516	14:0	12:0	H, 16:0
P. mirabilis	–	14:0	14:0	H, 16:0

Figure 3. Chemical structure of the lipid A component of Escherichia coli, Salmonella minnesota and Proteus mirabilis (Rietschel et al., 1984). The number in circles indicate the number of carbon atoms in acyl chains. In natural lipid A of S. minnesota, the substitution with palmitic acid is not quantitative, whereas the synthetic counterpart (compound 516) contains this fatty acid in stoichiometric amounts.

As Fig. 3 demonstrates, these lipid A's are made up by a ß-D-glucosaminyl-(1-6)-α-D-glucosamine disaccharide which carries two phosphoryl groups: one in position 4' (of the distal glucosaminyl residue, GlcN II) and one in position 1 (of the reducing glucosaminyl residue, GlcN I). This hydrophilic lipid A backbone is, in all three cases, acylated by four residues of (R)-3-hydroxytetradecanoic acid [14:0(3-OH)] at positions 2, 3, 2' and 3'. The hydroxyl groups of two (E. coli) or three (S. minnesota) of these 3-hydroxy groups are acylated by nonhydroxylated fatty acids. The distribution of normal fatty acids (chain length 12, 14 and 16 carbon atoms) over available hydroxyl groups is highly specific and characteristic for the bacterial genus. Thus, in both Escherichia and Salmonella lipid A, dodecanoic acid (12:0) is ester-linked to the 14:0(3-OH) group at position 2' (GlcN II) and tetradecanoic acid (14:0) to that at position 3' (GlcN II). In P. mirabilis lipid A, 12:0 is absent and positionally replaced by 14:0. S. minnesota and P. mirabilis lipid A contain, in addition, hexadecanoic acid (16:0) which is linked to 14:0(3-OH) at position 2 of GlcN I and which is lacking in E. coli. While the substitution by 12:0 and 14:0 is, to a large extent, stoichiometric that by 16:0 is not. Thus, e.g. within various batches of S. minnesota Re mutant lipopolysaccharide, values between 0.2 and 0.8 moles of 16:0 per mole D-glucosamine disaccharide have been determined. This suggests that in a S. minnesota lipid A preparation a heptaacyl species but also a structure corresponding to E. coli lipid A (lacking 16:0) is present. The nonmolar presence of 16:0 is only one facet of the intrinsic heterogeneity of lipid A (for literature see Nowotny et al., 1984; Qureshi et al., 1985). It is, therefore, emphasized that the structures shown in Fig. 3 represent that molecular species which contains all known acyl components of a free lipid A preparation and that this structure is only one (but major) among several others present in bacterial lipid A. The nonstoichiometric substitution of the backbone also extends to charged substituents of the phosphoryl groups in positions 1 and 4' of lipid A. These polar headgroups (4-amino-4-deoxy-L-arabino-pyranose, phosphorylethanolamine, phosphate) are not essential for endotoxic activity and are not shown in Fig. 3 (for literature see Rietschel et al., 1982). As Fig. 3 demonstrates, the backbone of enterobacterial lipid A's contains two free hydroxyl groups at positions 4 and 6'. The latter primary hydroxyl group is only free in polysaccharide-deprived lipid A as obtained on acid treatment of lipopolysaccharide. In lipopolysaccharide, it represents the attachment site of KDO I, i.e. the polysaccharide component (Zähringer et al., 1985).

LIPID A, THE ENDOTOXIC PRINCIPLE OF LIPOPOLYSACCHARIDES

More than 30 years ago, Westphal and Lüderitz postulated that the endotoxic principle, i.e. that lipopolysaccharide structure which is responsible for the induction of fever and shock, is located in the lipid A component (Westphal and Lüderitz, 1954). During the last 15 years an overwhelming body of data confirming this idea has been accumulated, mainly by the work of Galanos and colleagues (Galanos et al., 1977). Thus, it was shown that serologically different wild type and also rough mutant lipopolysaccharides express comparable pyrogenic and lethal toxicity proving that the O-specific chain is not involved in endotoxicity. Also, lipopolysaccharides of core deficient mutants (see below) were biologically highly active, indicating that the core oligosaccharide does not directly participate in endotoxicity. Further, it was demonstrated that polysaccharide-deprived (water-soluble) free lipid A being prepared from enterobacterial lipopolysaccharide by mild acid-catalyzed hydrolysis followed by electrodialysis and neutralization was of comparable activity as lipopolysaccharide in typical endotoxin test systems. Final proof could be provided after Shiba, Kusumoto and their colleagues at the Osaka University (Japan) had succeeded in the total chemical synthesis of *E. coli* lipid A (Imoto et al., 1985). The biological analysis of synthetic *E. coli* type lipid A (Fig. 4; preparation 506 or LA-15-PP) performed at the Max-Planck-Institut für Immunbiologie in Freiburg (C. Galanos, M. Freudenberg, O. Lüderitz) and at the Forschungsinstitut Borstel by our group showed that synthetic lipid A exhibited qualitatively and quantitatively identical activity as bacterial free *E. coli* lipid A (Galanos et al., 1985). This was found for lethal toxicity, pyrogenicity, the local Shwartzman phenomenon, leukopenia, macrophage activation (PGE_2-release), B-lymphocyte mitogenicity and induction of endotoxin cross-tolerance, some of the results being shown in Table 1. In addition, synthetic lipid A exhibited identical immunogenic and antigenic properties as bacterial lipid A (Brade et al., 1986), and both preparations were equally effective in inducing production and release of interleukin 1 from human mononuclear cells (Loppnow et al., 1986). These findings which were confirmed and extended by other laboratories (Kotani et al., 1985; Homma et al., 1985; Kanegasaki et al., 1986) provided final proof for the postulate of Westphal and Lüderitz that lipid A represents the endotoxic principle of lipopolysaccharides and they further showed that the structure derived from analytical work is correct.

STRUCTURE-ACTIVITY-RELATIONSHIPS OF LIPOPOLYSACCHARIDES

Endotoxicity

Lipid A represents a structurally complex molecule which harbours a hydrophilic (bisphosphorylated D-glucosamine disaccharide) and a lipophilic region (fatty acids). Its amphipathic nature and the presence of charged groups certainly plays an important role for the physicochemical behaviour and also the biological activity of lipid A. The recent availablity of synthetic lipid A partial structures and analogues allowed first studies concerning the minimal structural requirements of endotoxicity. The synthetic lipid A related structures used include glucosamine monosaccharide structures and ß(1-6)-linked D-glucosamine disaccharide compounds with a different phosphorylation and acylation pattern.

In our studies we have concentrated on the role of fatty acids for lipid A activity. Previous findings had indicated that fatty acids are essential for the expression of endotoxicity. It was not known, however, whether the number of acyl groups present, their nature, their type of linkage and their distribution, or all these factors together were of importance. In order to answer these questions, synthetic preparations were biologically analyzed which belong to a series of synthetic compounds sharing a bisphosphorylated ß(1-6)-linked D-glucosamine disaccharide and carrying four moles of (R)-14:0(3-OH) residues at positions 2, 3, 2' and 3', but differing in the number and type of acyl groups bound to 3-hydroxy fatty acids (Fig. 4). These molecules include the tetraacyl precursor Ia preparation (406 or LA-14-PP), the pentaacyl precursor Ib (LA-20-PP), hexaacyl *E. coli* lipid A (506 or LA-15-PP), and the recently described heptaacyl species of *S. minnesota* lipid A (516 or LA-16-PP) (Galanos et al., 1984, 1985, 1986; Hansen-Hagge et al., 1985).

The results of some biological analyses of these compounds are summarized in Table 1. Both bacterial and synthetic precursor Ia lacked Shwartzman reactivity and were of comparatively low pyrogenicity (Table 1). Precursor Ia differs from enterobacterial lipid A's in that it is devoid of the nonhydroxylated fatty acids do-, tetra- and hexadecanoic acid and, thus, of 3-acyloxyacyl residues (Fig. 4).

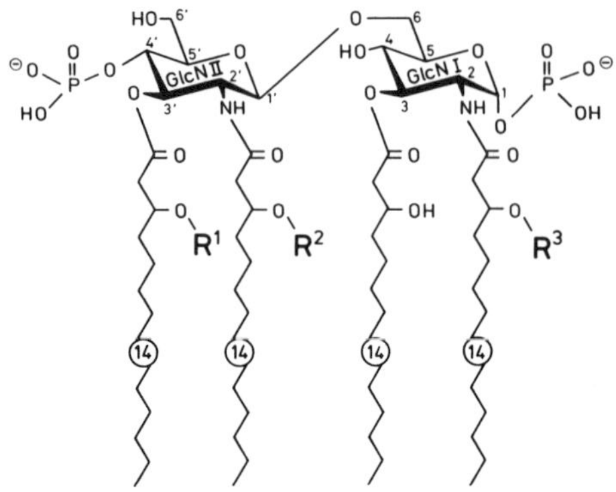

Designation of Preparation		Nature of		
Bacterial	Synthetic	R^1	R^2	R^3
Precursor Ia	LA-14-PP (406)	H	H	H
Precursor Ib	LA-20-PP	H	H	16:0
Isomer of Precursor Ib	LA-21-PP	H	16:0	H
E. coli Lipid A	LA-15-PP (506)	14:0	12:0	H
S. minnesota heptaacyl Lipid A	LA-16-PP (516)	14:0	12:0	16:0

12:0, dodecanoic ; 14:0, tetradecanoic ;
16:0, hexadecanoic acid

Figure 4. Chemical structure of synthetic enterobacterial lipid A's and partial structures (Imoto et al., 1984; Imoto et al., 1985; Galanos et al., 1986; Rietschel et al., 1987).

On the other hand, precursor Ib which structurally differs from precursor Ia by containing one mole of hexadecanoic acid bound to one of the amide-bound 3-hydroxytetradecanoyl groups is significantly more pyrogenic than precursor Ia and active in preparing for or in provoking the local Shwartzman reaction (Table 1). The only common structural feature of precursor Ib and E. coli lipid A [apart from the tetra-(3-hydroxy)acyl backbone] is the presence of a 3-acyloxyacyl residue. Considering that precursor Ia is of low endotoxic potency, while lacking this type of structure, it appears that 3-acyloxyacyl groups are essential for the expression of dermal Shwartzman reactivity and endotoxicity in general, and that it is irrelevant whether the (amide-bound) 3-acyloxyacyl group is linked to GlcN I or GlcN II of the lipid A backbone. Synthetic heptaacyl lipid A expressed, as compared to E. coli lipid A, very low pyrogenicity and Shwartzman reactivity (Table 1).

TABLE 1. Biological activity of synthetic precursor Ia, precursor Ib, isomer of precursor Ib, E. coli lipid A and S. minnesota heptaacyl lipid A

Endotoxin Test System	Dose required to induce biological effect with				
	Synthetic tetraacyl Precursor Ia P-[2]-[2]-P	Synthetic isomer of pentaacyl Precursor Ib P-[3]-[2]-P	Synthetic pentaacyl Precursor Ib P-[2]-[3]-P	Synthetic hexaacyl E.coli lipid A P-[4]-[2]-P	Synthetic heptaacyl S.minnes. lipid A P-[4]-[3]-P
	μg				
Pyrogenicity (MPD-3)	0.2	0.05	0.01	0.004	0.4
Local Shwartzman Reaction (Preparation)	(50)[a]	25	12.5	3.1	(50)[a]

a, no effect with the dose indicated

This molecule structurally corresponds to E. coli lipid A but contains, in addition, hexadecanoic acid linked to the amide-bound 14:0(3-OH) residue of GlcN I (Fig. 4). Therefore, heptaacyl lipid A contains three 3-acyloxyacyl residues, E. coli lipid A containing two, precursor Ib containing one mole of this structure, and precursor Ia being devoid of it. The low in vivo activity of synthetic S. minnesota heptaacyl lipid A shows that the endotoxic potency of a preparation does not simply increase with the number of 3-acyloxyacyl residues present. Rather, the total number of acyl chains (in a defined distribution) appears to be of importance for the expression of endotoxicity. According to present knowledge, the full spectrum of endotoxic effects is maximally expressed by a molecule containing six acyl residues including at least one 3-acyloxyacyl group (E. coli lipid A). Molecules with five fatty acids including one 3-acyloxyacyl group (precursor Ib) are endotoxically active in all parameters but exhibit, in general, a somewhat lower endotoxic potency (factor of 5 to 10). Beyond these merely quantitative differences more dramatic differences, are noted in the biological activity of compounds containing less than five or more than six acyl groups.

In other experiments the role of the glucosamine disaccharide and the phosphoryl groups was elucidated. Employing lipid A partial structures, it was generally found that acylated and phosphorylated glucosamine monosaccharides exhibit *in vitro* activity (such as B-lymphocyte mitogenicity and macrophage activation) but that they are incapable of inducing typical *in vivo* endotoxin effects such as fever and the local Shwartzman reaction. It was also demonstrated that the monodephospho partial structures of the D-glucosamine-disaccharide preparations (Fig. 4 and 5) are less active than the bisphosphorylated parent compound as pyrogens, lethal toxins and inducers of the local Shwartzman phenomenon. Many of the monodephospho partial structures were also less active in *in vitro* tests.

The combined data indicate that full endotoxicity is expressed by a molecule containing two ß(1-6)-interlinked D-glucosamine residues, two phosphoryl groups and minimally five or maximally six fatty acids including at least one 3-acyloxyacyl group in a defined distribution. Molecules lacking only one component, irrespective of its chemical nature, or molecules with a different location of components are endotoxically less or not active. This shows that slight modifications of the lipid A architecture result in a significant change of biological activity, suggesting that endotoxicity is not dependent on one single lipid A constituent, i.e. on a toxophore group, but rather that it is determined by several factors. Thus, a defined balance of hydrophilic and hydrophobic regions or charged and neutral groups could generate a unique molecular (or supramolecular) conformation of lipid A which allows the expression of endotoxic activity (Rietschel et al., 1987b). Since the conformational state of lipid A appears also to be of great importance for its immunoreactive properties (see below), our laboratory persues at present the analysis of the three-dimensional organization of lipid A aiming at the characterization of the nature and properties of an "endotoxic conformation".

Immunogenicity and Antigenicity

Besides their endotoxic activities, the immunogenic and antigenic properties of lipopolysaccharide are well known since the early studies on structure-activity relationships of endotoxins. Actually, immunogenicity and antigenicity were the first biological parameters to be analyzed on a molecular

level (Lüderitz et al., 1966a).
As the major surface antigens (O-antigens) of gram-negative bacteria, lipopolysaccharides serve as diagnostic markers in clinical bacteriology. At the same time they represent structures being involved in the very early events of infection. Moreover, as discussed above, lipopolysaccharides are the endotoxins of gram-negatives. Thus, it is expected that investigations on the immunogenicity and antigenicity of endotoxins i) will contribute to improved diagnosis of gram-negative infections, ii) should help to reveal and understand mechanisms of prevention of infection by gram-negative bacteria and iii) may lead to the recognition and characterization of antibodies (polyclonal or monoclonal) which are able to neutralize the deleterious effects of endotoxin *in vivo* and which could be used as therapeutic agents in clinical medicine. In the following, the particular immunoreactive properties of the O-specific side chain, the core oligosaccharide and the lipid A component will be summarized and discussed.

O-specific chain

As discussed above the O-specific chain is a heteropolysaccharide consisting of repeating oligosaccharide units which harbours the O-antigenic determinants. Distinct carbohydrate residues within a given repeating unit may exhibit

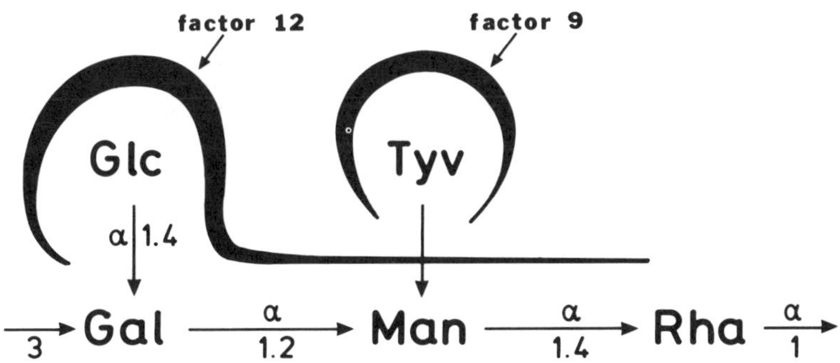

Figure 5. Chemical and antigenic structure of the repeating unit in the lipopolysaccharide of *Salmonella typhi* indicating the location of antigenic determinants (O-factors). (Modified according to Lüderitz et al., 1966a, 1966b)

distinct antigenic properties and give rise to the production of specific antibodies during experimental or natural infection. These antigenic determinants are named O-factors. In a single repeating unit, several such O-factors can be present which together determine the serotype of a given bacterial species. There exist as many different structures of repeating units as different serotypes. Thus, the structural diversity of O-side chains is paralleled by the antigenic diversity of this region. An example of the serology of the genus Salmonella is explained in Fig. 5 showing the chemical structure of the repeating unit of S. typhi. It consists of a pentasaccharide in which a linear trisaccharide of D-galactose, D-mannose, and L-rhamnose is substituted with D-glucose and D-tyvelose (3,6-dideoxy-D-mannose) as branches. Tyvelose determines the specificity of factor 9 and is characteristic for Salmonella serogroup D_1. The α(1-4)-linked terminal glucose residue, together with the backbone trisaccharide, determines the specificity of factor 12. Therefore, the complete serological formula for S. typhi is Salmonella group D_1 [9, 12]. The glucose residue contributes most significantly to the binding of anti-factor 12 antibodies (determined by inhibition assays with partial structures). To indicate this particular importance of a single sugar for the antigenicity of a complex oligosaccharide, the terms immunodominant sugar and immunodominance have been introduced (Lüderitz et al., 1966a).

This sophisticated serology resulted in a serotyping scheme which, in the case of the genus Salmonella is well known to clinical and general microbiologists as the Kauffmann-White scheme. The principles for typing bacteria based on their O-specific chains (either chemically or serologically) also holds true for other genera of the Enterobactereacae and even bacteria outside this family. It is noteworthy that these studies were performed with polyclonal antisera (and not with monoclonal antibodies) and at a time when the determination of carbohydrate structures posed enormous experimental problems.

In recent years, many of these results have been confirmed by the methodology available today. Several repeating units have been chemically synthesized, and it has been shown that synthetic antigens provide a useful basis for the production of artificial vaccines (e.g. Lindberg et al., 1983). However, the high degree of structural and antigenic diversity of the O-antigens has so far limited their usefulness in biomedical applications. This is regrettable since antibodies directed against the O-side chain are opsonic, bactericidal, and protective against infection and endotoxicity in the homologous serotyp-specific system. Immunization with live vaccines has been successfully applied to prevent typhoid and paratyphoid fever and is still useful for people travelling to countries with low standards in hygiene. It has also been shown that calves can be protected from Salmonellosis by vaccination with pure haptenic oligosaccharides (representative for a certain serotyp) coupled to immunogenic protein carriers (Lindberg et al., 1983). The high efficiency of anti-O antibodies in infection is impressively documented by results reported by Mäkelä and co-workers. These investigators presented data showing that less than 0.01 μl of a rabbit antiserum against _S. typhimurium_ (with an ELISA titer of 650 in the IgG and 400 in the IgM fraction) protected mice against infection initiated with a 10 LD_{50} challenge dose (Saxen et al., 1986).

Core oligosaccharide

The immunogenic and antigenic properties of the core region could be studied after the detection of rough (R)-mutants (Lüderitz et al., 1966a). These mutants possessing a defect in the biosynthesis of the core oligosaccharide and which are also called core-defective mutants are labelled Ra to Re depending on the depth of the biosynthetic defect (Fig. 6). The most defective mutant known is the Re chemotype the lipopolysaccharide of which contains only a disaccharide of KDO (in addition to lipid A). Rough mutant bacteria lack the O-chain and part of the core oligosaccharide, and consequently, they have lost the serological O-specificity. However, on experimental immunization the core oligosaccharide also gives rise to the production of antibodies which are specific for each of the different core-chemotypes. Therefore, the chemotype classification is paralleled by a serotype scheme which uses the same designations Ra to Re. The corresponding antigenic specificities are termed core- or R-specificities. Interestingly, anti-R antibodies are not observed (or only in very low titers) after immunization with wild-type (S-form) bacteria. Thus, the R-specificities appear to be shielded in S-form lipopolysaccharide. Likewise, R-antibodies obtained after immunization with rough mutant bacteria do neither exhibit significant cross reaction within the different chemotypes nor with wild-type S-form lipopolysaccharide. It has been proposed that for a given rough serotype the terminal sugar of the core oligosaccharide constitutes the immunodominant sugar.

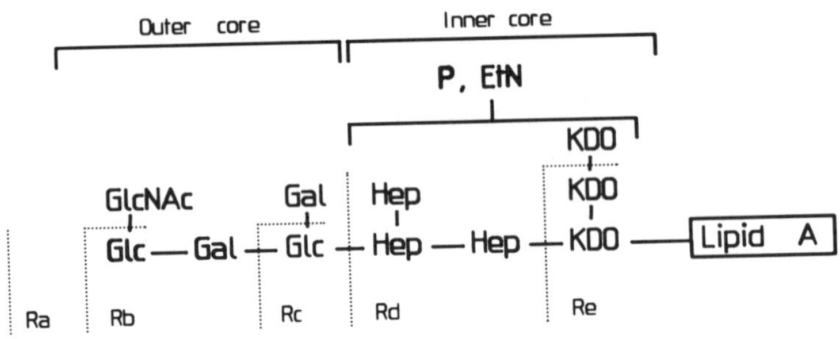

Figure 6. Schematic structure of the core oligosaccharide in the lipopolysaccharide of Salmonella minnesota (Ra-core). Dotted lines indicate partial structures as they are found in core-defective mutants labeled Ra to Re. GlcN = N-acetyl-D-glucosamine, Gal = D-galactose, Glc = D-glucose, Hep = L-glycero-D-mannoheptose, KDO = 3-deoxy-D-mannooctulosonic acid, P = phosphate, EtN = ethanolamine.

The fact that the core is in many different gram-negative bacteria similarly structured has nevertheless initiated studies concerning the potential cross-protective effects of antibodies against this region. Actually, two research groups have described independently the cross-protective capacity of antisera obtained after immunization with either the Re mutant strain R595 of S. minnesota (McCabe et al., 1977) or the UDP-galactose epimerase-less mutant strain J5 of E. coli O111 (Braude et al., 1977). These observations have stimulated many investigators to study this phenomenon in more detail, and in the meantime an overwhelming body of data has been reported in the literature. However, the results obtained could not be reproduced in some laboratories and a clearcut decision as to the usefulness of such antisera in clinical medicine is not yet possible (see eg. Greisman et al., 1978; Ng et al., 1976). A major difficulty in assessing this approach relates to our poor knowledge on the structural organization of the core region of clinically relevant pathogens and to the fact that the epitope specificities of the above antibodies has never been clearly defined.

With the recent advances made in the understanding of the structural organization of the inner core region, we have started to charcterize polyclonal antisera and monoclonal antibodies against common core structures, with emphasis on

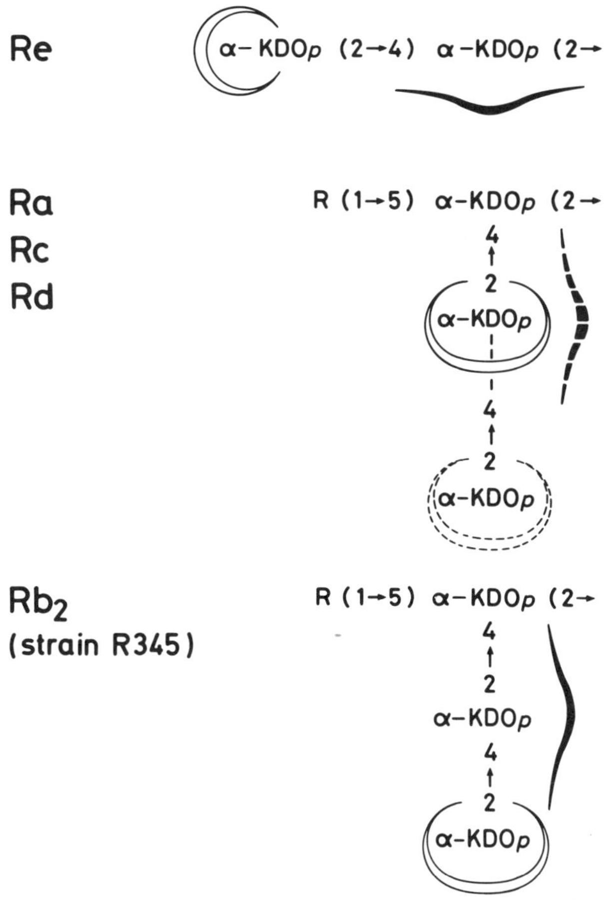

Figure 7. Schematic representation of epitope specificities of monoclonal antibodies recognizing terminal α-linked KDO (open symbol) and α(2-4)-linked KDO disaccharide (closed symbol). Broken symbols indicate the possible reaction pattern of these antibodies depending on the degree of substitution with KDO III (see Fig. 2). The lower panel displays the reactivity with lipopolysaccharide of Salmonella minnesota chemotype Rb_2 (strain R345) in which the substitution is quantitative (Brade et al., 1987a).

the KDO region. Murine monoclonal antibodies (prepared with Re mutant bacteria by B.J. Appelmelk, Amsterdam, The Netherlands) were analyzed for their epitope specificity. Using chemically synthesized antigens (copolymerization products of acrylamide with allylglycosides of different KDO derivatives, prepared by P. Kosma, Vienna, Austria), the epitope specificity of two monoclonal antibodies could be determined (Brade et al., 1987a). One antibody recognizes, in enterobacterial lipopolysaccharide, a terminal α-pyranosidic KDO residue whereas the second reacts with an α(2-4)-linked KDO disaccharide. The cross-reaction of these two antibodies with natural lipopolysaccharide is explained in Fig. 7 and can be correlated with the structural heterogeneity of the KDO region illustrated in Fig. 2. The monosaccharide antibody reacts with all lipopolysaccharides since they all share a terminal KDO residue whereas the disaccharide antibody reacts only with lipopolysaccharides expressing quantitatively a KDO disaccharide such as those of S. minnesota Re (strain R595) and Rb_2 (strain R345). Present investigations are performed (in cooperation with B. J. Appelmelk) to evaluate the protective capacity of these monoclonal antibodies with well-defined epitope specificity.

Lipid A

As the O-chain and the core region, also the lipid A component is an antigen and immunogen. Lipid A immunoreactivity however, exhibits some particular features which explain that lipid A immunogenicity and antigenicity were only discovered in the early seventies (Galanos et al., 1971). Galanos et al. found that this antigen is not exposed in lipopolysaccharides of either S- or R-form bacteria unless the poly- or oligosaccharide is removed. The same held true for lipid A immunogenicity, and only acid-treated bacteria (exposing free lipid A on their surface) or free lipid A complexed to suitable carriers were found to be good immunogens for the preparation of anti-lipid A antibodies. Such antisera could be raised with lipid A preparations derived from various gram-negative bacteria provided they contained the general structure described above. Moreover, these antibodies cross-reacted with all lipid A's. Therefore, it appeared that anti-lipid A antisera consisted of only one antibody specificity.

The rapidly developing field of research on lipid A antibodies is of great biomedical interest since lipid A is a common structure present in all endotoxically active lipopolysaccharides which, at the same time, harbour the endotoxic principle. Thus, lipid A antibodies have been shown to be cross-reactive and lipid A antisera to prevent some biological effects of lipid A or endotoxins (see e.g. Rietschel and Galanos, 1977). Second, the reaction of antibodies with lipid A can be used to study the interaction of proteins with lipid A which we regard as a model of how a putative humoral or cellular receptor may recognize lipid A

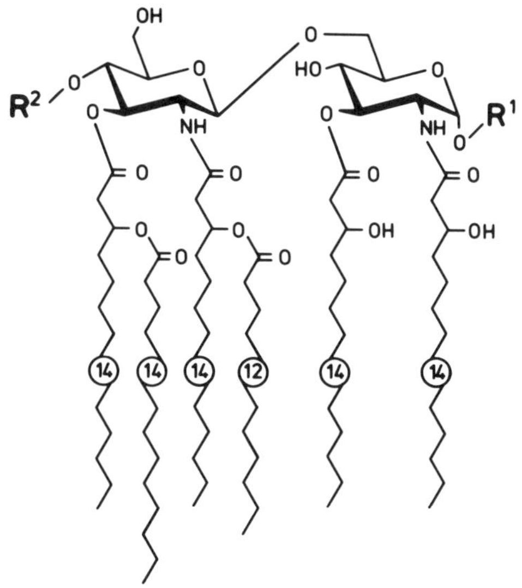

Preparation	Nature of substituents	
	R^1	R^2
504	H	$PO(OH)_2$
505	$PO(OH)_2$	H
506	$PO(OH)_2$	$PO(OH)_2$

Figure 8. Chemical structure of synthetic Escherichia coli-type lipid A (preparation 506) and its 1- and 4'-monodephospho derivatives (preparation 505 and 504, respectively) (Imoto et al., 1985; Kusumoto et al., 1985). The numbers in circles indicate the number of carbon atoms in acyl chains.

or endotoxin. Since in this model only two chemically defined molecular species are involved, the system appears to be ideally suited to investigate structure-function or structure-activity relationships of lipid A in molecular terms. In first studies we have characterized three different antibody specificities which were obtained after immunization of rabbits with synthetic E. coli lipid A (compound 506, Imoto et al., 1985) and its 1- and 4'- dephospho partial structures (compounds 504 and 505, respectively, Kusumoto et al. 1985) being incorporated into liposomes (Fig. 8). In the passive hemolysis assay, using sheep erythrocytes (SRBC) coated with the respective synthetic antigen and guinea pig complement, the following results were obtained (Brade et al., 1986; Brade et al., 1987b). Anti-504 and anti-505 antisera reacted with the homologous antigen and cross-reacted with the bisphosphorylated compound 506, whereas anti-506 antiserum was specific for the homologous antigen and did not cross-react with the monodephospho structures. Therefore, anti-504 and anti-505 antibodies recognize an

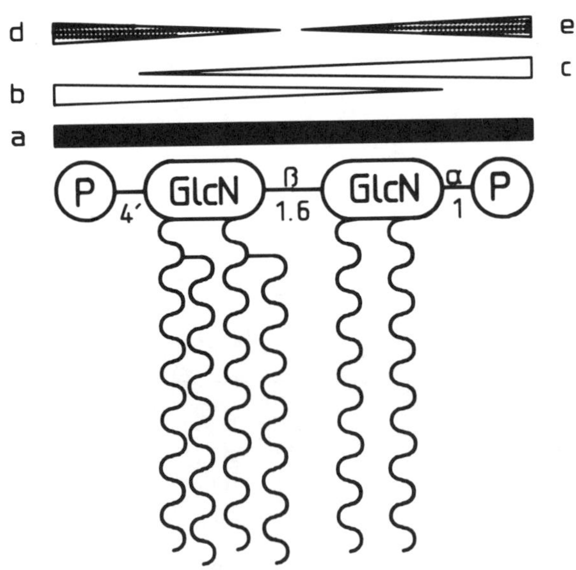

Figure 9. Schematic representation of different antibody specificities recognizing lipid A and partial structures thereof (Brade et al., 1987b).

epitope which requires the presence of the 4'- or the 1-phosphoryl group, respectively, as it is present in the corresponding homologous antigens and in the bisphosphorylated compound 506. On the other hand, anti-506 antibodies only react with the structure containing both phosphoryl residues, and thus, exhibits the most restricted specificity. Recently, in addition to disaccharide-specific antibodies, antibodies reacting with acylated and phosphorylated monosaccharide partial structures were detected (Brade et al., 1987b). Again, the position of the phosphate group determined their specificity. Therefore, a total of five different antibody specificities against lipid A are now known so far, their reactivity pattern being schematically displayed in Fig. 9.

From these results the following general conclusion can be drawn. The lipid A antibodies described recognize epitopes residing in the hydrophilic backbone of lipid A. These antibodies are reactive with monosaccharide (specificities d and e) or disaccharide antigens (specificities a, b and c, Fig. 9) carrying one or two phosphate groups. Antibodies against a defined partial structure cross-react with structurally more complex antigens containing this partial structure (Fig. 9). Thus, compound 506 has the broadest reactivity since all five antibodies are detected with this antigen. The main fraction of natural free lipid A prepared from E. coli Re lipopolysaccharide by mild acid-catalyzed hydrolysis corresponds to compound 506. Upon immunization with bacterial lipid A, high levels of antibodies against compound 506 (specificity a, Fig. 9) are obtained but also antibodies against other lipid A determinants can be detected in such antisera (Brade and Brade, 1985a).
The diversity of the immune response to lipid A could help to explain some of the not well understood results with regard to the protective effects of anti-lipid A antibodies. Today, with the availability of defined, i.e. synthetic lipid A antigens the different antibody specificities described above can be differentiated from each other and their potential biological potency can be studied in the future.

CONCLUDING REMARKS

During the last decades an overwhelming body of data has been elaborated and published concerning the chemical, physicochemical, serological and biological properties of lipopolysaccharides as well as their significance as bacterial cell wall components, surface antigens, phage receptors, virulence factors, and immunostimulators. By this work some of the basic questions in endotoxin research (e.g. general architecture, genetics and biosynthesis, structure and synthesis of the endotoxic principle, involvement of macrophages and endogenous mediators in endotoxic manifestation) have been answered. Many other problems, however, remain to be solved. Our work in the field of endotoxin research focusses on the structural elucidation of conserved lipopolysaccharide domains such as the lipid A component and the inner core oligosaccharide. Knowledge on the fine structure of these regions is essential for the identification of ubiquitous antigens and, thus, for the synthesis of nontoxic vaccines with broad serological reactivity and the preparation of cross-protecting polyclonal antisera or monoclonal antibodies. In view of the high mortality associated with gram-negative sepsis such immunoprophylactic and immunotherapeutic strategies appear highly indicated. The chemical characterization of conserved lipopolysaccharide domains is also a prerequisite for the elucidation of the molecular mechanisms underlying endotoxic activity in vivo. It is known today that endotoxin effects represent mediated phenomena. They are the result of the interaction of endotoxin with, notably, mononuclear host cells, the secretion by these cells of endogenous mediators such as tumor necrosis factor, interleukin 1, metabolites of arachidonic acid (including prostaglandins and leukotrienes), colony stimulating factor, interferon, and the subsequent action of these mediators on different organs and regulatory systems of the host (for literature see Proctor, 1984-86). Clinically, the characterization of these mediators is of utmost importance, since knowledge as to their nature and action opens the possibility to suppress endotoxin-induced toxic phenomena by means of antisera against mediators or drugs interfering with the synthesis or action of mediators. Promising results in this direction have recently been published (Beutler et al., 1985; Hagmann et al., 1984).

The chemical synthesis of new partial structures of lipid A and the inner core region obviously also presupposes knowledge on the fine architecture of the corresponding natural structures. A great number of partial structures and analogues are presently being synthesized by various laboratories in Europe and Japan. It is to be expected that such homogenous synthetic molecules will significantly contribute to the elucidation of the basic mechanisms of endotoxin action in vivo. By using radio-labelled synthetic preparations, the in vivo fate and biochemical dynamics of endotoxin can now be followed with higher sensitivity and precision. Such studies may reveal deeper insight into the primary humoral and cellular targets of endotoxin, its degradation (detoxification) and ways of its elimination from the host organism (Freudenberg et al., 1985; Peterson and Munford, 1987). Synthetic partial structures will also be indispensible for the establishment of quantitative structure-activity relationships of endotoxins. Further, previous studies suggest that certain lipid A analogues, derivatives and partial structures which are less toxic than native endotoxin still express biological effects considered beneficial for the host. Such effects include the induction of nonspecific resistance to irradiation and microbial infections, adjuvant activity and tumor necrosis. So far, its toxic properties have hampered the clinical application of endotoxin. Synthetic lipid A-related preparations which are largely devoid of toxicity but still exhibit wanted activities may, in the future, find their prophylactic or therapeutic application in clinical medicine.

Finally, the physiological significance of lipopolysaccharides for the bacterial cell merits consideration. So far, viable gram-negative bacteria which lack lipopolysaccharide have not been isolated pointing to a vital role of lipopolysaccharide for bacterial growth and division. Since bacterial Re mutants (which express a lipopolysaccharide consisting only of lipid A and KDO) do multiply, it appears that the O-specific chain, the outer core and the heptose region are dispensible for growth and division. Thus, the innermost (KDO-containing) region of the core and lipid A are essential for the structural and functional integrity of the bacterial cell. Presently, however, it is not known which structural and physical properties endow the KDO-lipid A domain with its vital significance (Labischinski et al., 1985). In view of the physiological importance of KDO, attempts have been made to develop structural analogues interfering with the biosynthesis of the lipid A-proximal inner core (Hammond et al., 1987). Such studies are performed with the hope that KDO analogues are active as antibiotics and that they can be used against pathogenic gram-negative bacteria in general. Lipopolysaccharides represent today, as they did in the past, a most rewarding target for interdisciplinary research notably in the field of infection biology. It is to be expected that lipopolysaccharides will continue to fascinate scientists of various research areas including chemistry, physics, microbiology, genetics, cell and molecular biology, immunology, and clinical medicine. The present worldwide interest in lipopolysaccharides and the joint efforts of different disciplines justify the hope that many of the still open problems in the field of endotoxin research will be solved in the near future.

Acknowledgments

The financial support of the Deutsche Forschungsgemeinschaft (Br 731-4/1, Scha 402-1/1) and the Fonds der Chemischen Industrie (EThR) is greatly acknowledged. We also thank Mrs. M. Lohs, G. Stegelmann, and B. Köhler for illustrations and photographic work and Mrs. I. Bendt for typing this manuscript.

REFERENCES

Beutler B, Milsark IW, Cerami AC (1985). Passive immunization against cachectin/tumor necrosis factor protects mice from lethal effect of endotoxin. Science 229:869-871.

Brade H, Galanos C, Lüderitz O (1983). Isolation of 3-deoxy-D-mannooctulosonic acid disaccharide from Salmonella minnesota rough-form lipopolysaccharides. Eur J Biochem 131:201-203.

Brade H, Rietschel ETh (1984). α-2.4-Interlinked 2-keto-3-deoxy-D-mannooctulosonic acid disaccharide: a common constituent of enterobacterial lipopolysaccharides. Eur J Biochem 145:231-236.

Brade H, Zähringer U, Rietschel ETh, Christian R, Schulz G, Unger FM (1984). Spectroscopic analysis of a 3-deoxy-D-manno-2-octulosonic acid (KDO) disaccharide from the lipopolysaccharide of a Salmonella godesberg Re-mutant. Carbohydr Res 134:157-166.

Brade H (1985). Occurrence of 2-keto-3-deoxyoctonic acid 5-phosphate in lipopolysaccharides of Vibrio cholerae Ogawa and Inaba. J Bacteriol 161:795-798.

Brade L, Brade H (1985a). Characterization of two different antibody specificities recognizing distinct antigenic determinants in free lipid A of Escherichia coli. Infect Immun 48:776-781.

Brade H, Rietschel ETh (1985). Identification of 2-keto-3-deoxy-1,7-dicarboxyheptonic acid as a constituent of the

lipopolysaccharide of Acinetobacter calcoaceticus NCTC 10305. Eur J Biochem 153:249-254.
Brade L, Rietschel ETh, Kusumoto S, Shiba T, Brade H (1986). Immunogenicity and antigenicity of synthetic Escherichia coli lipid A. Infect Immun 51:110-114.
Brade L, Kosma P, Appelmelk BJ, Paulsen H, Brade H (1987a). Use of synthetic antigens to determine the epitope specificity of monoclonal antibodies against the 3-deoxy-D-manno-2-octulosonate region of bacterial lipopolysaccharide. Infect Immun 55:462-466.
Brade L, Rietschel ETh, Kusumoto S, Shiba T, Brade H (1987b). Immunogenicity and antigenicity of natural and synthetic Escherichia coli lipid A. In Watson SW, Levin J, Novitsky T (ed.), Detection of bacterial endotoxins with the Limulus amebocyte lysate test. Prog Clin Biol Res 231:75-97. Alan R. Liss Inc., New York.
Braude AI, Ziegler EJ, Douglas H, McCutchan JA (1977). Antibody to cell wall glycolipid of gram-negative bacteria: induction of immunity to bacteremia and endotoxemia. J Infect Dis 136S:167-172.
Chaby R, Szabo L (1975). 3-Deoxy-2-octulosonic acid 5-phosphate: a component of the endotoxin of Bordetella pertussis. Eur J Biochem 59:277-280.
Freudenberg MA, Galanos C (1985). Alterations in rats in vivo of the chemical structure of lipopolysaccharides from Salmonella abortus equi. Eur J Biochem 152:353-359.
Galanos C, Lüderitz O, Westphal O (1971). Preparation and properties of antisera against the lipid-A component of bacterial lipopolysaccharides. Eur J Biochem 24:116-122.
Galanos C, Lüderitz O, Rietschel ETh, Westphal O (1977). Newer aspects of the chemistry and biology of bacterial lipopolysaccharides, with special reference to their lipid A component. In: Goodwin TW (ed.): International review of biochemistry, Biochemistry of lipids II. University Park Press, Baltimore, Vol. 14, pp. 239-335.
Galanos C, Lehmann V, Lüderitz O, Rietschel ETh, Westphal O, Brade H, Brade L, Freudenberg MA, Hansen-Hagge T, Lüderitz Th, McKenzie G, Schade U, Strittmatter W, Tanamoto K, Zähringer U, Imoto M, Yoshimura H, Yamamoto M, Shimamoto T, Kusumoto S, Shiba T (1984). Endotoxic properties of chemically synthesized lipid A part structures. Comparison of synthetic lipid A precursor and synthetic analogues with biosynthetic lipid A precursor and free lipid A. Eur J Biochem 140:221-227.
Galanos C, Lüderitz O, Rietschel ETh, Westphal O, Brade H, Brade L, Freudenberg MA, Schade U, Imoto M, Yoshimura H,

Kusumoto S, Shiba T (1985). Synthetic and natural Escherichia coli free lipid A express identical endotoxic activities. Eur J Biochem 148:1-5.

Galanos C, Lüderitz O, Freudenberg M, Brade L, Schade U, Rietschel ETh, Kusumoto S, Shiba T (1986). Biological activity of synthetic heptaacyl lipid A representing a component of Salmonella minnesota R595 lipid A. Eur J Biochem 160:55-59.

Greisman SE, DuBuy JB, Woodward CL (1978). Experimental gram-negative bacterial sepsis: reevaluation of the ability of rough mutant antisera to protect mice. Proc Soc Exp Biol Med 158:482-490.

Hagmann W, Denzlinger C, Keppler D (1984). Role of peptide leukotrienes and their hepatobiliary elimination in endotoxin action. Circ Shock 14:223-235.

Hammond SM, Claesson A, Jansson AM, Larsson L-G, Pring BG, Town CM, Ekström B (1987). A new class of synthetic antibacterials acting on lipopolysaccharide biosynthesis. Nature 327:730-732.

Hansen-Hagge Th, Lehmann V, Seydel U, Lindner B, Zähringer U (1985). Isolation and structural analysis of two lipid A precursors from a KDO deficient mutant of Salmonella typhimurium differing in their hexadecanoic acid content. Arch Microbiol 141:353-358.

Homma JY, Matsuura M, Kanegasaki S, Kawakubo Y, Kojima Y, Shibukawa N, Kumazawa Y, Yamamoto A, Tanamoto K, Yasuda T, Imoto M, Yoshimura H, Kusumoto S, Shiba T (1985). Structural requirements of lipid A responsible for the functions: a study with chemically synthesized lipid A and its analogues. J Biochem (Tokyo) 98:395-406.

Imoto M, Yoshimura H, Yamamoto M, Shimamoto T, Kusumoto S, Shiba T (1984). Chemical synthesis of phosphorylated tetraacyl disaccharide corresponding to a biosynthetic precursor of lipid A. Tetrahedron Lett 25:2667-2670.

Imoto M, Yoshimura H, Sakaguchi N, Kusumoto S, Shiba T (1985). Total synthesis of Escherichia coli lipid A. Tetrahedron Lett 26:1545-1548.

Jann K, Jann B (1984). Structure and biosynthesis of O-antigens. In: Proctor RA (ed.): Handbook of Endotoxin, Vol. I. Rietschel ETh (ed.): Chemistry of Endotoxin. Elsevier, Amsterdam, New York, Oxford, pp. 138-186.

Jansson P-E, Lindberg AA, Lindberg B, Wollin R (1981). Structural studies on the hexose region of the core in lipopolysaccharide from Enterobactereaceae. Eur J Biochem 115:571-577.

Kanegasaki S, Tanamoto K, Yasuda T, Homma JY, Matsuura M,

Nakatsuka M, Kumazawa Y, Yamamoto A, Shiba T, Kusumoto S, Imoto M, Yoshimura A, Shimamoto T (1986). Structure-activity relationship of lipid A: Comparison of biological activities of natural and synthetic lipid A's with different fatty acid compositions. J Biochem (Tokyo) 99:1203-1210.

Kawahara K, Brade H, Rietschel ETh, Zähringer U (1987). Studies on the chemical structure of the core-lipid A region of the lipopolysaccharide of Acinetobacter calcoaceticus NCTC 10305. Detection of a new 2-octulosonic acid interlinking the core oligosaccharide and lipid A component. Eur J Biochem 163:489-495.

Kotani S, Takada H, Tsujimoto M, Ogawa T, Takahashi I, Ikeda T, Otsuka K, Shimanchi H, Kasai N, Mashimo J, Nagao S, Tanaka S, Harada K, Nagaki K, Kitamura H, Shiba T, Kusumoto S, Imoto M, Yoshimura H (1985). Synthetic lipid A with endotoxic and related biological activities comparable to those of a natural lipid A from an Escherichia coli Re-mutant. Infect Immun 49:225-237.

Labischinski H, Barnickel G, Bradaczek H, Naumann D, Rietschel ETh, Giesbrecht P (1985). High state of order of isolated bacterial lipopolysaccharide and its possible contribution to the permeation barrier property of the outer membrane. J Bacteriol 162:9-20.

Lindberg AA, Wollin R, Bruse C, Ekwall E, Svensson SB (1983). Immunology and immunochemistry of synthetic and semi-synthetic Salmonella O-antigen-specific glyco-conjugates, p. 195-218. In Anderson L, Unger FM (ed), Bacterial lipopolysaccharides: structure, synthesis, biological activities. Am Chem Soc Symp Series 231. Washington D.C.

Loppnow H, Brade L, Brade H, Rietschel ETh, Kusumoto S, Shiba T, Flad H-D (1986). Lipid A-induced production of inter-leukin 1 in human monocytes. Eur J Immunol 16:1263-1267.

Lüderitz O, Staub AM, Westphal O (1966a). Immunochemistry of O and R antigens. Bacteriol Rev 30:192-255.

Lüderitz O, Galanos C, Risse HJ, Ruschmann E, Schlecht S, Schmidt G, Schulte-Holthausen H, Wheat R, Westphal O, Schlosshardt J (1966b). Structural relationships of Salmonella O and R antigens. Ann N Y Acad Sci 133:349-374.

Lüderitz O, Freudenberg MA, Galanos C, Lehmann V, Rietschel ETh, Shaw DH (1982). Lipopolysaccharides of gram-negative bacteria. In: Razin S, Rottem S (eds.): Membrane lipids of prokaryotes. Current topics in membranes and transport. Academic Press Inc., New York, pp. 79-151.

McCabe WR, Bruins SC, Craven DE, Johns M (1977). Cross-reactive antigens: their potential for immunization-induced immunity to gram-negative bacteria. J Infect Dis 136S:161-166.

Ng A-K, Chen C-LH, Chang C-M, Nowotny A (1976). Relationship of structure to function in bacterial endotoxins: serologically cross-reactive components and their effect on protection of mice against some gram-negative infections. J Gen Microbiol 94:107-116.

Nowotny A (1984). Heterogeneity of endotoxins. p. 308-338. In Proctor R (ed.), Handbook of endotoxins. Vol. 1, Rietschel ETh (ed.), Chemistry of endotoxin. Elsevier/North-Holland Biomedical Press, Amsterdam.

Peterson AA, Munford RS (1987). Dephosphorylation of the lipid A moiety of Escherichia coli lipopolysaccharide by mouse macrophages. Infect Immun 55:974-978.

Qureshi N, Mascagni P, Ribi E, Takayama K (1985). Monophosphoryl lipid A obtained from the lipopolysaccharides of Salmonella minnesota R595. Purification of the dimethyl derivative by high performance liquid chromatography and complete structural determination. J Biol Chem 260:5271-5278.

Rietschel ETh, Galanos C (1977). Lipid A antiserum-mediated protection against lipopolysaccharide- and lipid A-induced fever and skin necrosis. Infect Immun 15:34-49.

Rietschel ETh, Wollenweber HW, Brade H, Zähringer U, Lindner B, Seydel U, Bradaczek H, Barnickel G, Labischinski H, Giesbrecht P (1984). Structure and conformation of the lipid A component of lipopolysaccharides. In: Proctor RA (Ed.): Handbook of endotoxin, Vol. 1, Rietschel ETh (ed.): Chemistry of endotoxin. Elsevier, Amsterdam, New York, Oxford, pp. 187-220.

Rietschel ETh, Brade H, Brade L, Brandenburg K, Schade U, Seydel U, Zähringer U, Galanos C, Lüderitz O, Westphal O, Labischinski H, Kusumoto S, Shiba T (1987b). Lipid A, the endotoxic center of bacterial lipopolysaccharide: relation of chemical structure to biological activity. In: Watson SW, Levin J, Novitsky T (ed.): Detection of bacterial endotoxins with the Limulus Amebocyte Lysate. Progr Clin Biol Res Vol. 231, Alan R Liss Inc., New York, pp. 25-53.

Rietschel ETh, Brade L, Schade U, Galanos C, Freudenberg M, Lüderitz O, Kusumoto S, Shiba T (1987a). Endotoxic properties of synthetic pentaacyl lipid A precursor Ib and a structural isomer. Eur J Biochem, submitted.

Saxen H, Nurminen M, Kuusi N, Svenson SB, Mäkelä PH (1986. Evidence for the importance of O antigen specific antibodies in mouse-protective Salmonella outer membrane protein (porin) antisera. Microbial Pathog 1:433-441.

Strain SM, Fesik SW, Armitage IM (1983). Structure and metal-binding properties of lipopolysaccharides from heptose-less mutants of Escherichia coli studies by ^{13}C and ^{31}P nuclear magnetic resonance. J Biol Chem 258:13466-13477.
Tacken A, Rietschel ETh, Brade H (1986). Methylation analysis of the heptose/3-deoxy-D-manno-2-octulosonic acid region (inner core) of lipopolysaccharide from Salmonella minnesota rough mutants. Carbohydr Res 149:279-291.
Westphal O, Lüderitz O (1954). Chemische Erforschung von Lipopolysacchariden gram-negativer Bakterien. Angew Chem 66:407-417.
Westphal O, Lüderitz O, Galanos C, Mayer H, Rietschel ETh (1986). The story of bacterial endotoxin. Adv Immunopharmacol Vol. 3 (Chedid L, Hadden JW, Ipreafico F, Dukor P, Willoghby D, eds), Pergamon Press, Oxford pp. 13-34.
Zähringer U, Lindner B, Seydel U, Rietschel ETh, Naoki H, Unger FM, Imoto M, Kusumoto S, Shiba T (1985). Structure of de-O-acylated lipopolysaccharide from the Escherichia coli Re mutant F 515. Tetrahedron Lett 26:6321-6324.

Bacterial Endotoxins: Pathophysiological Effects, Clinical Significance, and Pharmacological Control, pages 47–62

ENDOTOXIN INTERACTIONS WITH SERUM PROTEINS RELATIONSHIP TO BIOLOGICAL ACTIVITY

Vernon L. Tesh, Stanley W. Vukajlovich, and David C. Morrison
Department of Microbiology and Immunology, Emory University School of Medicine, Atlanta GA 30322 (V.L.T.) and Department of Microbiology, Kansas University Medical Center, Kansas City, KS 66103 (S.W.V. and D.C.M.)

INTRODUCTION

Lipopolysaccharides (LPS) are one of the major constituents of the outer membrane of Gram-negative bacteria. Because of their exterior location in the bacterial membrane, LPS molecules are readily available to interact with a variety of host biologic and immunologic systems (reviewed in Morrison and Ulevitch, 1978). When injected into experimental animals, purified LPS preparations can reproduce many of the pathophysiologic symptoms which are characteristic of patients with Gram-negative septicemia, including fever, coagulopathies, changes in hemodynamics and activation of the complement cascade. Extrapolation of these observations to infection with Gram-negative bacteria in man or animals would suggest that endotoxins may be of considerable relevance to the pathophysiology of disease in Gram-negative bacteremia and septic shock. It is pertinent, therefore, to ask how these potent immunomodulatory and potentially deleterious molecules are derived and/or processed *in vivo*.

The majority of *in vivo* studies carried out to date have employed bacterial endotoxins extracted and purified from bacteria grown *in vitro* and subsequently injected into experimental animals. Implicit in this experimental protocol is the assumption that the chemical structures and the biological activity of endotoxins generated by chemical extraction procedures will parallel those found in endotoxins generated during infection. Endotoxins may be released from the intact bacterium in clinical infection by natural death of the organism, by bacteriolysis mediated via antibiotics to which the organism is sensitive, or by the action of host defense systems. Primary systems which may be operative in the latter include serum complement at the bacterial cell surface and phagocytic cell mediated bacterial digestion and release of bacterial constituents.

Endotoxin interactions with serum have been extensively investigated over the last forty years and, in this respect, there exists a large body of experimental evidence for various serum proteins which can bind lipopolysaccharides. The seminal observation made by several investigators in the mid-1950's that LPS preparations that have been exposed to serum are reduced in the ability to induce fever in rabbits, led to the development of the concept of serum-mediated LPS "detoxification" (reviewed in Skarnes and Rosen, 1971). Prominent among normal serum proteins which can bind and neutralize endotoxin is natural antibody directed against LPS. In addition, however, Rudbach et al. suggested that a mechanism of endotoxin detoxification might involve the non-enzymatic disaggregation of LPS micelles due to a reversible interaction with some serum factor(s). Although the serum factor(s) involved were not defined, a detergent-like effect, possibly mediated by bile salts, was implicated (Rudbach et al., 1966). Schultz and Becker suggested that a serum lipoprotein lipase was involved (Schultz and Becker, 1967), while Skarnes suggested that a heat-stable serum esterase first initiated LPS disaggregation followed by a second enzymatic "detoxification" reaction involving a heat-labile alpha-globulin (Skarnes, 1966). Johnson et al. isolated a single protein from human serum which they termed LPS inactivator which inhibited LPS toxicity in actinomycin D-treated mice and reduced reactivity in the Limulus amoebocyte lysate assay (Johnson et al., 1977). Serum factors which bind specifically to determinants characterized by Ra chemotype LPS and some which are cryptically contained in many LPS but expressed following mild acid hydrolysis have also been described (Ihara, et al., 1982; Brade and Brade, 1985). More recently, several investigators have demonstrated that the incubation of LPS preparations in normal mammalian sera resulted in the alteration of both the characteristic buoyant densities of these preparations and their biologic properties (Ulevitch and Johnston, 1978; Freudenberg, et al., 1980; Munford et al., 1981). In intact bacteria, Joiner et al. demonstrated that in 10% NHS, 80% of the radiolabelled C3 was associated with the LPS or outer membrane proteins of an E. coli Rc mutant (Joiner et al., 1984). A summary of serum proteins which have been shown to interact with LPS is presented in Table 1.

Table 1

LPS BINDING FACTORS PRESENT IN PLASMA/SERA

1. NATURAL ANTI-LPS ANTIBODIES (IgM and IgG)
2. HIGH AND LOW DENSITY LIPOPROTEINS
3. COMPLEMENT COMPONENTS C1 and C3
4. LPS INACTIVATION FACTOR IN α-GLOBULIN FRACTION
5. LIPOPROTEIN ASSOCIATED ESTERASE
6. ANTI-Ra LPS BACTERICIDAL FACTOR (28 Kda)
7. KDO REGION BINDING FACTOR (28 KDa)
8. SERUM ALBUMIN
9. ACUTE PHASE PROTEIN IN LEPORINE PLASMA (60.5 KDa)

These combined studies suggest that the interaction patterns of LPS with serum may be complex and involve multiple serum protein components. Thus while many serum proteins may bind to endotoxin, it is more difficult to evaluate the relative importance of such interactions. Further the demonstration that serum proteins bind to purified LPS does not guarantee that LPS released from a bacterium will necessarily bind the equivalent population of serum proteins. In this manuscript we address these two issues using recently developed technologies in our laboratory. In the first, we have employed a photoactivatable radioiodinated disulfide reducible LPS probe recently synthesized in our laboratory to characterize serum proteins with affinity for LPS (Wollenweber and Morrison, 1985). To address the second question, we have employed the J-5 mutant strain of E. coli 0111:B4 grown in the presence of radiolabelled galactose to label specifically the LPS component (Duncan and Morrison, 1984). Incubation of the labelled bacteria with normal human serum results in the complement dependent lysis and release of radiolabelled LPS (Tesh, Duncan and Morrison, 1986). In the studies to be reported here we have partially purified and characterized biologically the serum-released LPS.

RESULTS

As suggested above, we have recently synthesized a unique photoactivatable LPS derivative which has been designed to assess potential LPS binding substances (Wollenweber and Morrison, 1985). In addition to containing a photoactivatable crosslinking azido group, the crosslinking group (sulfosuccinimidyl-2-(p-azidosalicylamido)=1,3'-dithiopropionate, SASD) contains a reducible disulfide between the covalent binding site to the LPS and the activatable azido group. Importantly, this photoactivatable LPS derivative may also be radioiodinated by the chloramine T procedure. Since the site of iodination is distal to the disulfide group, post-crosslinking reduction effectively transfers radiolabel to LPS targets while eliminating covalent LPS attachment to the target molecules. Subsequent fractionation of crosslinked and reduced LPS-serum mixtures by SDS-PAGE and autoradiography allows positive identification of serum proteins with affinity for LPS.

For the studies to be reported here, both intact undiluted normal human serum (NHS) and serum fractionated by affigel blue column chromatography and reconstituted to approximate original serum concentrations have been studied. This latter chromatography procedure readily allows fractionation of serum into three major serum protein pools: i) an immunoglobulin containing pool, which elutes from the column at low ionic strength; ii) a complement protein pool which elutes at high ionic strength; (0.5 M NaCl); iii) and an albumin pool, which can be eluted with guanidine hydrochloride. The elution profiles for NHS fractionated as described are shown in Figure

1a. The use of these fractionated serum pools has facilitated the identification of LPS binding proteins in serum which are not readily detected in NHS.

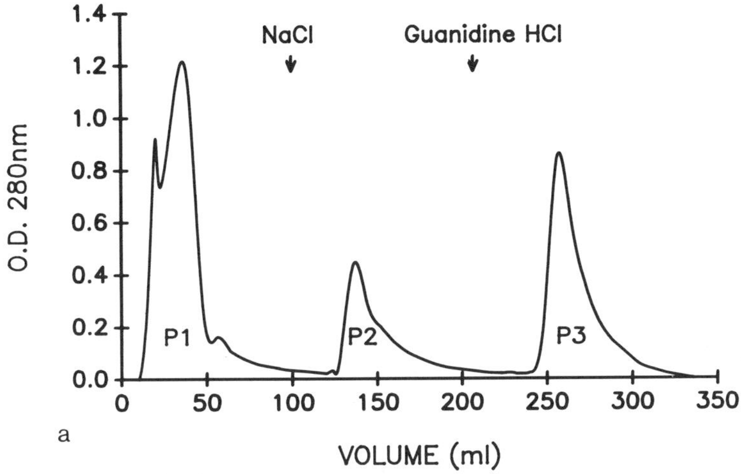

a

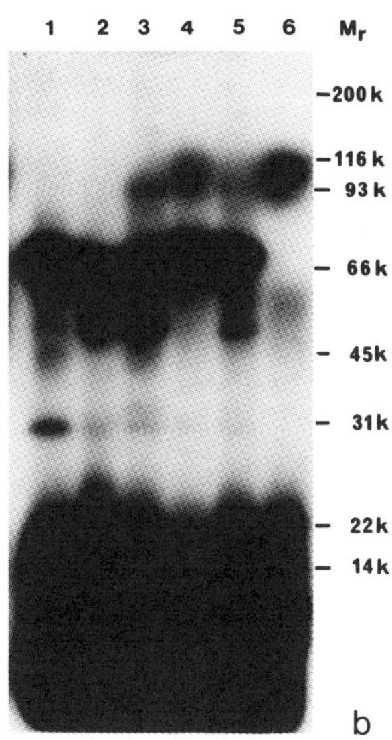

b

Figure 1

a) Fractionation of NHS chromatography on Affi-Gel Blue. 3.5 ml of NHS was applied to a 30 ml column of Affi-Gel Blue. The column was developed as indicated with VBS MgCa, pH 7.35, VBS MgCa with 0.5 M NaCl and 2.0 M Guanidine HCl to generate fractions P1, P2, and P3 respectively.

b) Analysis of LPS binding proteins specifically labeled using a photoaffinity probe. NHS or NHS fractions were treated with 0.2 ug of ^{125}I-ASD-LPS and analyzed as described in the text. Samples are: lane 1, NHS (1:400 final); lane 2, P1 (1:8 final); lane 3, P2 (1:8 final); lane 4, P3 (1:400 final); lane 5, P1, P2, and P3 (1:1:1 mixture of the above dilutions); lane 6, ^{125}I-ASD-LPS alone.

For these experiments NHS or the NHS fraction pools indicated in Figure 1a were incubated with ^{125}I-LPS-ASD for 15 minutes at 37 C and then irradiated with ultraviolet light as described earlier (Wollenweber and Morrison, 1985). Samples were then reduced with dithioerythritol and analyzed on 5-25 percent gradient acrylamide gels under reducing conditions. The gels were then subjected to autoradiography using exposure conditions determined to be optimal for each experiment (approximately 0.5 - 3 days).

The results of one such experiment are shown in Figure 1b. In lane 1 binding of LPS to proteins in intact NHS as reflected by autoradiography illustrates a major binding protein at about 66 Kda, a second at approximately 28 Kda, and a number of additional proteins. We have confirmed by Western blot that the major 66 Kda binding protein contains serum albumin and that binding to purified human serum albumin can also be demonstrated. Further, binding to albumin can be competitively inhibited by unsubstituted LPS and by S. minnesota R595 LPS, indicating that the binding specificity is for the lipid A-KDO region of the LPS and not for the crosslinking group. These data suggest that serum albumin may be a dominant target in NHS for LPS binding.

Since serum albumin constitutes the dominant protein present in NHS (approximately 40 mg/ml) we considered the possibility that the observed albumin binding might be obtained primarily due to excess albumin concentration and/or of mass action rather than a high binding affinity of LPS for this serum protein target. To obviate this problem, serum was fractionated as described above and in Figure 1 and individual serum pools assessed by LPS binding. Specific albumin concentrations in pools P1, 2, and 3 were 0.09 mg/ml , 0.2 mg/ml , and 70.5 mg/ml respectively (as determined by RIA). LPS binding profiles as determined by autoradiography are shown in Figure 1b (lanes 2, 3, and 4). Of significance, specific binding of LPS to albumin can be detected in all three serum pools in spite of the marked concentration differences of albumin in these fractions. In addition, specific binding to proteins of 96 KDa, 55 KDa, 28 KDa, and several additional species can be detected in the various serum pools. By specific immunoprecipitations and/or Western blot analysis, we have identified LPS binding to IgG (heavy and light chain, lane 3), complement component C3 (α chain, lane 3) and several apolipoproteins.

A summary of the serum proteins which bind LPS as detected by photocrosslinking is shown in Table 2.

Table 2

^{125}I-LPS-ASD Binding to Serum Proteins

Albumin
Immunoglobulin (IgG)
Complement Component C3
Apolipoproteins (212 , 26 , 18 , 6 KDa)
Unidentified (Major) (111 , 84 , 28 , 17 , 14 KDa)
Unidentified (Minor) (159 , 148 , 74 , 56 , 12 KDa)

In addition to examining the cross-linking of serum proteins to iodinated derivatized LPS preparations, we have also carried out experiments to examine the serum protein binding capacities of LPS released from the outer membrane of intact gram negative bacteria by the action of serum complement. E. coli J5, grown to mid-logarithmic phase in basic salts medium containing ^{3}H-galactose, were briefly pulsed with excess unlabelled galactose and harvested by centrifugation. The washed, labelled E. coli were then incubated in NHS at 37 C. At varying time intervals, aliquots of the sera were collected and assayed for bacterial viability by plating onto trypticase soy agar plates, and for serum-mediated release of radiolabelled LPS by liquid scintillation counting after brief centrifugation (9000 x G, 1 min.). In earlier published studies, we have demonstrated that E. coli J5 exposed to NHS were susceptible to the bactericidal action of NHS (> 99%

reduction in viability after 15 min.) and coincident release of ^{3}H-LPS into the serum. LPS release also occurred rapidly, with maximal release occurring in 30 min. Of importance, these studies documented that only a finite fraction (approximately 30%) of the LPS molecules were "releasable" from the outer membrane, even following multiple or prolonged exposure to NHS. Bacterial killing and LPS release were shown to be complement mediated as evidenced by a lack of killing and detectable LPS release in heat inactivated serum. Both the alternative and classical pathways of complement contributed to bacterial killing and LPS release. Furthermore, the intact membrane attack complex appeared to be necessary for the bactericidal process, while LPS release did not occur in C7-depleted serum, but LPS release comparable to release seen in NHS was detected in C9-deficient serum (Tesh, et al., 1986). These results suggested that although complement mediated LPS release from the outer membrane of gram negative bacteria accompanies the serum bactericidal event, it is not, in and of itself, sufficient to mediate bacterial killing.

In the studies to be reported here, we have initiated experiments to characterize the nature of the LPS released from the bacteria following the complement dependent lysis of the organism. Initial purification of the serum released LPS (S-LPS) was by gel permeation chromatography on Sepharose 4B. Approximately 6.0 ml. of S-LPS was chromatographed on a 5 cm. X 54 cm. Sepharose 4B column equilibrated with 0.1M ammonium bicarbonate buffer (pH = 7.2). One-hundred and fifty 10 ml. fractions were collected and assayed for both ^{3}H-LPS by liquid scintillation counting and protein content by either the Lowry or Bradford assays. Passage of S-LPS over Sepharose 4B (Figure 2) partitions the S-LPS into three ^{3}H-LPS containing fractions: i) peak 1 (P1), a high molecular weight, relatively protein-free component; ii) a broad peak (P2) which elutes slightly ahead of the major protein peak of the crude S-LPS; and iii) peak 3 (P3), a lower molecular weight component. Fractions containing peak radioactivity were pooled. P1 and P3 were lyophilized and resuspended to a final volume of 10 ml. in ammonium bicarbonate buffer. P2 was concentrated to a final volume of 20 ml. by drying with repeated changes of powdered Bio-Gel P-300 (Biorad Corp. San Francisco, CA).

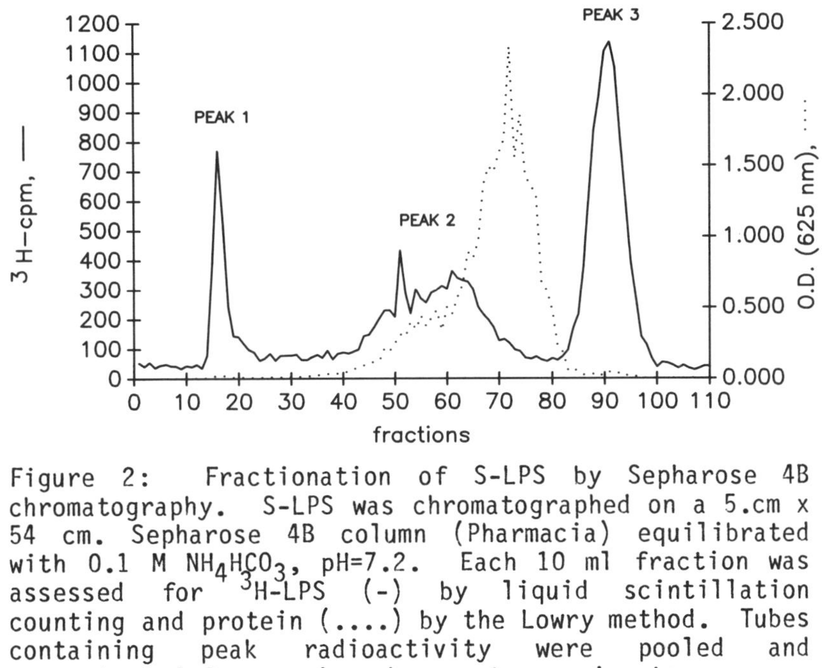

Figure 2: Fractionation of S-LPS by Sepharose 4B chromatography. S-LPS was chromatographed on a 5.cm x 54 cm. Sepharose 4B column (Pharmacia) equilibrated with 0.1 M NH_4HCO_3, pH=7.2. Each 10 ml fraction was assessed for ^{3}H-LPS (-) by liquid scintillation counting and protein (....) by the Lowry method. Tubes containing peak radioactivity were pooled and concentrated for use in subsequent experiments.

To characterize further the proteins associated with each of these three major ^{3}H-LPS containing fractions, 200 ul. samples of each pooled fraction were analyzed by Superose 12 chromatography (Pharmacia Inc. Piscataway, NJ). Eighty 0.5 ml. fractions were collected. ^{3}H-LPS was detected by liquid scintillation counting, and protein estimated by absorption at 280 nm. using a Pharmacia UV-1 monitor with a 10 mm. optical path length. Fraction P1 partitioned into two tritium-labelled components: a fraction which eluted in the void volume (300 kDa) and a second fraction with the major peak eluting at fraction 42 (Figure 3a). By analyzing molecular weight standards (Biorad) on the Superose 12 column, we have estimated that a protein

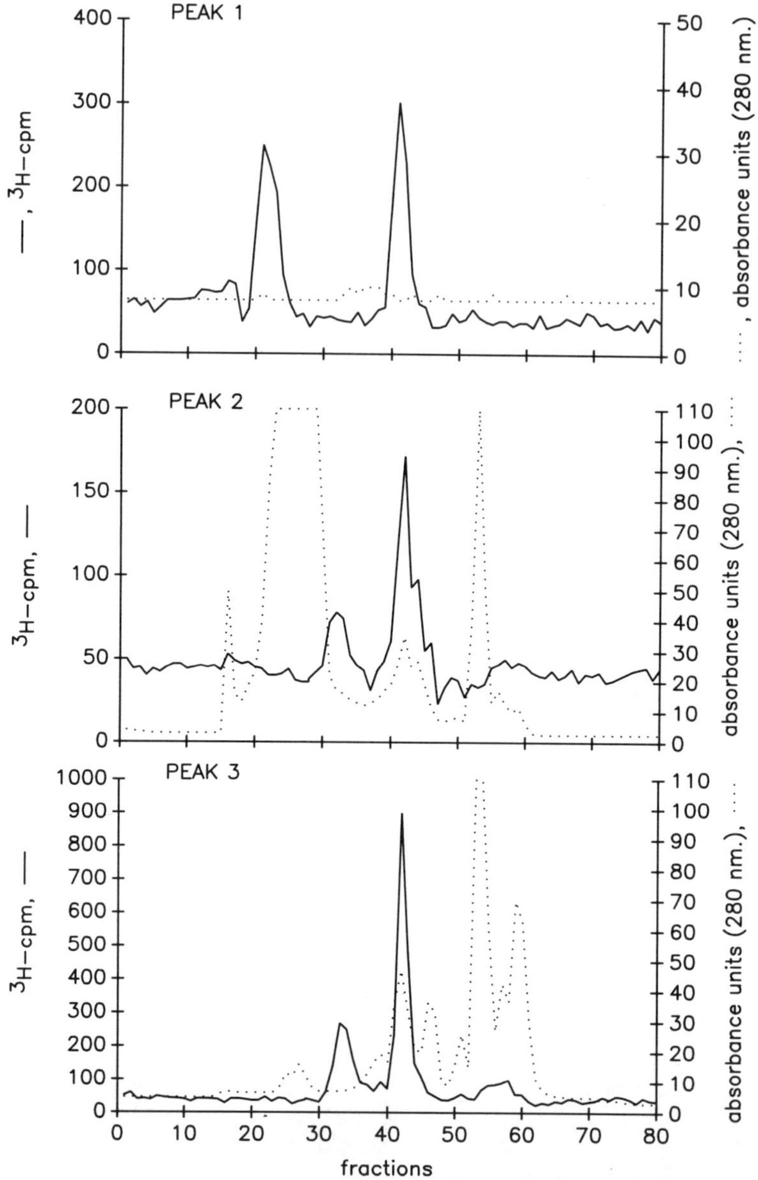

Figure 3: Analysis of partially purified S-LPS by Superose 12 chromatography. Samples (200 ul) of the concentrated peaks eluted from the Sepharose 4B column were chromatographed on a 10 mm x 300 mm Superose 12 column (Pharmacia) equilibrated with 0.1 M NH_4HCO_3 buffer, pH=7.2. Each 0.5 ml fraction was assessed for ^{3}H-LPS (-) by liquid scintillation counting and protein (...) by absorbance at 280 mm.

of approximately 60 kDa would elute at fraction 42. The protein elution profile of P1 was unremarkable due to the absence of significant amounts of protein in this fraction. The protein elution pattern of P2 was more complex, demonstrating at least 4 major protein peaks. Of significance, however, the ^{3}H-LPS was detected in fractions eluting in the region of tube 43, with a minor radioactive peak at tube 32 (Figure 3b). The protein elution pattern of P3 was also complex, but different when compared to P2. Again, the ^{3}H-LPS containing peaks were similar to P2, with a major peak at tube 43 and a secondary peak at tube 32 (Figure 3c).

These combined data thus suggest that a substantial portion of LPS released from bacteria by the action of NHS exist in a molecular form significantly lower in molecular weight than native phenol water extracted LPS. As estimated by gel permeation chromatography, the LPS in all three fractions chromatograph at an approximate molecular weight of about 60-70 KDa. We have shown the presence of serum albumin in each of these peak fractions from the superose 12 columns (tube 42); whether the LPS is in reversible association with this or some other serum protein remains to be established.

We have also examined several of the biological activities of S-LPS and its partially purified fractions. Experiments to assess lethality were carried out in outbred CF1 mice treated simultaneously with 20 ug of actinomycin D. The S-LPS, phenol water purified LPS (PW -LPS) and phenol water purified LPS added to NHS, all prepared from E. coli 0111:B4-J5 grown in ^{3}H-galactose, were normalized to equivalent concentrations of ^{3}H and graded doses injected intraperitoneally into 20 gm female CF1 mice. Deaths were recorded daily and LD_{50} values calculated by the method of Reed and Muench (1938). The results of this experiment are shown in Figure 4a and indicate that, in contrast to control LPS or control LPS treated with serum, the S-LPS is remarkably less toxic to mice.

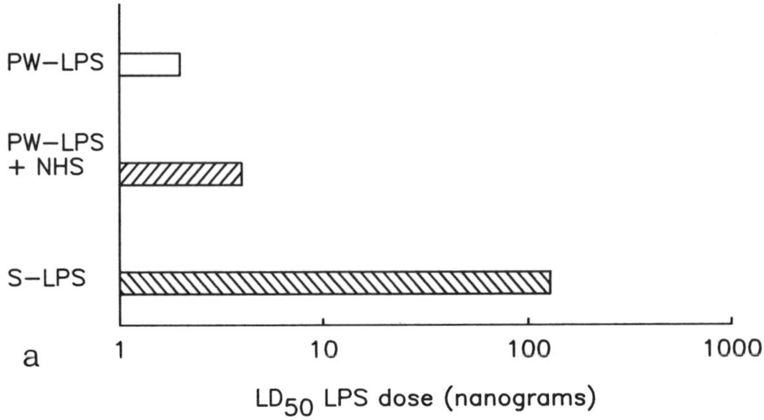

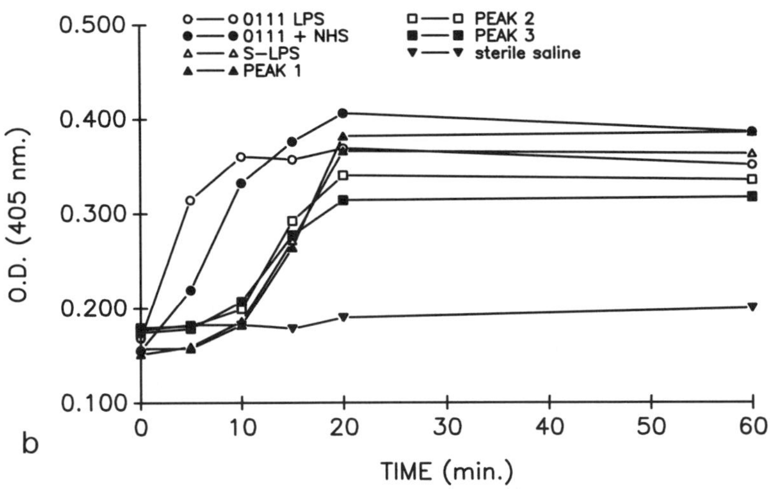

Figure 4a: Toxicity of S-LPS in actinomycin D-treated mice. CF-1 Mice (8-12 weeks old) were simultaneously injected i.p. with 20 ug. Actinomycin D (Cosmegen, Merck) and serial two-fold dilutions of LPS preparations. Lethality was monitored over 72 hours and LD_{50} values computed by the method of Reed and Muench.

Figure 4b: Kinetics of LAL activation by S-LPS. Equal volumes of LPS preparations and a non-clotting LAL (Associates of Cape Cod) were incubated in sterile 96-well microtiter plates. At each time point, turbidity was measured in Multiskan microtiter plates reader (Flow Laboratories) using a 405 nm. filter.

As a second assay for endotoxicity, we have employed a LAL kinetics assay utilizing a modification of the turbidimetric assay described by Novitsky et al. (1985). In this assay, serial two-fold dilutions of LPS preparations were incubated with a non-clotting LAL (Associates of Cape Cod) in sterile 96-well microtiter plates. The turbidity of the LAL were then monitored (405 nm.) at room temperature over a 60 min. period using a Multiskan microtiter plate reader (Flow Laboratories). PW-LPS, in nanogram quantities, was a potent activator of LAL (Figure 4b). When PW-LPS was mixed with NHS at room temperature and tested in the LAL kinetics assay, we noted a reproducible dimunition in LAL activity which may be due to the presence of an inhibitor in serum (Levin, et al., 1970). The crude S-LPS and all the partially purified S-LPS preparations were markedly less active in the LAL kinetics assay, with the time necessary to reach one-half maximal turbidity prolonged four-fold when compared to PW-LPS.

DISCUSSION

In this manuscript we have provided evidence for specific interaction of purified bacterial lipopolysaccharide with a variety of serum proteins. Further, we have shown that much of the LPS released from the surface of bacteria following bacteriolysis with serum complement may be in association with a serum protein and that the LPS protein complex has a combined molecular weight of 60-70 Kda. Finally our data suggest that, as normalized by galactose content, the serum released LPS is significantly reduced in its endotoxic activity at least as assessed by mouse lethality and Limulus amoebocyte lysate gelation.

In studies to assess binding of purified LPS to serum proteins, our results have confirmed that LPS will bind to natural antibody present in NHS, as well as several of the serum lipoproteins previously suggested by several investigators to complex with LPS (Ulevitch, et al., 1979; Freudenberg, et al., 1980; Munford, et al., 1981). Further, since the LPS used in these studies is an activator of the alternative pathway, it is not particularly surprising to find LPS bound to C3. We have not detected binding of C4, in spite of the presence of natural anti-LPS antibody. We believe it to be noteworthy and significant that albumin is one of the major serum proteins binding to purified LPS and that this observed binding is not totally explained by the relative concentration of albumin present in NHS. We have also detected a variety of additional, as yet unidentified serum proteins which can also bind LPS. The significance of these latter findings remains to be defined. Nevertheless, these results confirm the complex nature of LPS-serum protein interactions and suggest that, at the least, any proposed scheme for such interactions should include albumin as an important component.

Results obtained with LPS released from bacteria provide evidence for the concept that a majority of this LPS is in association with a single species of serum protein. The combined results of Figures 2 and 3 would suggest that reversible aggregation may occur between individual LPS-protein complexes. We have shown that each of the various species of S-LPS isolated and partially purified contains serum albumin and it is possible that LPS is released from the bacteria as single LPS subunit-albumin complexes. Assuming an average LPS monomer molecular weight of 8-10 KDa and 66 KDa for human serum albumin, an aggregate molecular weight of about 75 KDa would be obtained for the complex, which is well within the experimental estimates obtained by gel permeation chromatography.

Our results also indicate that S-LPS is significantly reduced in biological activity in comparison to purified LPS suspended in NHS. Since our data suggest that a majority of the LPS may be in a dissociated form, it may be that a requisite aggregate molecular weight for LPS, not obtained with S-LPS is necessary for biological activity. Rudbach and his colleagues (1966) have earlier suggested that an obligate step in serum-mediated detoxification of LPS is a reversible dissociation. Our data would not be inconsistant with this interpretation. Efforts are currently underway to identify unequivocally the serum protein binding LPS released from bacteria.

Acknowledgments

A portion of this research was carried out in partial fulfillment of the requirements for the Ph.D. degree from Emory University School of Medicine (V.L.T.). This research was supported by NIH grants AI 22948 (D.C.M.) and AI 24565 (S.W.V.). The authors would like to thank Ms. Janet Hollands for outstanding secretarial assistance in the preparation of this manuscript.

REFERENCES

Brade L, Brade H (1985). A 28,000-dalton protein of normal mouse serum binds specifically to the inner core region of bacterial lipopolysaccharide. Infect Immun 50:687-694.

Duncan RL, Morrison DC (1984). The fate of E. coli lipopolysaccharide after the uptake of E. coli by murine macrophages in vitro. J Immunol 132:1416-1424.

Freudenberg MA, Bog-Hansen TC, Back U, Galanos, C (1980). Interaction of lipopolysaccharides with plasma high density lipoproteins in rats. Infect Immun 28:373-380.

Ihara I, Hirada Y, Ihara S, Kawakami M (1982). A new complement-dependent bactericidal factor found in nonimmune mouse sera: specific binding to polysaccharide of Ra chemotype Salmonella. J Immunol 128:1256-1260.

Johnson KJ, Ward PA, Goralnick S, Osborn MJ (1977). Isolation from human serum of an inactivator of bacterial lipopolysaccharide. Am J Pathol 88:559-574.

Joiner KA, Goldman R, Schmetz M, Berger M, Hammer CH, Frank MM, Leive L (1984). A quantitative analysis of C3 binding to O-antigen capsule, lipopolysaccharide, and outer membrane proteins of E. coli 0111:B4. J Immunol 132:369-375.

Levin J, Tomasulo PA, Oser RS (1970). Detection of endotoxin in human blood and demonstration of an inhibitor. J Lab Clin Med 75:903-911.

Morrison DC, Ulevitch RJ (1978). The interaction of bacterial endotoxins with cellular and humoral mediation systems. Am J Pathol 93:527-618.

Munford RS, Hall CL, Dietschy JM (1981). Binding of Salmonella typhimurium lipopolysaccharides to rat high density lipoproteins. Infect Immun 34:835-843.

Novitsky TJ, Roslansky PF, Siber GR, Warren HS (1985). Turbidimetric method for quantifying serum inhibition of Limulus amoebocyte lysate. J Clin Microbiol 20:211-216.

Reed LJ, Muench H (1938). A simple method of estimating fifty percent endpoints. Am J Hyg 27:493-497.

Rudbach JA, Anacker RL, Haskins WT, Johnson AG, Milner KC, Ribi E (1966). Physical aspects of reversible inactivation of endotoxin. Ann NY Acad Sci 133:629-643.

Schultz DR, Becker EL (1967). The alteration of endotoxin by postheparin plasma and its purified fractions. II. Relationship of the endotoxin detoxifying activity of euglobulin from postheparin plasma to lipoprotein lipase. J Immunol 98:482-489.

Skarnes, RC (1966). The inactivation of endotoxin afterinteraction with certain proteins of normal serum. Ann NY Acad Sci 133:644-662.

Skarnes RC, Rosen FS (1971). Host dependent detoxification of bacterial endotoxins. In Ajl SJ, Kadis S, Weinbaum G (eds): Microbial Toxins, Vol 5. Academic Press, New York. pp 151-164.

Ulevitch RJ, Johnston, AR (1978). The modification of biophysical and endotoxic properties of bacterial lipopolysaccharides by serum. J Clin Invest 62:1313-1324.

Ulevitch RJ, Johnston AR, Weinstein DB (1979). New function for high density lipoproteins. Their participation in intravascular reactions of bacterial lipopolysaccharides. J Clin Invest 64:1516-1524.

Wollenweber H-W, Morrison DC (1985). Synthesis and biochemical characterization of a photoactivatable, iodinatable, cleavable bacterial lipopolysaccharide derivative. J Biol Chem 260:15068-15074.

Tesh VL, Duncan RL, Morrison DC (1986). The interaction of Escherichia coli with normal human serum: the kinetics of serum-mediated lipopolysaccharide release and its dissociation from bacterial killing. J Immunol 137:1329-1335.

Bacterial Endotoxins: Pathophysiological Effects, Clinical Significance, and Pharmacological Control, pages 63–75

THE METABOLIC FATE OF ENDOTOXINS

Marina A. Freudenberg and Chris Galanos

Max-Planck-Institut für Immunbiologie, Stübeweg 51, 7800 Freiburg, F.R.G.

INTRODUCTION

It has been generally believed in the past that the mammalian organism possesses mechanims which render endotoxin harmless and a number of such detoxifying mechanisms both humoral and cellular have been proposed. However, in former studies no precise knowledge on the fate of LPS in the organism was available mainly because only poorly defined LPS preparations were available at that time. The improvement in the last years of isolation and purification methods for LPS, which made pure well defined LPS preparations available, and the more complete knowledge of the chemical structure and composition of LPS provided a better starting point for studies on the distribution, degradation, and elimination of LPS *in vivo*.

Our own investigations on the fate of endotoxin were carried out mainly in rats with LPS of *S.abortus equi* (smooth form, S-LPS) and *S.minnesota* R595 or *E.coli* F515 (Re forms, Re-LPS). Results obtained are shown and discussed in the present communication.

RESULTS AND DISCUSSION

Clearance

Biosynthetically radiolabeled LPS-preparations were injected in rats intravenously and their removal from the blood measured. The data obtained revealed the dependency of the rate of clearance on the chemical and physicochemical properties of the LPS injected (Freudenberg et al.,1980; Galanos et al.,1984).

Re-LPS was cleared from the circulation much faster (half live $T_{1/2}$: minutes) than S-LPS ($T_{1/2}$: hours) (Fig. 1). This is in agreement with earlier studies (Chedid et al., 1966; Hofman and Dlabac, 1974) and shows that the difference in the rate of clearance between smooth and rough forms is a general phenomenon.

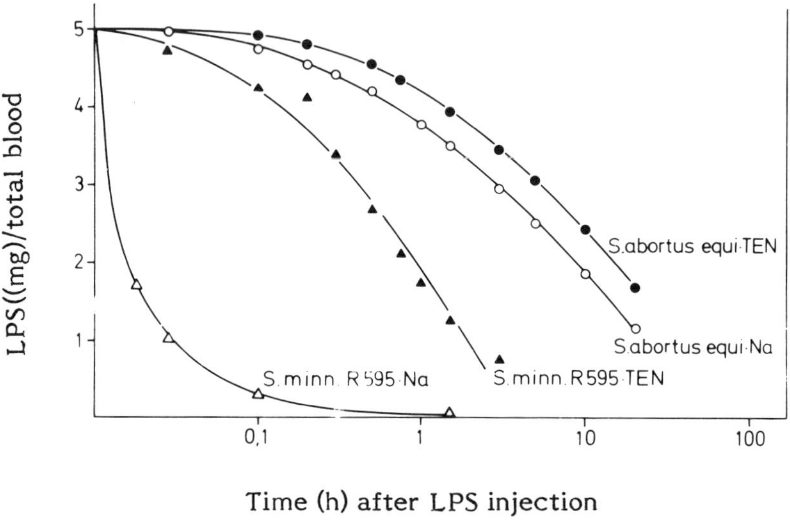

Figure 1: Blood clearance of LPS in low (triethylamine) and high (Na) molecular weight. ^{14}C uniformly labeled LPS preparations were injected in rats i.v., at different times thereafter blood samples were collected and remaining radioactivity measured.

For a given smooth or rough LPS the rate of clearance was dependent on the state of molecular aggregation. This could be demonstrated after the method of electrodialysis was introduced and one given LPS could be prepared in different uniform salt forms, each with characteristic physichochemical properties (Galanos and Lüderitz, 1975). The highly aggregated sodium salt forms of S- and Re-LPS injected in rats were cleared faster from the circulation than their less aggregated triethylamine forms (Fig. 1). In conventional LPS preparations phosphate and carboxylic groups are neutralized by an unpredictable mixture of metal cations and polyamines that vary from one batch to another, thereby influencing differentially the aggregation properties of different preparations. Therefore different batches of the same LPS may exhibit different $T_{1/2}$ of clearance depending on which cations are present in the preparation. The state of molecular aggregation influences not only the clearance of LPS but also other biological properties such as toxicity, pyrogenicity, complement activation, etc. (Freudenberg MA and Galanos, 1978; Galanos and Lüderitz, 1976; Galanos et al., 1984). For this reason the conversion of LPS to a defined salt form results in preparations giving very reproducible results in different biological assays independent of batch of isolation.

In contrast to earlier reports, the rate of clearance of LPS was found to be independent of the injected dose (Freudenberg MA et al., 1984).

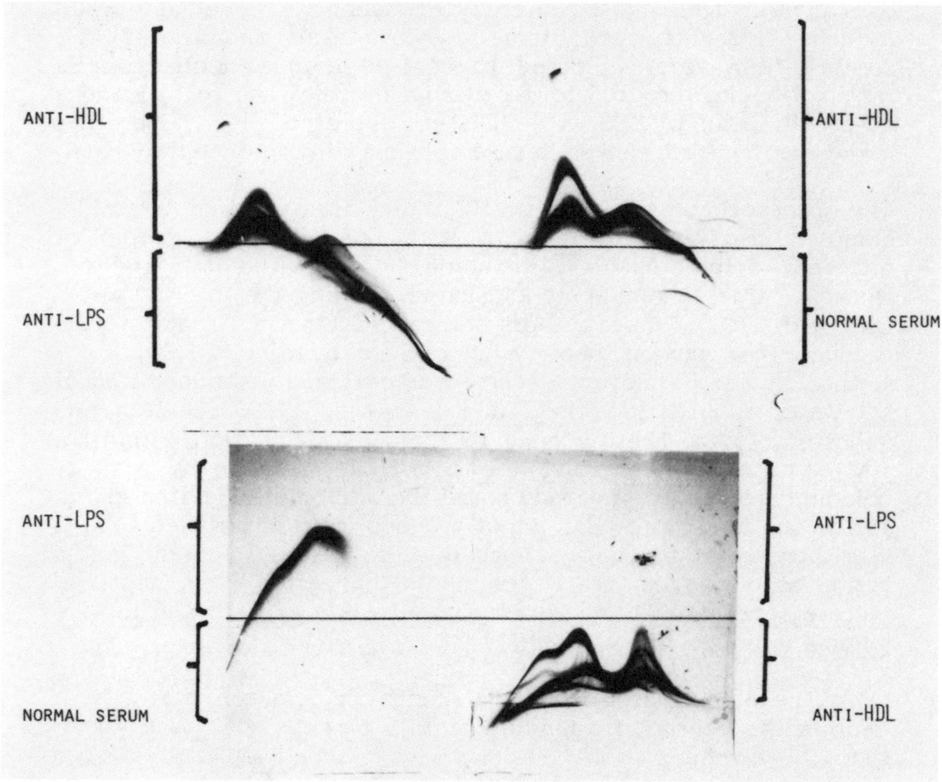

Figure 2: Crossed-immunoelectrophoresis with intermediate gel of rat plasma 1 h after i.v. injection of LPS (5 mg).

Analysis of blood samples obtained from rats from the second minute onwards after injection showed that LPS was present exclusively in plasma, the formed blood elements were free of LPS. This, however, does not exclude the possibility that in the first 1 or 2 min. after injection small amounts of LPS may associate with blood cells which thereafter disappear from the circulation and are thus not detectable. That this indeed may be the case is indicated by the presence of LPS-positive granulocytes in liver sinusoidal cells in the initial period after LPS injection (Freudenberg N et al., 1985).

In plasma, LPS undergoes a complex formation with high density lipoprotein (HDL) and remains in this complex form until it is cleared from the circulation (Freudenberg MA et al., 1980). Evidence for this was obtained when plasma samples of LPS treated rats were subjected to crossed immunoelectrophoresis with antibodies directed to rat plasma proteins and to LPS and a co-precipitation of LPS with HDL was detected (Fig. 2). No other major carrier for LPS was found in plasma of normal healthy rats.

The interaction of LPS with HDL and formation of a stabile complex also proceeds in vitro with the plasma or serum of different animal species, including humans. Ulevitch and co-workers came to similar conclusions regarding the interaction of LPS with HDL. These authors showed that normal rabbit serum reduces the buoyant density of LPS to a value < $1,2/cm^3$. A similar change in buoyant density was obtained after injection of LPS into rabbits (Mathison and Ulevitch, 1979; Ulevitch and Johnston, 1978). The evidence for complex formation with HDL was obtained by isolating the LPS/HDL complex by affinity column chromatography using anti-LPS antibodies (Ulevitch et al., 1981). The binding to HDL is not property only of purified LPS. It was shown that the native form of LPS, i.e. LPS present in the bacterial outer membrane fragments (harvested from bacterial culture supernatant) undergoes a comparable in vitro and in vivo binding (Munford et al., 1982).

The binding of LPS to HDL has biological consequences. It inhibits the nonspecific binding of LPS to cells such as erythrocytes (Freudenberg MA et al., 1980) and lymphocytes (own unpublished results). This makes understandable why, as mentioned before, we were not able to detect LPS on circulating blood cells despite the fact that LPS binds very easily in the absence of serum to erythrocytes and lymphocytes in vitro.

In their clearance studies in rabbits Mathison and Ulevitch, (1979) provided evidence that binding to HDL retards the clearance of LPS. It was shown in the past that cells of RES are the main cells participating in the clearance of LPS. Therefore, we have investigated whether the binding to HDL has an influence on the uptake of LPS by macrophages. Bone marrow-derived macrophages were cultivated with different amounts of free and HDL-bound LPS for 24 h. As shown in Fig. 3 free LPS was incorporated in cells much more effectively than HDL-bound LPS, which is in agreement with the results on LPS clearance obtained in vivo.

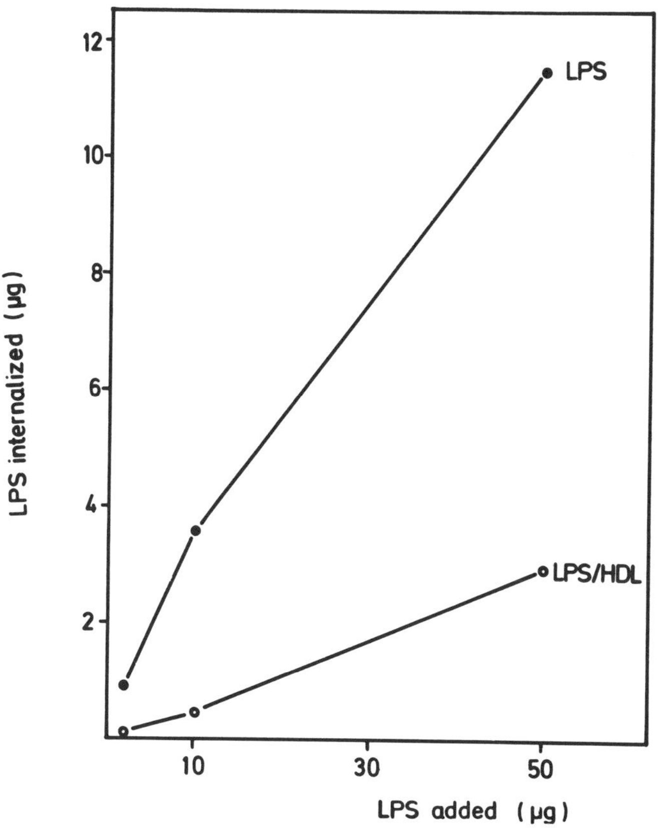

Figure 3: Uptake of free and HDL-bound S-LPS by mouse macrophages. 10^7 C3H/Tif F macrophages were incubated with different amounts of ^{14}C labeled S-LPS for 24 h at 37°C in a humidified atmosphere with 8% CO_2, thereafter washed and the cell associated radioactivity measured.

The strong effect of binding to HDL on the rate of LPS clearance stimulated us to investigate whether the difference in the speed of clearance between smooth and rough LPS preparations described above, has something to do with this binding. Therefore, we have compared in rats the clearance of radiolabeled Re- and S-LPS with and without precomplexation with HDL (unpublished results). There was practically no difference in the clearance time, when the S-LPS was injected in the free or HDL-bound form ($T_{1/2}$ = 7,5 h versus 8 h). This demonstrates that after its injection the bulk of S-LPS is converted to the LPS/HDL complex and explains the long time necessary for clearance of S-LPS from the circulation. In contrast, Re-LPS injected in the free form disappeared from the blood rapidly with a $T_{1/2}$ of 2 min, whereas the HDL-bound Re-LPS was cleared slowly with a $T_{1/2}$ of 6,5 h. This shows that after injection the largest part of the Re-LPS is removed from the circulation before it can interact with plasma HDL and that Re-LPS/HDL- as the S-LPS/HDL complex needs for its clearance many hours.

In our experiments in rats factors known to influence LPS clearance such as an activated reticuloendothelial system and specific LPS antibodies were eliminated as far as possible by using healthy animals grown under specific pathogen-free conditions in which LPS antibodies were not detectable. Munford and Dietschy (1985) studied the effect of antibodies to LPS on LPS-HDL binding and showed that high titers of homologous IgG-antibodies inhibit the binding of LPS to HDL and thereby augment the removal of LPS from the blood. Tobias and Ulevitch identified an LPS binding protein (LBP) which is inducible in mice, rabbits and humans by different agents causing acute phase response. LBP was shown to decrease the rate of LPS/HDL complex formation and as consequence to increase the rate of LPS clearance (Tobias and Ulevitch, 1983; Tobias and Ulevitch, 1986).

Distribution

Measurement of the content of LPS in various rat organs using radiolabeled preparations revealed differences in the distribution of Re- and S-LPS. The Re-LPS accumulated rapidly in the liver, 80% of injected LPS appearing in this organ after 5 h (Fig. 4). In the liver also the highest tissue concentration of LPS was measured. A slightly lower tissue concentration was found in the spleen (Fig. 5), in other organs the LPS concentration was very low or not detectable (unpublished results). The S-LPS accumulated in the liver more slowly than the Re-LPS and to a significantly lesser extent (on average 43% of injected on day 3, Fig. 4). The highest tissue concentration of S-LPS was found in the spleen, followed by the liver and adrenal gland (Fig. 5). The S-LPS content in other organs was low however in general higher than that found after injection of Re-LPS (Freudenberg MA and Galanos, 1985).

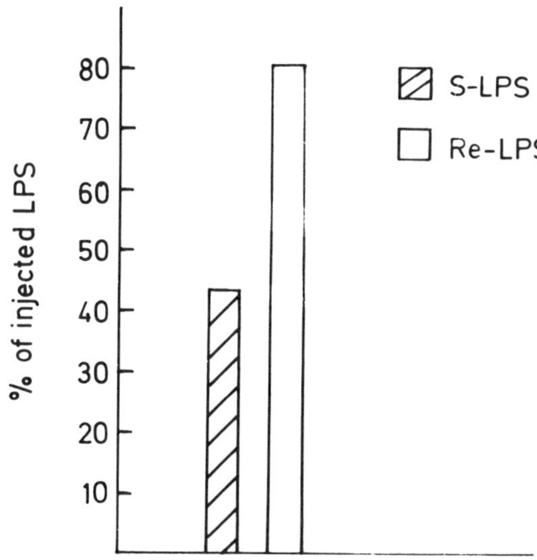

Figure 4: Amounts of LPS in the liver at the time of highest accumulation. The highest accumulation of Re-LPS in tissue was found at 5 h, that of S-LPS on day 3 after i.v. injection and was independent of the dose of LPS.

As shown above at the time of its uptake S-LPS is present in the circulation as LPS-/HDL complex while Re-LPS mostly in unbound form. It is suggested that this difference is the reason for the difference in their organ distribution.
It was shown earlier that only HDL-bound LPS enters the adrenal gland and that this uptake is effected via HDL receptors (Mathison and Ulevitch, 1979; Munford et al., 1981). Therefore the fact that S-LPS but not Re-LPS is concentrated in the adrenal gland is explained by the higher efficiency of S-LPS to bind to HDL in vivo.

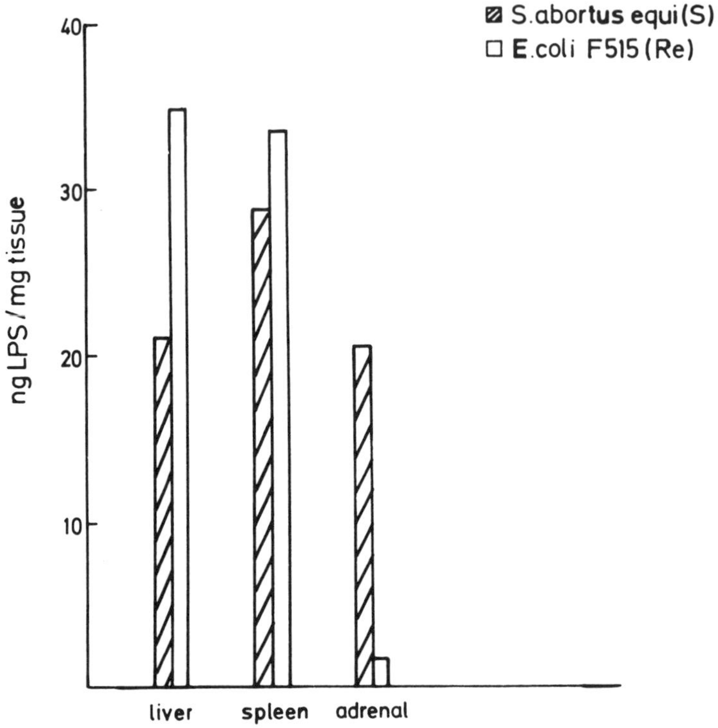

Figure 5: Concentration of tissue-bound LPS after i.v. injection (100 µg/rat) at the time of highest accumulation.

A study on the cellular distribution of LPS in the liver revealed that there is a difference in the cellular uptake between S- and Re-LPS (Freudenberg MA et al., 1982). By immunoperoxidase staining of liver sections it could be shown that Re-LPS is primarily taken up by sinusoidal cells, hepatocytes and granulocytes, whereas S-LPS only by sinusoidal cells and granulocytes. Both LPS pass ultimately from sinusoidal cells to hepatocytes.

In organs other than liver again macrophages and granulocytes are the cells involved in the uptake of LPS (S- and Re-). In the adrenal gland S-LPS is concentrated in the phagocytic vacuoles of the cortical epithelium (Freudenberg N et al., 1985).

Degradation

Although LPS may circulate in experimental animals for a long time, no evidence for chemical degradation or detoxification in plasma was found (Freudenberg MA et al., 1984). Despite the differences in the distribution, the liver is the main clearance organ for both S- and Re-LPS. When the persistence of LPS in the liver was tested by two different methods, i.e., by immunohistochemistry using anti-LPS antibodies directed against the O-polysaccharide, and by radioactivity measurements using ^{14}C uniformly labeled LPS, different results were obtained. By the former method the detectability of antigenically active LPS was lost already 5-9 days after injection (Freudenberg MA et al., 1982). In contrast, by the latter method, the amount of radioactivity in the liver decreased only slowly with time, the S-LPS persisting more than 5 weeks (Freudenberg MA et al., 1984) and the Re-LPS more than 8 weeks (unpublished). The discrepancy in the data obtained indicated that chemical changes in the O-polysaccharide part of the LPS molecule resulting in loss of antigenic activity had occured. Evidence for chemical degradation of LPS was also obtained in studies with S- and Re-LPS preparations labeled in sugars and fatty acids with two different isotopes. In comparison to the starting LPS, that present in the liver and other organs contained less fatty acids.

Chemical degradation of LPS in vivo was confirmed by chemical analysis of S-LPS reisolated from rat liver 3 days after its injection (Freudenberg MA and Galanos, 1985). It could be shown that the macromolecular structure of LPS remained grossly unaltered, however a significant reduction in the amount of abequose, the branched sugar of the O-chain of S.abortus equi and a partial loss of fatty acids was detected. As expected, the chemical alteration of the O-chain resulted in a reduction of the antigenic activity. The fatty acid loss included the 3-hydroxytetradecanoic acid (reduced by 40%), dodecanoic acid (reduced by 50%) hexadecanoic acid (reduced by 60%) and 2-hydroxytetradecanoic acid (almost absent). The amount of tetradecanoic acid however remained similar to that of starting LPS. It is known for many years that the partial cleavage of fatty acids from LPS by, for example, alkaline hydrolysis leads to an extensive loss of biological activity. It was therefore surprising that the considerable loss of fatty acids from LPS, which occurred in rat liver had no influence on the endotoxic activity (pyrogenicity, lethal toxicity, local Shwartzman reaction, Limulus amoebocyte lysate gelation).

Experiments with double labeled LPS revealed that the cleavage of fatty acids in vivo occurs in all LPS containing organs. The vast majority of the injected LPS is present here in macrophages and granulocytes. Recently, enzymatic deacylation of LPS by human neutrophils and mouse macrophages in vitro was demonstrated (Munford and Hall, 1985; Munford and Hall, 1986). It was claimed however that the deacylation of LPS by human neutrophils results in detoxification of LPS. The obvious absence of chemical detoxification seen in our study in rat liver raises the question of species differences, which may be answered only by further investigations.

It is not known at present whether hepatocytes, the third cell type through which large amounts of LPS in vivo have to pass contribute to the chemical degradation of LPS.

Elimination

The partially degraded LPS is slowly excreted from the body mainly through the gut (Kleine et al., 1985). Fig. 6 shows the 14 days excretion of radioactivity in faeces after injection of double labeled S- and Re-LPS. Faeces collected during this period contained 32% of S-LPS and 22% of Re-LPS. With both LPS, the quantitative difference in excretion of ^{14}C (sugar) and ^{3}H (fatty acids) was an evidence of deacylation of LPS.

At least, a part of the degraded LPS reaches the intestine via bile as has been detected by canulating the bile duct of injected rats (Komuro, Freudenberg MA and Galanos, unpublished). Using immunohistochemistry it was recognized however that another route of LPS excretion through the lung exists (Freudenberg MA et al., 1982; Freudenberg N et al., 1984). Macrophages carrying LPS migrate to the lung and pass into the alveolar and bronchiolar space. During the first hours after injection of S-LPS alveolar macrophages were free of LPS. Thereafter the number of LPS-positive macrophages increased and from day 3 up to day 14 all alveolar macrophages obtained by broncho-alveolar lavage carried LPS. A significant number of alveolar macrophages obtained between the second and eighth day after LPS was morphologically identified as activated macrophages. There was a striking correlation between the appearance of LPS-positive macrophages and histological signs of lung tissue injury. The cell-bound LPS excreted through the lung is also chemically altered (partially deacylated) and we assume that in rats it later forms part of the LPS eliminated in faeces.

A small amount of S- and Re-LPS radioactivity (4-7% in 14 days) is excreted in urine. It could be shown that it represents mainly small molecular weight metabolic products (Freudenberg MA et al., 1984; Kleine et al., 1985).

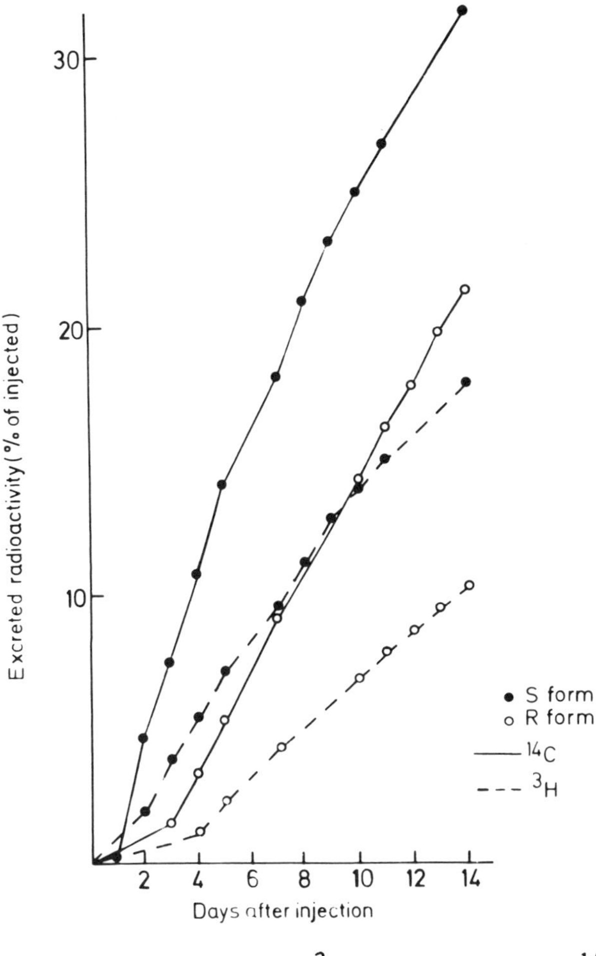

Figure 6: Excretion of ^{3}H (fatty acids) and ^{14}C (sugars) in faeces of rats injected with ^{3}H, ^{14}C-LPS preparations.

REFERENCES

Chedid L, Parant F, Parant M, Boyer F (1966) Localization and

fate of ^{51}Cr-labeled somatic antigens of smooth and rough Salmonellae. Ann NY Acad Sci 133: 712-726.
Freudenberg MA, Galanos C (1978) Interaction of lipopolysaccharides and lipid A with complement in rats and its relation to endotoxicity. Infect Immun 19: 875-882.
Freudenberg MA, Bøg-Hansen TC, Back U, Galanos C (1980) Interaction of lipopolysaccharides with plasma high-density lipoprotein in rats. Infect Immun 28: 373-380.
Freudenberg MA, Bøg-Hansen TC, Back U, Jirillo E, Galanos C (1980) Interaction of lipopolysaccharides with plasma high density lipoprotein in rats. p. 349-354. In Eaker D, Wadstrom T (ed), Natural Toxins, Pergamon Press, New York.
Freudenberg MA, Freudenberg N, Galanos C (1982) Time course of cellular distribution of endotoxin in liver, lungs and kidneys of rats. Br J Exp Path 63: 56-65.
Freudenberg MA, Kleine B, Galanos C (1984) The fate of lipopolysaccharide in rats: evidence for chemical alteration in the molecule. Rev Infect Dis 6: 483-487.
Freudenberg MA, Galanos C (1985) Alterations in rats in vivo of the chemical structure of lipopolysaccharide from Salmonella abortus equi. Eur J Biochem 152: 353-359.
Freudenberg MA, Galanos C (1986) The fate of endotoxin in the host. Eos VI-n.3/Supplement: 66-67.
Freudenberg N, Freudenberg MA, Guzman J, Mittermayer CH, Bandara K, Galanos C (1984) Identification of endotoxin-positive cells in the rat lung during shock. Virchows Arch [Pathol Anat] 404: 197-211.
Freudenberg N, Freudenberg MA, Bandara K, Galanos C (1985) Distribution and localization of endotoxin in the reticulo-endothelial system (RES) and in the main vessels of the rat during shock. Path Res Pract 179: 517-527.
Galanos C, Lüderitz O (1975) Electrodialysis of lipopolysaccharides and their conversion to uniform salt forms. Eur J Biochem 54: 603-610.
Galanos C, Lüderitz O (1976) The role of the physical state of lipopolysaccharides in their interaction with complement: high molecular weight as prerequisite for the expression of anticomplementary activity. Eur J Biochem 65: 403-408.
Galanos C, Freudenberg MA, Joh J, Freudenberg N, Lüderitz O (1984) Bacterial endotoxins: chemical and biological properties. p. 203-213. In Keppler D, Popper H, Bianchi L, Reutter W (ed), Mechanisms of hepatocyte injury and death, MTP Press Limited, Lancaster, Boston, The Hague, Dordrecht.
Hofman J, Dlabac V (1974) The role of lipid A in phagocytosis of gram-negative bacteria and their lipopolysaccharides. J Hyg Epidemiol Microbiol Immunol 18: 447-453.

Kleine B, Freudenberg MA, Galanos C (1985) Excretion of radioactivity in faeces and urine of rats injected with ^{3}H, ^{14}C-lipopolysaccharide. Br J Exp Path 66: 303-308.

Mathison JC, Ulevitch RJ (1979) The clearance, tissue distribution, and cellular localization of intravenously injected lipopolysaccharide in rabbits. J Immunol 123: 2133-2143.

Munford RS, Andersen JM, Dietschy JM (1981) Sites of tissue binding and uptake in vivo of Bacterial lipopolysaccharide-high density lipoprotein complexes. Studies in the rat and squirrel monkey. J Clin Invest 68: 1503-1513.

Munford RS, CL Hall, JM Lipton, JM Dietschy (1982) Biological activity, lipoprotein-binding behavior, and in vivo disposition of extracted and native forms of Salmonella typhimurium lipopolysaccharides. J Clin Invest 70: 877-888.

Munford RS, Dietschy, JM (1985) Effects of specific antibodies, hormones, and lipoproteins on bacterial lipopolysaccharides injected into the rat. J Infect Dis 152: 177-184.

Munford RS, Hall CL (1985) Uptake and Deacylation of bacterial lipopolysaccharides by macrophages from normal and endotoxin-hyporesponsive mice. Infect Immun 48: 464-473.

Munford RS, Hall CL (1986) Detoxification of bacterial lipopolysaccharides (endotoxins) by a human neutrophil enzyme. Science 234: 203-205.

Tobias PS, Ulevitch RJ (1983) Control of lipopolysaccharide-high density lipoprotein binding by acute phase protein(s). J Immunol 131: 1913-1916.

Tobias PS, Ulevitch RJ (1986) Regulation by acute phase proteins of LPS binding to lipoproteins. Eos VI.-n.3/Supplement: 68-70.

Ulevitch RJ, Johnston AR (1978) The modification of biophysical and endotoxic properties of bacterial lipopolysaccharides by serum. J Clin Invest 62: 1313-1324.

Ulevitch RJ, Johnston AR, Weinstein DB (1981) New function for high density lipoproteins. Isolation and characterization of a bacterial lipopolysaccharide-high density lipoprotein complex formed in rabbit plasma 67: 827-837.

Bacterial Endotoxins: Pathophysiological Effects, Clinical Significance, and Pharmacological Control, pages 77–88

Cachectin/TNF Mediates the Pathophysiological Effects of Bacterial Endotoxin/Lipopolysaccharide (LPS)

Kevin J. Tracey*#, Stephen F. Lowry*, and Anthony Cerami#

*Laboratory of Surgical Metabolism,
New YorkHospital-Cornell University Medical Center,
New York, New York, 10021

#Laboratory of Medical Biochemistry,
Rockefeller University, New York, New York, 10021

Introduction

The isolation and characterization of bacterial endotoxin/lipo-polysaccharide (LPS) has led to the subsequent delineation of pathophysiological responses attributable to LPS (reviewed in Aggarwal, ed., 1980). Wide derangements of organ function and cellular homeostasis characterized by acute inflammatory responses, altered energy metabolism, and lethal tissue injury are observed after LPS administration. It is now known that many (if not all) of these effects are mediated by humoral factors released in response to endotoxemia. In this review we will focus on the role of the macrophage-secreted protein cachectin as an early mediator of the deleterious events induced by LPS. In addition, the potential therapeutic role of inhibiting cachectin/TNF-mediated toxicity during sepsis will be discussed.

The cardiovascular, hematologic, and metabolic responses during lethal endotoxemia have been extensively studied since the 1940's. In general, the infusion of a lethal dose of LPS in experimental animals is followed by hypotension, shock, critical organ injury, and death by respiratory arrest (Franke, 1944; Gilbert, 1960; MacLean and Weil,1956). The principal histopathological abnormalities include acute pulmonary inflammation with margination of neutrophils, acute renal tubular necrosis, and diffuse hemorrhagic necrosis secondary to intravascular thrombosis in a variety of tissues (Franke, 1944; Balis et. al., 1978).

Cachectin was first isolated during studies directed towards characterization of humoral mediators present in the serum of rabbits afflicted with a lethal infection caused by Trypanosoma brucei (Rouzer and Cerami, 1980). Termed "cachectin" because of its suspected involvement in the wasting diathesis (cachexia) accompanying advancing infection, the protein was purified to homogeneity from the supernatant of murine macrophages stimulated with LPS (Beutler et. al., 1985b). The polypeptide was found to have a relative molecular weight of 17,000 D and a P.I. of 4.7. Subsequent molecular cloning of the cDNA encoding cachectin confirmed that cachectin and tumor necrosis factor (TNF) (Fransen et. al., 1985; Pennica et. al., 1985; Wang et. al., 1985) are identical (Beutler et. al., 1985a; Caput et. al., 1986). The availability of endotoxin-free, recombinant human cachectin/-TNF has enabled recent studies of the biological activities and systemic toxicity of this hormone.

Pathophysiology induced by cachectin

The biodistribution and clearance of exogenously administered cachectin/TNF has been determined in studies using radio-labeled peptide in experimental animals (Beutler et. al., 1985c; Palladino et. al., 1987). The plasma half-life of cachectin is relatively short (6-20 minutes) and the highest tissue levels of radio-labeled protein occur in kidneys, lungs, skin, and liver. Electrophoretic analysis of tissues after cachectin exposure suggest that the protein is rapidly degraded after tissue binding (Beutler et. al., 1985c). High affinity plasma membrane receptors are present on many normal tissue cell types, but the post-receptor events leading to intracellular responses remain obscure. Cachectin-induced expression of a 26 kDa protein mRNA may be related to the resistance of some transformed cells to TNF-cytotoxicity (Defilippi et. al., 1987), but additional studies of post-receptor mechanisms and second messengers are needed to better understand the wide variety of cellular responses induced by cachectin in normal and transformed cells.

The administration of cachectin to normal animals is capable of inducing a series of acute inflammatory and systemic responses which are perhaps indistinguishable from endotoxemia (Tracey et. al., 1986a; Kettelhut et. al., 1987). Cachectin is an endogenous pyrogen; a biphasic fever response is observed following infusion of the cytokine in rabbits (Dinarello et. al., 1986). The first fever spike is observed within 45-55 min and is due to a direct effect of cachectin-stimulated hypothalamic PGE_2 synthesis. This is followed by a second, interleukin-1 mediated fever peak (3 hours later), resulting from the cachectin-induced release of IL-1 from monocytes (Bachwich et. al., 1986) and endothelial cells (Nawroth et. al., 1986). As with the other pyrogenic cytokines, cachectin is capable of inducing important immunological responses during endotoxemia, including the activation of leukocytes and stimulation of neutrophil phagocytosis, degranulation, and superoxide release (Djeu et. al., 1986; Shalaby et. al., 1985; Tsujimoto et. al., 1986).

During infection with Gram-negative organisms, plasma concentrations of several proteins, the acute phase proteins, increase significantly whereas albumin levels typically decrease. These responses, considered a hallmark of endotoxemia, are frequently associated with increased net peripheral tissue protein catabolism. Cachectin exposure to human hepatoma cells is capable of inducing increased biosynthesis of several acute phase proteins, and decreased synthesis of albumin, and these responses occur in a reversible and cachectin dose-dependent manner (Perlmutter et. al., 1986). IL-1 also induces similar effects on the regulation of acute phase protein biosynthesis in this model, and the administration of either monokine is associated with elevated plasma levels of alpha-1 glycoprotein. Only cachectin however, has been reported to induce significantly decreased levels of plasma albumin as typically occurs during prolonged acute phase responses (Flores et. al., 1987). Thus, the appearance of cachectin and the subsequent induction of IL-1 during endotoxemia likely participate in the regulation of altered acute phase protein levels.

We first reported the severe systemic toxicity of cachectin/TNF following studies of rats infused with recombinant human cachectin. The cachectin-infused animals acutely developed hypotension, hyperglycemia, lactic acidosis, shock, and after a cachectin dose-dependent period of time, abrubtly succumbed to respiratory arrest (Tracey et. al., 1986a). Hemoconcentration, which occurs within minutes after cachectin-induced hypotension, is similar to the intravascular depletion typically observed in endotoxemic shock, and appears to be secondary to fluid sequestration in both interstitial and intracellular spaces. During cachectin-induced shock, very high respiratory rates may represent either a compensatory response to metabolic acidosis, or the effect of direct hypothalamic respiratory stimulation by cachectin. Prolonged hyperventilation cannot be maintained in the face of progressive shock and abrupt respiratory arrest ensues.

The rat is relatively resistant to the biological effects of recombinant human cachectin; the $L.D._{50}$ has been estimated to be 700 μg cachectin/kg body weight (Tracey et. al., 1986a). Recent evidence suggests that this may be in part due to inter-species variability of decreased receptor binding to human cachectin in the rodent (Smith et. al., 1986). The dog however, is exquisitely sensitive to human cachectin and becomes acutely hypotensive after doses (10-100 μg cachectin/kg body weight) that produce only nanomolar serum concentrations (Tracey et. al., 1987b). The mechanisms of cachectin-induced shock are incompletely characterized, but as generally occurs during lethal endotoxemia, dogs receiving a lethal dose of cachectin develop hypotension and shock in association with a decreased cardiac output.

Cardiac output falls after cachectin infusion even though the serum adrenaline levels are significantly increased, suggesting that cachectin-induced shock is in part dependent on depressed cardiac function (Tracey et. al., 1987b). Whether this response is due to a direct effect of cachectin on cardiac muscle remains uncertain, but this monokine has been implicated as a mediator of skeletal muscle membrane function and metabolism. Cachectin induces a diminished resting transmembrane potential difference in rat skeletal muscle in vitro, and L-6 myoblasts are induced to express increased hexose transporters, accelerated glucose uptake, glycogen breakdown, and lactate efflux (Tracey et. al., 1986c, Lee et. al., 1987). The significance of these myoblast metabolic responses in cardiac muscle is unknown, but recently we observed that chronic administration of sublethal cachectin doses was associated with sub-endocardial infarct necrosis (Tracey et. al., unpublished). We have hypothesized that infection-induced cardiac depression is in part mediated by cachectin, but further investigations are required to discern the humoral mediators and cellular mechanisms involved.

Cachectin infusion in dogs also induces significant alterations of circulating glucose counter-regulatory hormones including epinephrine, norepinephrine, glucagon and cortisol (Tracey et. al., 1987b). The release of these hormones is characteristic of the "stress hormone" response observed after lethal endotoxemia, and has previously been attributed to a direct effect of LPS on both the hypothalamus and adrenal cortex. Vascular baroreceptors are stimulated by systemic hypotension and may effectively stimulate stress hormone release, but cachectin alone is sufficient to induce marked perturbations in systemic hormone milieu, even in the absence of significant hypotension or endotoxemia. Since these catabolic hormones exert a dominant role in the metabolic control of carbohydrate and lipid utilization after acute injury or infection, endogenous cachectin appearance may be important during the mobilization of host energy stores as necessary to meet the increased energy demands of the infected host. We and others (Tracey et. al., 1986a; Kettelhut et. al., 1987) have also observed acute hyperglycemia following cachectin administration, after which a profound hypoglycemia develops (within 2-3 hours). The role of insulin, the counter-regulatory hormones, and altered peripheral tissue glucose utilization (Lee et. al., 1987) during these metabolic changes requires further study.

Necropsy of animals succumbing to the lethal effects of cachectin reveals gross evidence of hemorrhagic necrosis and acute inflammation in many vital organs (Tracey et. al., 1986a; Tracey et. al, 1987b; Remick et. al., 1987). These lesions are strikingly similar to those observed after endotoxic shock. Typically the lungs of cachectin-treated rats or dogs are grossly congested with hemorrhage and pulmonary edema. Microscopic hematuria is observed, the kidneys appear distended, and the renal cortex is diffusely pale. In the rat, the cecum is particularly sensitive to cachectin-induced hemorrhagic necrosis, patchy ischemia and segmental infarction. Hemorrhagic necrosis of the adrenal medulla frequently occurs after overwhelming infection or lethal endotoxemia (reviewed by Gilbert, 1960) and has also been observed after cachectin-induced shock in dogs (Tracey et. al., 1987b).

Histopathological analysis of vital organ sections obtained after cachectin-induced shock are also strikingly similar to those observed after endotoxic shock. The pulmonary capillaries are congested with leukocytes, primarily polymorphonuclear cells adherent to the endothelial surface, and acute interstitial and alveolar inflammation is diffuse. Moreover, the alveolar walls appear somewhat thickened and the histopathological appearance is essentially indistinguishable from the acute inflammation observed after endotoxemia. The kidneys undergo acute tubular necrosis, proteinaceous deposits are present in the tubular lumens and pyknotic changes are observed in tubular epithelium. The cecum of the rat frequently exhibits hemorrhagic tissue necrosis with sloughing of the epithelium, loss of normal glandular structure, and acute submucosal inflammation. Ultrastructural analysis of small intestine from mice treated with even very low doses of cachectin has revealed significant endothelial damage with extravasation of red blood cells into the interstitium, and sequestration of neutrophils in the submucosa (Remick et. al., 1987).

The pathophysiologic cellular mechanisms underlying cachectin-induced shock, multiple organ failure and death are partially understood. It is clear that cachectin enhances endothelial procoagulant activity, and suppresses cofactor activity for the anticoagulant protein C pathway, resulting in an increased endothelial propensity towards clot formation (Stern and Nawroth, 1986; Bevilacqua et. al., 1986). These responses have been implicated in both infection-associated disseminated intravascular coagulation, and intra-vascular thrombosis causing hemorrhagic necrosis in normal tissues. Moreover, cachectin induces increased expression of endothelial activation antigens which participate in the enhanced binding of leucocytes to the endothelial surface (Gamble et. al., 1985). Since cachectin also activates neutrophils (Klebanoff et. al., 1986) resulting in degranulation and superoxide anion release, significant subendothelial tissue injury may be mediated via accelerated coagulation and inflammation. Kettelhut et. al. inhibited cachectin/TNF toxicity by pretreatment with cyclo-oxygenase inhibitors, suggesting that PGE_2 synthesis mediates some cachectin-induced responses (altered blood glucose, acidosis, and lethargy) (1987). Clearly the activation of other humoral mediators (e.g. interferon-γ, IL-1, IL-2, platelet activating factor, products of complement activation) would contribute to these deleterious and injurious responses, but the role of these mediators has not been fully characterized during cachectin-induced shock.

Cachectin production in endotoxemia and disease

Cachectin appearance in serum has been studied under a variety of experimental and clinical conditions including: animals injected with endotoxin (LPS) or live E. coli; patients with parasitic and Gram-negative infections; and normal volunteers injected with small quantities of LPS. Although a variety of bacterial, viral and parasitic products are capable of inducing macrophages to release cachectin, LPS appears to be the most potent stimulus capable of inducing a rapid translation of nascent cachectin mRNA normally present in resting macrophages (Beutler et. al., 1986). Cachectin gene transcription, which may be under the control of short-lived suppressor molecules (Collart et. al., 1986), is also stimulated by LPS, leading to the release of relatively large quantities of this potent hormone. Persistent exposure of macrophages to LPS in vitro, however, does not lead to continuous production of cachectin, and biosynthesis is discontinued within several hours (Gifford and Lohmann-Matthes, 1986).

Infusion of a lethal dose of LPS in rabbits, mice, or baboons stimulates the appearance of serum cachectin levels in the nanomolar range as measured by either ELISA or bioassay (Beutler et. al., 1985c; Tracey et. al., 1987a; Hesse et. al., 1987). Maximal levels are detected between 1.5-2 hours following endotoxin bolus. Since the serum half-life of cachectin is 6-20 minutes, circulating cachectin is cleared within 4-5 hours and serum levels normalize to baseline (nondetectable) values. The appearance of nanomolar serum cachectin/TNF concentrations, however, is invariably associated with progressive shock, multiple system organ failure, and death within minutes to hours. Maximal serum levels are related to the dose of LPS administered, such that sub-lethal doses of LPS induce lesser circulating cachectin levels. When human subjects are infused with sub-lethal doses of LPS, maximal cachectin/TNF bioactivity is achieved within 1.5-2 hours, and the appearance of headache and fever coincides with the peak serum concentration (Hesse et. al., 1987). The magnitude of the serum cachectin response, however, is only one-tenth to one-one hundredth that measured in baboons given a lethal intravenous dose of live E. coli.

The highest serum concentrations of endogenous cachectin reported in man were measured in samples obtained from patients with mennigococcal meninigitis and/or septicemia (Waage et. al., 1987). In that study, all patients with serum cachectin/TNF concentrations greater than 0.1 ng/ml died; the magnitude of this circulating cachectin response agrees closely with estimates of lethality in animals. Earlier observations from this group of investigators also demonstrated the appearance of cachectin/TNF in the serum of cancer patients with sepsis caused by a variety of pathogens (Waage et. al., 1986). We have detected cachectin bioactivity in 40% of patients with critical illness and sepsis using an assay based on cachectin-induced depolarization of skeletal muscle (Tracey et. al., 1986). Finally, Scuderi et. al. have observed significant elevations of cachectin/TNF in 68% of patients with parasitic disease caused by leishmania or P. falciparum (Scuderi et. al., 1987). These measurements of serum cachectin appearance during septicemia and parasitic infection in man suggest that this toxic hormone is normally produced in response to invasion.

Clinical significance and treatment

Septic shock is associated with a mortality rate of 50-70%, and represents a serious complication of invasive infection. While cachectin has previously been implicated as a proximal mediator of this syndrome, the significance of systemic cachectin toxicity is now better appreciated from the results of experimental studies in animals. Moreover, systemic toxicity has been observed (fever, headache, hypotension, pulmonary edema) after recombinant TNF administration to cancer patients. While significant toxicity to TNF may prevent the drug from becoming a panacea in the treatment of malignancy, extensive investigation has also focused on the development of antibodies to cachectin/TNF for use in the treatment of septicemia.

The administration of polyclonal anti-murine cachectin antibodies to mice confers significant survival advantages during subsequent injection with LPS (Beutler et. al., 1985d). Moreover, monoclonal antibodies against recombinant human cachectin/TNF also protected against shock and death in rats subsequently infused with cachectin (Tracey et. al., 1986a). We have recently observed that monoclonal anti-cachectin antibodies protect baboons from septic shock when administered two hours before the infusion of a lethal dose of E. coli (Tracey et. al., 1987a). Thus, a large body of evidence suggests that cachectin/TNF is a crucial mediator in the development of lethal septic shock, and may be both a.) sufficient to induce a sepsis-like syndrome, and b.) necessary during the pathophysiological responses observed in Gram-negative septicemia.

The discovery and isolation of cachectin as a pivotal mediator of the deleterious consequences of endotoxemia has led to a new appreciation of infection-associated pathophysiology. Septic shock and multiple organ failure, which can progress even after the invading pathogen has been eradicated, may be secondary to systemic toxicity induced by cachectin. The syndrome of clinical sepsis may actually occur as an example of immune system "over-responsiveness" to invading pathogens, leading to the eventual demise of the host (Tracey et. al, 1987a). Thus, while immune responsiveness to LPS may be benefical and protective, the detrimental pathobiology induced by cachectin may be analagous to the life-threatening anaphylactic response induced by simple allergens. Future investigations are needed to determine whether inhibition of cachectin toxicity will improve survival during human infections and septic shock.

References

Aggarwal MK, ed. (1980). Bacterial endotoxins and host response. Elsevier/North-Holland Biomedical Press, Amsterdam.

Bachwich PR, Chensue SW, Larrick JW, Kunkel SL (1986). Tumor necrosis factor stimulates IL-1 and PGE_2 production in resting macrophages. Biochem. Biophys. Res. Comm. 136: 94-101.
Balis JU, Rappaport ES, Gerber L, Fareed J, Buddingh F, Messmore HL (1978). A primate model for prolonged endotoxin shock: Blood-vascular reactions and effects of glucocorticoids. Lab. invest. 38: 511-523.
Beutler B, Greenwald D, Hulmes JD, Chang M, Pan Y-CE, Mathison J, Ulevitch R, Cerami A (1985a). Identity of tumor necrosis factor and the macrophage-secreted factor cachectin. Nature 316: 552-554.
Beutler B, Mahoney J, LeTrang N, Pekala P, Cerami A(1985b). Purification of cachectin, a lipoprotein lipase-suppressing hormone secreted by endotoxin-induced RAW 264.7 cells. J. Exp. Med. 161: 984-995.
Beutler B, Milsark IW, Cerami A (1985c). Cachectin/tumor necrosis factor: production, distribution, and metabolic fate in vivo. J. Immunol. 135: 3972-3977.
Beutler B, Milsark IW, Cerami A (1985d). Passive immunization against cachectin/tumor necrosis factor (TNF) protects mice from the lethal effects of endotoxin. Science 229: 869-871.
Beutler B, Krochin N, Milsark IW, Leudke C, Cerami A (1986). Control of cachectin (tumor necrosis factor) synthesis: mechanism of endotoxin resistance. Science 232: 977-980.
Bevilacqua MP, Pober JS, Majeau GR, Fiers W, Cotran RS, Gimbrone Jr MA (1986). Recombinant tumor necrosis factor induces procoagulant activity in cultured human vascular endothelium: characterization and comparison with the effects of interleukin-1. Proc. Natl. Acad. Sci. (USA) 83: 4533-4537.
Caput D, Beutler B, Hartog K, Thayer R, Brown-Shimer S, Cerami A (1986). Identification of a common nucleotide sequence in the 3'untranslated region of mRNA molecules specifying inflammatory mediators. Proc. Natl. Acad. Sci. (USA) 83: 1670-1674.
Collart MA, Belin D, Vassalli J-D, DeKossodo S, VassalliP (1986). Gamma-interferon enhances macrophage transcription of the tumor necrosis factor/cachectin,interleukin-1, and urokinase genes, which are controlled by short lived repressors. J. Exp. Med. 164: 2113-2118.
Defilippi P, Poupart P, Tavernier J, Fiers W, Content J (1987). Induction and regulation of mRNA encoding 26-kDa protein in human cell lines treated with recombinant human tumor necrosis factor. Proc. Natl. Acad. Sci. (USA) 84: 4557-4561.
Dinarello CA, Cannon JG, Wolff SM, Bernheim HA, Beutler B, Cerami A, Palladino MA, O'Connor JV (1986). Tumor necrosis factor (cachectin) is an endogenous pyrogen and induces production of interleukin-1. J. Exp. Med. 163: 1433-1450.

Djeu JY, Blanchard DK, Halkais D, Friedman H (1986). Growth inhibition of Candida albicans by human polymorphonuclear neutrophils: activation by interferon-gamma and tumor necrosis factor. J. Immunol. 137:2980-2984.

Flores E, Drabik M, Bistrian BR, Istfan N, Blackburn GL, Dinarello CA (1987). Acute phase response to recombinant mediators. Surgical Forum 37: 28-30.

Franke FE (1944). Action of toxic doses of the polysaccharide from Serratia marcescens (Bacillus prodigiosis) on the dog and guinea pig. J. Natl. Cancer Inst.5: 185-193.

Fransen L, Muller R, Marmenout ATavernier J, Van der Heyden J, Kawashima E, Chollet A, Tizard R, Van Heuverswyn H, Van Vliet A, Ruysschaert M-R, Fiers W (1985). Molecular cloning of mouse tumor necrosis factor cDNA and its eukaryotic expression. Nucleic Acids Res. 13: 4417-4429.

Gamble JR, Harlan JM, Klebanoff SJ, Vadas MA (1985). Stimulation of the adherence of neutrophils to umbillical vein endothelium by human recombinant tumor necrosis factor. Proc. Natl. Acad. Sci. (USA) 82: 9667-8671.

Gifford GE, Lohmann-Matthes ML (1986). Requirement for the continual presence of lipopolysaccharide for production of tumor necrosis factor by thioglycollate-induced peritoneal murine macrophages. Int. J. Cancer 38: 135-137.

Gilbert RP (1960). Mechanism of the hemodynamic effects of endotoxin. Physiol. Rev. 40: 245-279.

Hesse DG, Tracey KJ, Fong Y, Manogue KR, Palladino Jr MA, Dinarello CA, Cerami A, Lowry SF (1987). Cytokine appearance in human endotoxemia and non-human primate bacteremia. Submitted.

Kettelhut IC, Fiers W, Goldberg AL (1987). The toxic effects of tumor necrosis factor in vivo and their prevention by cyclooxygenase inhibitors. Proc. Natl. Acad. Sci. 84: 4273-4277.

Klebanoff SJ, Vadas MA, Harlan JM, Sparks LH, Gamble JR, Agosti JM, Waltersdorph AM (1986). Stimulation of neutrophils by tumor necrosis factor. J. Immunol. 136: 4220-4225.

Lee DM, Zentella A, Pekala PH, Cerami A (1987). Effect of endotoxin-induced monokines on glucose metabolism in the muscle cell line L-6. Proc. Natl. Acad. Sci. (USA) 84: 2590-2594.

MacLean LD, Weil MH (1956). Hypotension (shock) in dogs produced by Escherichia coli endotoxin. Circul. Res. 4: 546-556.

Nawroth PP, Bank I, Handley D, Cassimeris J, Chess L, Stern D (1986). Tumor necrosis factor/cachectin inter-acts with endothelial cell receptors to induce release of interleukin-1. J. Exp. Med. 163: 1363-1375.

Palladino MA, Shalaby MR, Kramer SM, Ferraiolo BL, Baughman RA, Deleo AB, Crase D, Marafino B, Aggarwal BB, Figari IS, Liggitt

D, Patton JS (1987). Characterization of the antitumor activities of human tumor necrosis factor-alpha and the comparison with other cytokines: induction of tumor specific immunity. J. Immunol. 138:4023-4032.

Pennica D, Hayflick JS, Bringham TS, Palladino MA, Goedell DV (1985). Cloning and expression in Escherichia coli of the cDNA for murine tumor necrosis factor. Proc. Natl. Acad. Sci. (USA) 82: 6060-6064.

Perlmutter DH, Dinarello CA, Punsal PI, Colten HR (1986). Cachectin/tumor necrosis factor regulates hepatic acute phase gene expression. J. Clin. Invest. 78: 1349-1354.

Remick DG, Kunkel RG, Larrick JW, Kunkel SL (1987). Acute in vivo effects of human recombinant tumor necrosis factor. Lab. Invest. 56: 583-590.

Rouzer CA, Cerami A (1980). Hypertriglyceridemia associated with Trypanosoma brucei infections in rabbits: role of effective triglyceride removal. Mol. Biochem. Parasitol. 2: 31-38.

Scuderi P, Lam KS, Ryan KJ, Petersen E, Sterling KE, Finley PR, Ray CG, Slymen DJ, Salmon SE (1987). Raised serum levels of tumor necrosis factor in parasitic infections. Lancet ii: 1364-1365.

Shalaby MR, Aggarwal BB, Rinderknecht E, Svedersky LP, Finkle BS, Palladino MA (1985). Activation of human polymorphonuclear neutrophil functions by interferon-gamma and tumor necrosis factors. J. Immunol. 135: 2069-2075.

Smith RA, Kirstein M, Fiers W, Baglioni C (1986). Species specificity of human and murine tumor necrosis factor. J. Biol. Chem. 261: 14871-14874.

Stern DM, Nawroth PP (1986). Modulation of endothelial hemostatic properties by tumor necrosis factor. J. Exp. Med. 163: 740-745.

Tracey KJ, Beutler B, Lowry SF, Merryweather J, Wolpe S, Milsark IW, Hariri RJ, Fahey III TJ, Zentella A, Albert JD, Shires GT, Cerami A (1986a). Shock and tissue injury induced by recombinant human cachectin. Science 234: 470-474.

Tracey KJ, Lowry SF, Beutler B, Cerami A, Albert JD, Shires GT (1986b). Cachectin/tumor necrosis mediates changes of skeletal muscle plasma membrane potential. J. Exp. Med. 164: 1368-1373.

Tracey KJ, Lowry SF, Beutler B, Cerami A, Albert JD, Shires GT (1986c). Cachectin/TNF participates in reductions of skeletal muscle membrane potential: evidence for bioactivity in plasma during critical illness. Surgical Forum 37: 13-15.

Tracey KJ, Fong Y, Hesse DG, Manogue KR, Lee AT, Kuo GC, Lowry SF, Cerami A (1987a). Anti-cachectin/TNF monoclonal antibodies prevent septic shock during lethal bacteremia in baboons. Nature (London), in press.

Tracey KJ, Lowry SF, Fahey III TJ, Albert JD, Fong Y, Hesse DG,

Beutler B, Manogue KR, Calvano SE, Wei H, Cerami A, Shires GT (1987b). Cachectin/tumor necrosis factor induces lethal shock and stress hormone responses in the dog. Surg. Gynecol. & Obstet. 164: 415-422.

Tsujimoto M, Yokota S, Vilcek J, Weissmann G (1986). Tumor necrosis factor provokes superoxide anion generation from neutrophils. Biochem Biophys. Res. Commun. 137: 1094-1100.

Waage A, Espevik T, Lamvik J (1986). Detection of tumour necrosis factor-like cytotoxicity in serum from patients with septicaemia, but not from untreated cancer patients. Scand. J. Immunol. 24: 739-743.

Waage A, Halstensen A, Espevik T (1987). Association between tumour necrosis factor in serum and fatal outcome in patients with meningococcal disease. Lancet i: 355-357.

Wang AM, Creasey AA, Ladner MB, Lin LS, Strickler J, Van Arsdell JN, Yamamoto R, Mark DF (1985). Molecular cloning of the complementary DNA for human tumor necrosis factor. Science 228:149-154.

Bacterial Endotoxins: Pathophysiological Effects, Clinical Significance, and Pharmacological Control, pages 89–101

CHARACTERISTICS OF LIPOPOLYSACCHARIDE-INDUCED INTERLEUKIN-1 PRODUCTION BY HUMAN MONOCYTES. CLINICAL RELEVANCE IN PATIENTS UNDERGOING HEMODIALYSIS.

Nicole Haeffner-Cavaillon, Françoise Bacle, Martine Caroff, and Jean-Marc Cavaillon.

INSERM U28, Hôpital Broussais, 75014 Paris;
Inst. Biochimie, Université Paris XI, Orsay;
U. Immuno-Allergie, Inst. Pasteur, 75015 Paris.

INTRODUCTION

Lipopolysaccharides (LPS) from a variety of Gram-negative bacteria such Enterobacteriaceae and Neisseriaceae are potent endotoxins that elicit typical effects in humans including fever, hypotension and irreversible shock. Most toxic activities of LPS are inducible by the hydrophobic moiety of the molecule termed lipid A. The hydrophilic region of Enterobacteriaceae LPS consists of two different regions, the O-specific chain and the core oligosccharide. Whereas the former exhibit a high compositional variability by which the serological specificity of each strain is manifested, the core oligosaccharide constitutes a domain of limited structural diversity (Lüderitz & al. 1983). The lipid A may vary according to its fatty acid content although it expresses a common backbone consisting of a β D-glucosaminyl-(1'-6)α-D-glucosamine disaccharide which is phosphorylated in position 4' of the distal glucosaminyl residue and in position 1 of the reducing glucosaminyl residue. The phosphoryl groups may be either free or substituted by non-acylated substituents. These substituents contribute to the diversity of lipid A. Structural diversity of LPS is also present within the "hinge-like" region of the molecule, consisting of one, two or three 3-deoxy-D-manno-2-octulosonic acid (KDO) residue(s) (Rietschel et al. 1984). The biochemical heterogeneity of LPS within the same strain (Nowotny 1984) and the diversity within a species may contribute to the different reactivity of LPS in various biological systems, i.e. reactivity of LPS with polymyxin B (Vaara et al. 1979), detection of LPS using Limulus assay (Arend and Massoni 1986) and

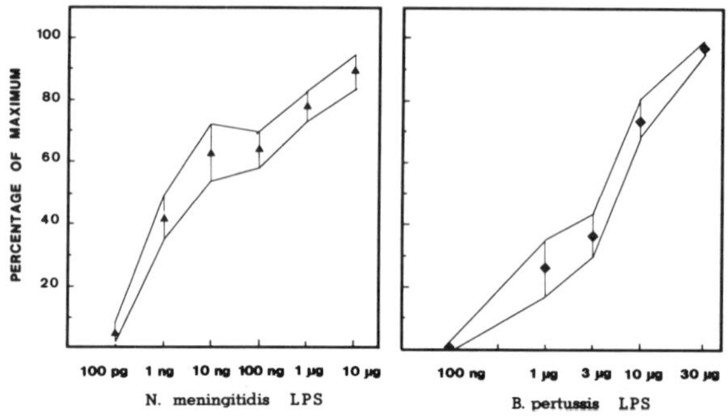

FIGURE 1. Dose dependent effect of Neisseria meningitidis and Bordetella pertussis endotoxins on the induction of IL-1 secretion by human monocytes. Each experimental point represents the mean value obtained from 10 to 22 different donors +/- SE. A suspension of peripheral blood mononuclear cells (0.5 ml) containing 10^6 NSE+ cells/ml were led to adhere in 24 wells plastic dishes in the absence of serum for 1h at 37 °C. After washing, the adherent cells are cultured in 0.5 ml RPMI for 24 h in the absence of serum, and in the absence or in the presence of IL-1 inducers. Cell supernatants were harvested, centrifuged and tested for their capacity to induce C3H/HeJ thymocytes proliferation in the presence of suboptimal doses of Con A. Diluted supernatants (1:10) were added to 0.75×10^6 thymocytes, in the presence of 4% FCS, 5×10^{-5} M 2-mercaptoethanol and 0.075 ug Con A per well. The cultures were maintained 72 h at 37 °C and 3H-thymidine was added 7 h before the end of the culture.

ability of LPS to induce IL-1 (Newton 1986).

I. INDUCTION OF IL-1 PRODUCTION BY LPS.

The capacity of LPS to induce IL-1 release by human monocytes depends upon the family of bacteria from which it originates (figure 1). Triggering of IL-1 production by human monocytes is a rapid process : the appearance of IL-1 mRNA (Auron et al. 1987) and of cell-associated IL-1 activity reach a maximum by 3 h . The release of IL-1 needs a further incubation period which does not require the presence of LPS throughout the culture period. After 1 h of stimulation, the culture medium can be replaced by fresh RPMI

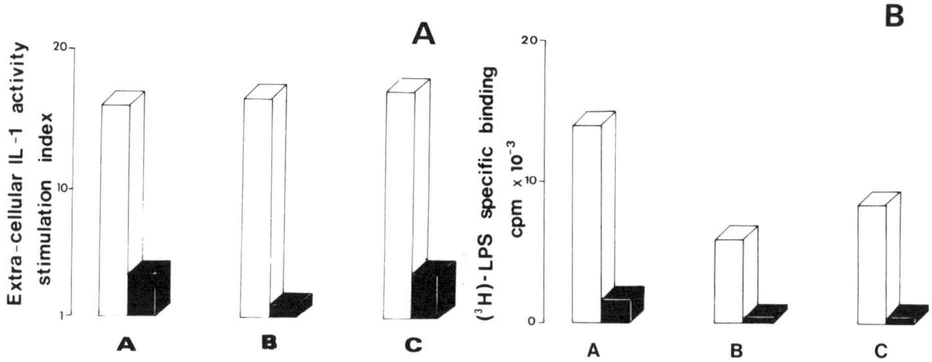

Figure 2. A. Extracellular IL-1 activity measured in supernatants of freshly isolated monocytes cultured 24 h in the presence of LPS (□), and of 24 h-cultured monocytes prior to LPS stimulation (10µg/ml)(■). Results obtained from three different donors (A,B,C) are expressed as stimulation index (cpm in the presence of supernatant from LPS-stimulated cells / cpm in the presence of supernatant from cells cultured in RPMI alone). B . Binding of 3H-LPS to freshly isolated human monocytes from three different donors (□) and to 24 h-cultured cells (■). Results are expressed as specific binding : cpm 3H-LPS (0.2µg/well) - cpm 3H-LPS (0.2µg/well) in the presence of 10µg unlabeled LPS.

without affecting the release of IL-1 by the cells. Human monocytes cultured in vitro over 24 h prior to addition of LPS lose the ability to release IL-1 (fig. 2A). However the cells are still able to respond to LPS since intracellular IL-1 activity can be detected . As shown by several authors (Arenzana-Seisdedos et al. 1985; Newton 1985) the capacity of 24 h-cultured monocytes to secrete IL-1 upon further LPS stimulation can be restored by the addition of interferon gamma (INFγ)(fig. 3). However, the total production of IL-1 (intracellular IL-1 + released IL-1) upon LPS stimulation is not decreased when the cells are cultured 24 h prior to LPS stimulation, rather only the pathway of IL-1 release is altered and upon stimulation with INFγ this release is partially restored (table 1). INF also modulates the expression of specific LPS-receptors shown to be present on human monocytes (Haeffner-Cavaillon et al. 1982). Specific binding of radiolabeled LPS is markedly decreased when cells are maintained in the culture medium for 24 h prior to the binding assay, indicating a loss or conformational change of LPS

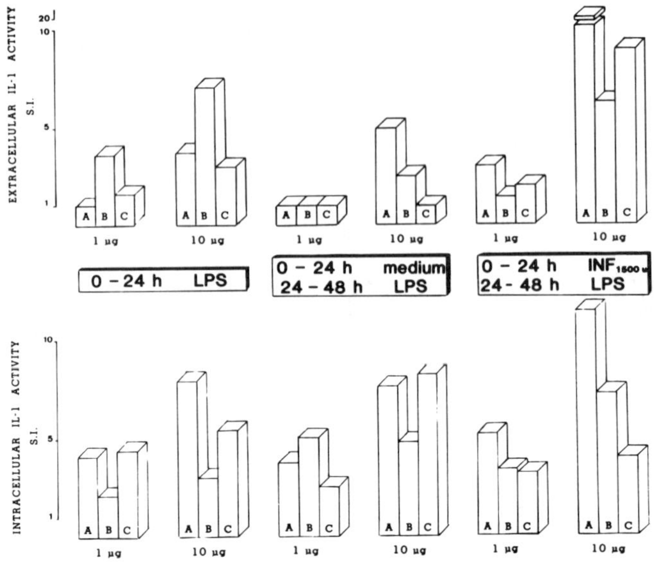

Figure 3. LPS-induced IL-1 production by freshly isolated human monocytes from 3 different donors (A,B,C) and 24 h -cultured cells in the absence and in the presence of INFγ (1500 u/well) prior to LPS stimulation.

receptors during the culture period (Fig. 2B). Addition of interferon gamma during the 24 h culture period prior to the binding assay partially maintained the expression of LPS-receptors (unpublished results). Furthermore, we have shown that gangliosides inhibit LPS-induced IL-1 release (Cavail-

TABLE 1. Percentage of total and released interleukin-1 activity produced by human monocytes cultured for 48 h, with or without INF for 24 h prior to LPS stimulation.

INF (1500 u/well)	-		+	
IL-1 activity	Total	Released	Total	Released
Donor A	100 %	7 %	116 %	39 %
Donor B	100 %	13 %	94 %	35 %
Donor C	100 %	17 %	128 %	21 %

lon and Fitting 1986) as well as the specific binding of LPS to human monocytes (Cavaillon & al. 1987). Altogether, these correlations between the specific binding of LPS and the IL-1 release suggest that the binding of LPS to its receptor is a prerequisit to IL-1 secretion as a signal provided by LPS to human monocytes. Certain serum components are IL-1 inducers including C5a, C3a (Goodman & al.1982; Haeffner-Cavaillon & al.1987), and thrombin (Lopreste and Dinarello 1987). Potent IL-1 inducers and inhibitors (Dinarello et al. 1981) are present in the serum, however the addition of 0.4% serum to cultures of monocytes potentiates the IL-1 release induced by LPS and PS (Cavaillon & Haeffner-Cavaillon 1985) and allows the detection of the specific binding of LPS to human monocytes (Haeffner-Cavaillon & al.1985). Serum components which interact with LPS such as C3 (Joiner et al. 1984), a 28 Kd protein (Brade and Brade 1985), or high density lipoproteins (Ulevitch et al. 1981), may be involved in the potentiating effect of serum. High density lipoproteins (HDL) are involved in the detection of specific binding. In the presence of sera which are partially depleted in HDL, the specific binding is markedly decreased (table 2). Fifty percent of the maximum specific binding obtained in the presence of normal human serum could be achieved by addition of purified HDL (data not shown). Lipoprotein-bound LPS is less potent than free LPS in causing complement activation, leukopenia and fever (Ulevitch & al. 1981); in contrast, the association of LPS to serum high density lipoprotein does not reduce LPS-induced IL-1 production (data not shown).

TABLE 2. N. meningitidis 3H-LPS specific binding to human monocytes in the absence or in the presence of lipoproteins.

Serum	A		B		C
concentration of Apolipoprotein A (a)	100 %		20 %		< 5 %
amount of serum used in the binding assay	10µl	2µl	10µl	2µl	10µl
SPECIFIC BINDING	100%	58%	76%	14%	2%

(a) expressed as percentage of Apo A in normal human serum

II. INDUCTION OF IL-1 PRODUCTION BY ISOLATED POLYSACCHARIDE (PS) AND LIPID A MOIETIES.

The isolated polysaccharide moieties from B. pertussis and N. meningitidis LPS are able to induce IL-1 release by human monocytes cultured in serum-free conditions (Lebbar & al. 1986). The structural determinants required for IL-1 induction by some PS are probably located in the "inner core like" region of the molecule which consists of unusual sugars such as L-glycero-D-manno-heptose and KDO. We observed that the activity of isolated PS was dependent upon the hydrolysis used to release it from the LPS. Thus PS isolated after hydrolysis of the LPS in acetate buffer (Rosner & al. 1979) was active, whereas those obtained after hydrolysis in acetic acid (11mM) or mineral acid (HCl 0.25N) were devoid of IL-1 inducing capacity. The lack of activity correlated with a modification within the KDO residue (unpublished results). Neither synthetic KDO residues alone (Lebbar & al. 1986), nor synthetic KDO residues substituted in position 5 (the position which is substituted by a heptose residue in almost all LPS) by a methyl group or a benzyl group were able to induce IL-1 secretion. The heptose-KDO disaccharide is the most conserved region within the core in LPS from Gram-negative bacteria. KDO-deficient LPS have been repor-

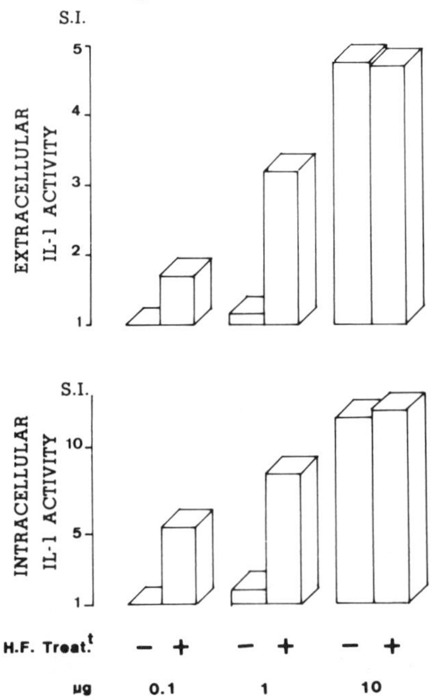

FIGURE 4. Intracellular IL-1 activity (in cell lysates), and extracellular IL-1 activity produced by human monocytes stimulated by untreated (−) Vibrio cholerae LPS or treated with aqueous hydrofluoric acid (50%) (+).

ted, based on a negative response in the thiobarbiturate assay, however, it has been shown that endotoxins devoid of KDO probably do not exist (Caroff et al. 1987). For example LPS from Bacteroides gingivalis shown to be an IL-1 inducer was reported to be devoid of KDO (Hamada 1986). However, after appropriate treatment, B.gingivalis LPS gives a positive reaction in the thiobarbiturate assay. Vibrio cholerae LPS, first reported as "KDO-free" has been shown to contain phosphorylated KDO (Brade & al. 1985) and to respond in the thiobarbiturate assay after treatment with aqueous hydrofluoric acid (50% HF, 4°C, 48h)(Caroff & al.1987). The capacity to induce IL-1 production by V. cholerae LPS treated with HF, in similar conditions to those required to improve the thiobarbiturate response, is increased compared to that

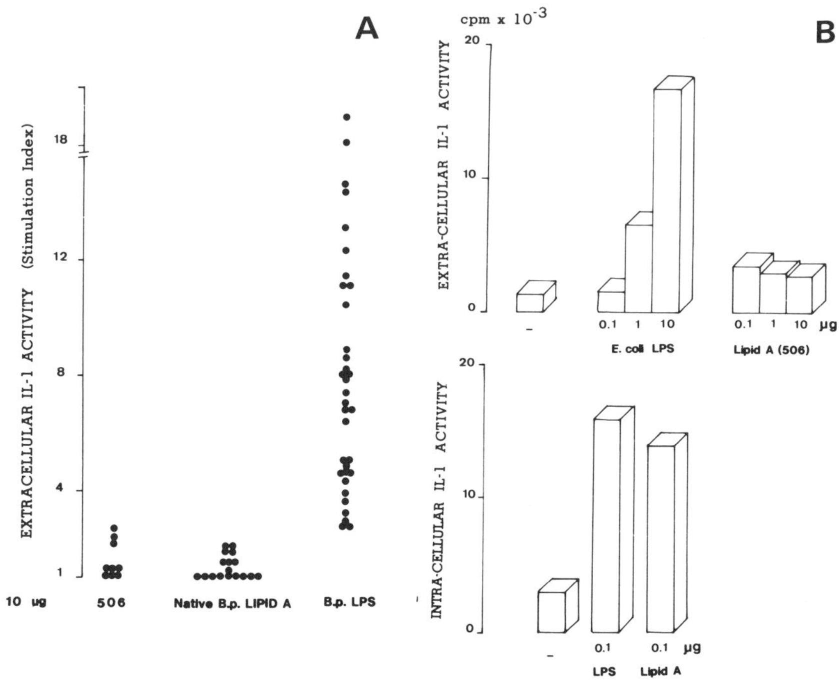

Figure 5. A. IL-1 release by human monocytes stimulated with synthetic lipid A (506), native purified lipid A from B.pertussis LPS and B.pertussis LPS. Each dot represent the results obtained with the monocytes from one individual donor. B. Intracellular IL-1 activity and IL-1 release (extracellular IL-1 activity) by monocytes from a given donor, cultured in the presence of synthetic lipid A (506) or E.coli LPS.

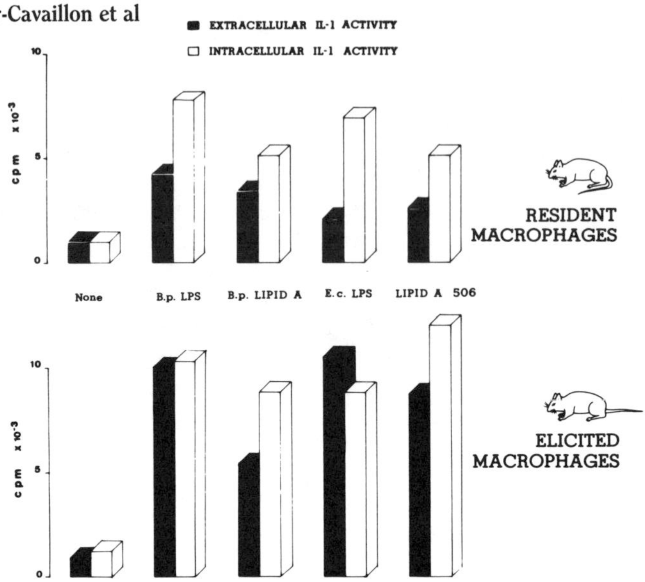

Figure 6. IL-1 production by resident or thioglycolate elicited Balb/c mouse macrophages stimulated with B. pertussis LPS, E. coli LPS, B. pertussis isolated lipid A and synthetic lipid A (compound 506) (10µg/well).

induced by untreated LPS (figure 4). To further investigate the determinants required for IL-1 induction we examined the role of the KDO-residue present at the reducing end of different LPS-derived oligosaccharides. A polysaccharide fragment, containing as a terminal reducing sugar a KDO residue was isolated from B. pertussis LPS, either after hydrolysis in acetate buffer (compound B) or after acidic methanolyis (B-OMe). Fragment B was as potent as the native PS whereas fragment B-OMe containing the methyl-ketoside-methylester-KDO residue as the reducing sugar was inactive, suggesting that the free carboxyl group of the KDO residue is involved for the induction of IL-1. This hypothesis was corroborated by the fact that after alkaline treatment of B-OMe, restoring the carboxyl group, the oligosaccharide was as active as fragment B (manuscript in preparation). Similar conclusions were reached with the trisaccharide (hep-hep-KDO) isolated from Salmonella minnesota which is representative of the inner-core region of many enterobacterial LPS. Isolated hep-hep-KDO was able to induce IL-1 secretion in a dose-dependent fashion (Lebbar & al. 1986). When the trisaccharide was treated with NaBH4, the carbonyl reduced trisaccharide was inactive, indicating a conformational requirement in the KDO molecule.

Human monocytes cultured in the presence of isolated B. pertussis and N. meningitidis lipid A are able to produce intracellular IL-1 activity but release little or no IL-1 in the culture medium (Caroff & al. 1986). Synthetic diphosphoryl hexaacyl compound (506) built up according to the native E. coli lipid A, also induces almost uniquely intracellular IL-1 production from human monocytes (fig. 5). It is noteworthy that the same compound triggers IL-1 release from thioglycolate elicited mouse peritoneal macrophages; lower levels of IL-1 release were obtained from resident mouse macrophages (fig. 6). These results favour the hypothesis that the IL-1 release pathway induced by lipid A may require an additional activation step in the cells. It has been reported that the 4'-dephospho-hexaacyl compound (505), which does not correspond to the native lipid A structure in most LPS, is a strong inducer of IL-1 release from human peripheral blood mononuclear cells (Loppnow & al. 1986). Thus, hydrophobic structures different from, but related to those present in LPS, trigger human monocytes to release IL-1. In contrast to the results obtained with 24 h-cultured monocytes in the presence of INF_{γ} prior to LPS stimulation, monocytes cultured in the presence of INF_{γ} prior to lipid A

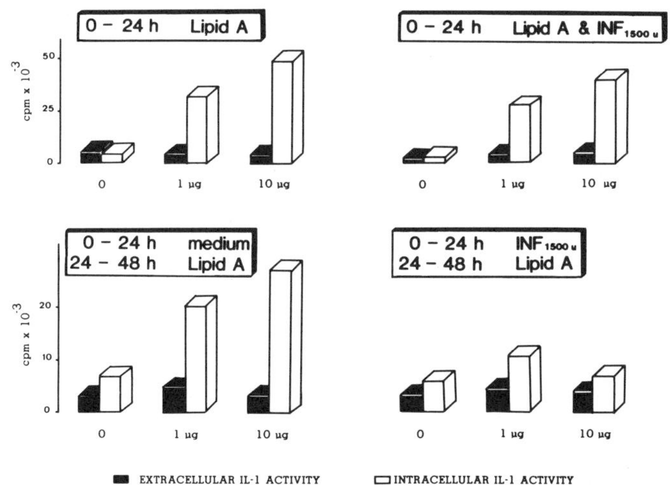

Figure 7. Natural B. pertussis lipid A -induced IL-1 production by freshly isolated human monocytes cultured in the presence or in the absence of interferon, and by 24 h-cultured cells in the absence or in the presence of interferon prior to lipid A stimulation.

stimulation did not release IL-1 and intracellular IL-1 activity was reduced (fig. 7). Altogether, these results suggest that lipid A and LPS do not trigger similar pathways of IL-1 production. It is noteworthy that different pathways of human pro-IL-1 genes regulation by PMA and LPS have been reported (Auron & al. 1987), suggesting that interactions of hydrophobic compounds with the cell membrane do not lead to the same gene activation in monocytes.

III. CLINICAL RELEVANCE OF LPS-INDUCED IL-1 PRODUCTION IN PATIENTS UNDERGOING HEMODIALYSIS

IL-1 is universally accepted as one of the most pluripotent proteins ever described (Dinarello 1984). Since IL-1 has been shown in relationship to the acute phase response, fever, decrease in circulating neutrophils, decrease in systemic arterial pressure, vascular resistance, etc... it has been postulated by Henderson & al. (1983) that IL-1 production in vivo occuring through interaction of blood with artificial membranes and dialysates may be responsible for the adverse effects associated with hemodialysis. The etiology of the phenomenon remains unresolved. However, pyrogens, such as muramyl dipeptide, staphylococcal toxic shock toxin and endotoxin, have been incriminated despite the fact that some authors were unable to demonstrate the passage of pyrogens accross the intact cellulosic dialysis membrane (Bernick & al. 1979). The recent use of high permeability membrane such as the polyacrylonitrile membrane, allowing the passage of molecules up to 10,000 daltons, raised again the question of whether LPS are involved in hemodialysis adverse reactions through the IL-1 induction. Indeed, rough LPS or PS moieties molecular weights between 2,000 and 10,000 d., may be transfered from the dialysis fluid to the patient. According to our study on the kinetics of the production of cell-associated IL-1 and release of IL-1 by human monocytes stimulated by LPS, we have investigated the production in vivo of intracellular IL-1 activity in monocytes from patients undergoing hemodialysis using cuprophane (CUP) membrane (which do not allow LPS passage) and in patients undergoing hemodialysis with polyacrylonitrile (PAN) devices. In 21 patients undergoing hemodialysis using CUP membrane, the mean value of monocyte-associated IL-1 activity at the beginning of the session (TO) was 12.5 +/- 3 arbitrary units (AU) /ml, a value which was higher than that measured in 14 healthy individuals (0.6 +/- 0.2 UA/ml). At the end of the dialysis, the mean value was 42.5 +/- 5 AU/ml. These data indicate for the first time that IL-1 was effectively produced in vivo during hemodialysis. In 22 patients undergoing hemodialysis with PAN membrane, a high value of monocyte-associated IL-1 activity of 33 +/- 5.6 AU/ml was found at T0, and a value of 59 +/- 7 AU /ml was found at the end of the dialysis session (manuscript submited for publication). The perdialytic increase of intracellular IL-1 activity in patients dialyzed with CUP membrane correlates with the perdialytic complement activation resulting in anaphylatoxin generation shown to induce IL-1 production (Haeffner-Cavaillon & al 1987). The PAN devices do not lead to complement activation, and the chronic induction of the production of intracellular IL-1 activity must result from other factors, such as a possibly high membrane permeability and pyrogenicity of the dialysate. The effective passage of LPS or derived fragments through such high permeability membranes is currently under investigation.

REFERENCES

Arend WP, Massoni RJ (1986). Characteristics of bacterial lipopolysaccharide induction of interleukin-1 synthesis and secretion by human monocytes. Clin. Exp. Immunol. 64: 656-664.

Auron PE, Webb AC (1987). The structure and regulation of the human prointerleukin-1 beta gene. Ann. Inst. Pasteur (Immunol.)(in press).

Arenzana-Seisdedos F, Virelizier JL, Fiers W (1985). Interferon as macrophage activating factors. III. Preferential effects of interferon on the interleukin-1 secretory effects of fresh or aged human monocytes. J. Immunol. 134: 2444-2448.

Bernick JJ, Port FK, Favero MS, Brown DG (1979). Bacterial and endotoxin permeability of hemodialysis membranes. Kidney Intern. 16: 491-496.

Brade H, Brade L (1985). A 28,000 d protein of normal mouse serum binds specifically to the inner core region of bacterial lipopolysaccharide. Infect. Immun. 50: 687-694.

Brade H. (1985). Occurence of 2-keto-deoxyoctonic acid 5-phosphate in lipopolysaccharide of Vibrio cholerae Ogawa and Inaba. J. Bacteriol. 61: 795-798

Caroff M, Cavaillon JM, Fitting C, Haeffner-Cavaillon N (1986). Inability of pyrogenic Bordetella pertussis purified Lipid A to induce interleukin-1 (IL-1/LAF) release by human monocytes. Infect. Immun. 54: 465-471.

Caroff M, Lebbar S, Szabo L (1987). Do endotoxins devoid of 3-deoxy-D-manno-2-octulosonic acid exist ? Bioch. Biophys Res. Comm. 143: 845-847.

Cavaillon JM, Haeffner-Cavaillon N (1985). The role of serum in interleukin-1 production by human monocytes activated by endotoxins and their polysaccharide moieties Immunol. Lett. 10: 34-41.

Cavaillon JM, Fitting C. (1986). Inhibition of lipopolysaccharide-induced monocyte interleukin-1 secretion by gangliosides. Eur. J. Immunol. 16: 1009-1012.

Cavaillon JM, Haeffner-Cavaillon N (1987). Characterization of the induction of human interleukin-1 by endotoxins. in Lipid Mediators in Immunology of Burn and Sepsis (P. Braquet Ed.), NATO ASI Series, Plenum Pub. Comp. Ltd. (London) (in press).

Dinarello CA (1984). Interleukin-1 and the pathogenesis of the acute phase response. N. Engl.J. Med. 311: 1413- 1418

Goodman MG, Chenoweth DE, Weigle WO (1982) Induction of interleukin-1 secretion and enhancement of humoral immunity by binding of huamn C5a to macrophage C5a receptors. J. Exp. Med. 156: 912-917.

Haeffner-Cavaillon N, Cavaillon JM, Szabo L (1982) Macrophage-dependent polyclonal activation of splenocytes by Bordetella pertussis endotoxin and its isolated polysaccharide and lipid A regions. Cell. Immunol. 74: 1-14.

Haeffner-Cavaillon N, Cavaillon JM, Szabo l (1985) cellular receptors for endotoxin in Handbook of Endotoxin Vol. 3, "The Cellular Biology of Endotoxin" (L.J. Berry Ed.) Elsevier Pub. Comp. (Amsterdam) p. 1-24.

Haeffner-Cavaillon N., Cavaillon JM, Laude M, Kazatchkine M (1987). C3a(C3adesArg) induces production and release of interleukin-1 (IL-1) by cultured human monocytes. J. Immunol. (in press).

Hamada S, Koga T, Nishihara T, Fujiwara T, Okahashi N (1986). Role of bacterial LPS in human periodontal disease. E.O.S. Riv. Immunol. Immunofarmacol. 6(Supl.3): 131.

Henderson LN, Koch KM, Dinarello CA, Shaldon S (1983). Hemodialysis hypotension : the interleukin hypothesis. Blood 1: 3-8.

Joiner KA, Goldman R, Schmetz M, Berger M, Hammer CH, Frank MM, Leive L (1984) A quantitative analysis of C3 binding to O-antigen capsule lipopolysaccharide, and outer membrane protein of E. coli 0111B4. J. Immunol. 132: 369-375

Lebbar S, Cavaillon JM, Caroff M, Ledur A, Brade H, Sarfati R., Haeffner-Cavaillon N (1986). Molecular requirement for interleukin-1 induction by lipopolysaccharide-stimulated human monocytes : involvement of the heptosyl-2-keto-3-deoxyoctulosonate region. Eur. J. Immunol. 16: 87-91.

Loppnow H, Brade L, Brade H, Rietschel ET, Kusomoto S, Shiba T, Flad HD (1986) Induction of interleukin-1 by bacterial and synthetic lipid A. Eur. J. Immunol. 16: 1263-1267.

Lüderitz O, Tanamoto K, Galanos C, Westphal O (1983). Structural principles of lipopolysaccharides and biological properties of synthetic partial structures in Bacterial lipopolysaccharides (Anderson L & Unger FM Eds) ACS Series p.1-17.

Newton RC (1985). Effect of interferon on the induction of human monocyte secretion of interleukin-1 activity. Immunol. 56 : 441-449.

Newton RC (1986). Human monocytes production of interleukin-1 : parameters of the induction of interleukin-1 secretion by lipopolysaccharides. J. Leuk. Biol. 39: 299-311.

Nowotny A (1984) Heterogeneity of endotoxins in Handbook of endotoxin, Vol.1, "Chemistry of endotoxin" (E.T. Rietschel Ed.), Elsevier Publ. Comp. (Amsterdam) p. 308-338.

Rietschel ET, Wollenweber HW, Brade H, Zahringer U, Linder B, Seydel U, Bradaczek H, Barnickel G, Labinchinski H, Griesbrecht P (1984). Chemistry of endotoxin in Handbook of endotoxin, Vol. 1, (E.T. Rietschel Ed.), Elsevier Publ. Comp. (Amsterdam) p. 187-214.

Rosner MR, Tang JY, Borzi-Lay J, Khorana MG (1979). Structure of the lipopolysaccharide from an E. coli heptose-less mutant. Chemical degradation and identification of products. J. Biol. Chem. 254: 5906-5917

Ulevitch RJ, Johnston AR, Weinstein DB (1981) New function for high density lipoporoteins. Isolation and characterization of a bacterial lipopolysaccharide-high density lipoprotein complex formed in rabbit plasma. J. Clin Invest 67: 827-837.

Vaara M, Vaara T, Sarvas M (1979). Decreased binding of polymyxin by polymyxin B-resistant mutants of Salmonella typhimurium. J. Bacteriol. 139: 64-67.

Bacterial Endotoxins: Pathophysiological Effects, Clinical Significance, and Pharmacological Control, pages 103–114

THE ROLE OF LIPOPOLYSACCHARIDE OF GRAM-NEGATIVE BACILLI IN SENSITIVITY AND RESISTANCE TO SERUM

Reuven Porat, Margaret A. Johns and
William R. McCabe
Departments of Medicine and Infectious Diseases
Ichilov Hospital, Tel-Aviv Medical Center, Israel.
Maxwell Finland Laboratory for Infectious Diseases
Boston University School of Medicine, Boston, MA.

INTRODUCTION

Mechanisms of the interaction of serum and complement with Gram-negative bacilli (GNB) have become an intriguing issue since the decription of the bactericidal activity of normal human serum (NHS) at the beginning of the century (Neisser and Wechsberg, 1901). Almost all species of GNB have been shown to be killed by sera (Neisser and Wechsberg 1901; Taylor, 1983). Research of bacterial clearance from the blood stream is concentrated in two major fields (a) phagocytosis and (b) complement mediated killing. Our work centered upon postulations made in the past concerning complement mediated killing of GNB and the role of lipopolysaccharide (LPS) as a target site in bactericidal activity of human serum.

NHS exhibits potent bactericidal activity against GNB comparable to or greater than that of any antibiotic agent. The bactericidal activity of serum can result in the killing of 99 to 99.9 per cent of an inoculum of sensitive bacteria within 15 minutes (Taylor, 1983). Thus, NHS provides a potent host defense mechanism against invasion by GNB but regrettably, many GNB have developed mechanisms for circumventing this mechanism. The majority of GNB from the gastrointestinal tract are sensitive to serum but most clinical isolates (85 per cent of blood cultures) are resistant to killing by NHS (McCabe et al., 1978; Roantree and Rantz, 1960: Vosti and Randall, 1970).

Gram-negative bacterial cell wall consists of an outer membrane, peptidoglycan layer and an inner membrane. The outer membrane is composed of variant proteins and LPS. LPS contains three components (a) lipid A (diglucosamine sub-unit with attached fatty acids), (b) core polysaccharide (2-keto 3-deoxyoctonate and heptoses) and (c) polysaccharide O-antigenic side chains (repeating sub-units of oligosaccharides). Cell wall components of GNB, especially LPS, K or capsular antigens and outer membrane proteins have been postulated as sites sensitive to the killing effect of serum and as mediators of resistance to killing by serum (Taylor, 1983; Wardlaw, 1963; Rowley, 1968; Moll et al., 1980; Ogata and Levine, 1980; Opal et al., 1982). Furthermore, LPS has been implied as the target site for complement action in complement and serum mediated killing (Galanos and Luderitz, 1976; Morrison and Kline, 1977; Pangburn et al., 1980; Joiner et al., 1984; Grossman and Leive, 1984). Killing of smooth strains of GNB is mediated through complement and antibody although rough bacilli may be killed by complement components alone (Neisser and Wechsberg, 1901; Vukajlovich and Morrison, 1983). LPS has been demonstrated as an alternative complement pathway activator (Morrison and Kline, 1977; Joiner et al., 1984) while both complement pathways participate in bacterial killing by serum. The classical pathway has been shown to induce more rapid bactericidal activity than the alternative pathway (Root et al., 1972). Furthermore, different complement system components have been found to form (C3 convertase) and to attach (C3b) to polysaccharide O-antigenic sub-units of LPS, forming an attack complex which damages the bacterial cell wall (Joiner 1985; Fearon, 1978; Kazatchkine et al., 1979). Based on the above mentioned data several mechanisms explaining serum resistance of GNB are suggested: (a) the membrane attack complex (MAC) is inactive and incapable of insertion into the bacterial membrane or (b) the MAC formation on the bacterial surface is blocked and therefore no damage of bacterial cell wall is possible. The latter is supported by studies of several investigators (Grossman et al., 1984; Joiner and Frank, 1985), in which they refer this formation blocking of the MAC to configurational or "steric" hindrance caused by the long polysaccharide O-antigenic side chains of the LPS.

Studies performed in our laboratory were directed towards assessment of the role of LPS and its polysaccharide O-antigenic sub-units in resistance to serum of GNB.

METHODS

(1) Analyses of LPS and its polysaccharide O-antigen sub-units: As previously described (Porat et al., 1987), LPS was extracted by the phenol-water method (Westphal et al., 1952), from clinical isolates of serum sensitive strains of smooth E. coli (#109=06 and #227=018ac:K1) and their serum resistant clones selected according to the method described by Fierer et al. (1972). Bactericidal activity was determined by using a standard microtiter plate assay technique (McCabe et al., 1978), demonstrating sensitivity or resistance to serum of parent strains and the derived clones respectively and is discussed in a previous publication (Porat et al., 1987). Polyacrylamide Gel Electrophoresis (PAGE) of LPS from either serum sensitive or derived resistant strain was performed, silver stained and analysed by scanning densitometry. We proceeded by disaggregating LPS molecules in buffer containing sodium deoxycholate (DOC) and fractionated on a Sephadex G 200 column into high and low molecular weight fractions using a method described by Vukajlovich and Morrison (1983). Hexose content of LPS fractions from either serum sensitive or resistant organisms was determined by the anthrone (Scott and Melvin, 1953) or by the cysteine sulfuric acid method (Wright and Rebers, 1972), while 2-keto 3-deoxyoctonate (KDO) was determined by the thiobarbituric acid method (Osborn, 1963).

(2) Utilization of liposomes in lytic assays simulating bacterial cell wall: Multilamellar liposomes prepared by mixing dimyristyl phosphatidyl choline, cholesterol and dicetylphosphate in a molar ratio of 2:1.5:0.22 were engaged in these studies according to methods described by Kinsky et al. (1969) and Kataoka et al. (1971). This enables us to simulate bacterial membrane by incorporating various bacterial cell wall components, in this case LPS or its high or low molecular weight polysaccharide O-antigenic fractions. As a marker we used 51Cr in the aqueous phase which could have been released from lysed liposomes representing bacteriolysis. The radioactive chromium was then counted on a gamma counter and percentage of liposomal lysis calculated.

RESULTS

(1) Figure 1 demonstrates PAGE patterns of LPS from serum sensitive parent strains and their serum resistant clones. The sub-unit structure of LPS from smooth GNB is

represented by the typical ladder formation and is seen in all lanes. The major differences between LPS from the serum sensitive and the resistant strains occur in the banding pattern at the top of the gel representing the longer chains of the polysaccharide O-antigenic sub-units. LPS from resistant clones contain more densely staining bands in this area. Two dense bands (c) are seen at the top of the gel of LPS from serum resistant 109 which are barely visible in LPS from serum sensitive 109 (lanes 1 and 2).

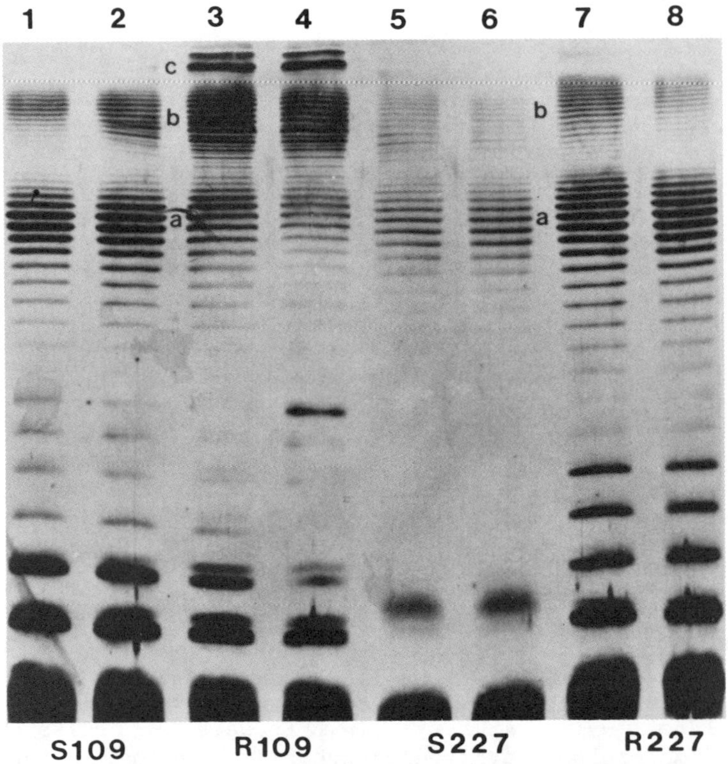

Figure 1. Polyacrylamide Gel Electrophoresis (PAGE) of LPS from two serum sensitive clinical isolates of E. coli and from two serum resistant clones derived from them. The densely staining broad bands at the bottom of the gel represent core components and account for at least 50% of the LPS present. No differences are observed in the number and density of the higher molecular weight bands (upper part of the gel) between heated (lanes 2,4,6,8) and unheated (lanes 1,3,5,7) samples.

Scanning densitometry pattern of the upper half of the gel is summarized in Table 1. There appears to be 3 distinct clusters of bands in this region in 109 LPS (a,b,c) and 2 in 227 LPS (a,b). The cumulative recorded density of bands in these areas, of LPS from serum sensitive 227 is 273 for area a and 143 for area b while those of the serum resistant 227 are three and two times higher respectively. In addition, the c area (highest molecular weight) of serum resistant 109 is approximately 20 times greater than that of the same area of serum sensitive 109. These results confirm the observation that LPS from the serum resistant clones contain more higher molecular weight polysaccharide O-antigenic sub-units, than LPS from their serum sensitive parents.

TABLE 1. Comparison of intensity of bands in the higher molecular weight regions of the PAGE gel.

		a	b	c	TOTAL
109 LPS	S*	580**	390	2.2	972
	R***	468	584	40	1192
227 LPS	S	273	143		416
	R	790	315		1105

* Serum sensitive ** Numbers representing absolute density
*** Serum resistant

Table 2 lists the weighed amounts and the results of chemical analyses of the recovered fractions after chromatography of LPS from the serum sensitive strains and the resistant clones on Sephadex G 200 after disaggregation with DOC, and of the LPS from which they were obtained. The weights of F1 fractions (higher molecular weight) from LPS of serum sensitive strains were 2.5-3 times less than the F2 fractions (lower molecular weight), while the weights of F1 fractions of LPS from serum resistant clones were 2 times greater than F2 fractions. Absolute amounts of hexose and KDO were determined and the ratio of hexose to KDO was calculated. The ratios of hexose:KDO among the LPS, among the F1 fractions, or among the F2 fractions from LPS of serum sensitive or resistant strains did not differ materially. However, F1 fractions of all LPS preparations had hexose: KDO ratios of approximately 20:1 while F2 fractions had approximately 4:1 ratios. F1 hexose: KDO ratios were consistently 4-6 times higher than F2 ratios.

TABLE 2. Hexose and KDO analyses of LPSs and their deoxycholate column fractions.

Smple	Dried weight (MGM)	Hexose (UG/MG)	KDO (UG/MG)	Ratio hexose:KDO
S 109		160	16	10:1
F1	14	230	11	21:1
F2	36	170	36	5:1
R 109		180	23	8:1
F1	32	278	14	20:1
F2	17	115	29	4:1
S 227		123	24	5:1
F1	13	190	10	18:1
F2	37	70	21	3:1
R 227		170	23	7:1
F1	34	210	10	20:1
F2	16	125	39	3:1

(2) The demonstration of differences in the sub-unit composition of LPS from serum sensitive strains and serum resistant clones suggested but did not establish that LPS was the site of killing by NHS. The actual role of LPS as the site of the bactericidal activity of NHS was determined using liposomes into which LPS from either serum resistant or sensitive GNB was incorporated along with 51Cr used as a "release marker" for lysis. Radiolabelled liposomes were separated from free 51Cr by passage through a 1.6x30 cm Sephadex G 25 column; the first peak contains radiolabelled liposomes while the second peak contains free 51Cr. After washing with Gelatin Veronal Buffered Saline (GVBS) to remove excess LPS, the liposome preparations were resuspended in GVBS containing magnesium chloride and calcium chloride (GVBS++) and 0.2ml amounts were incubated in NHS, heat decomplemented NHS or GVBS++ for 30 min. at 35 degrees centigrade. Table 3 lists percentages of free 51Cr which was released from the lysed liposomes after incubation and separation on a 1.6x30 cm Sephadex G 25 column.

The results clearly indicate that release of 51Cr occurred only from liposomes incorporated with LPS from the serum sensitive strains and incubated with NHS. Liposomes containing LPS from serum resistant clones and incubated in the same conditions were not affected and were not lysed. Furthermore, absence of serum or even heat inactivation of complement alone, proved to be crucial for the liposomes to remain intact and not to undergo lysis.

TABLE 3. Gamma counting of column fractions; intact radio-labelled liposomes (first peak) and released 51Cr from lysed liposomes (second peak).

Liposomes with LPS from	First peak	Second peak	Total	Percentage of 51Cr release
109 S *	4150	3850	8000	48
**	7800	0	7800	0
109 R *	9200	0	9200	0
**	9400	0	9400	0
227 S *	4250	4450	8700	51
**	9000	0	9000	0
227 R *	8950	0	8950	0
**	9350	0	9350	0

* Incubation with NHS.
** Incubation with decomplemented NHS or buffer alone.

After the demonstration of these striking differences in lysis, we examined the role of the higher (F1) and the lower (F2) molecular weight LPS fractions in determining susceptibility to lysis by NHS. F1 and F2 fractions obtained from LPS of either serum sensitive or resistant strains were incorporated into liposomes in the same manner as parent LPS and treated with NHS. As shown in table 4, similar concentrations of F1 fractions (longer polysaccharide O-antigen side chains) from both serum sensitive or resistant of either strain protected liposomes from lysis by NHS.

No release of 51Cr occurred in liposomes containing F1 fractions from serum sensitive or serum resistant strains. In contrast, liposomes containing F2 fractions (shorter polysaccharide O-antigen side chains) from LPS of serum sensitive or resistant strains were lysed and 51Cr was released. Incubation with heat inactivated NHS or with buffer alone did not result in lysis of liposomes and 51Cr was not released. These observations provide convincing evidence that serum resistance in clinical isolates of GNB is mediated by the longer side chains of polysaccharide O-antigen sub-units of LPS and that these are present in greater amounts in serum resistant GNB than in serum sensitive GNB.

TABLE 4. Gamma counting of column fractions after incubation with NHS; intact radiolabelled liposomes (first peak) and released 51Cr from lysed liposomes (second peak).

Liposomes with LPS	First peak	Second peak	Total	Percentage of 51Cr release
109 S F1	6500	0	6500	0
F2	2950	4250	7200	59
109 R F1	7000	0	7000	0
F2	2750	4150	6900	60
227 S F1	7500	0	7500	0
F2	2850	4450	7300	61
227 R F1	6800	0	6800	0
F2	2950	4050	7000	58

DISCUSSION

Resistance to killing by NHS is often considered a virulence factor since GNB in the gastrointestinal tract are usually sensitive to serum while those isolated from clinical infections are usually resistant (McCabe et al., 1978; Roantree and Rantz, 1960). Several investigators have hypothesized that the difference between serum sensitive and resistant bacilli is associated with the length of O-antigen side chains of LPS molecules (Joiner, 1985; Wardlaw, 1963; Rowley, 1968; Grossman et al., 1984; Goldman et al., 1984). Many investigators have compared serum resistant strains with rough mutants, lacking polysaccharide O-antigen side chains (Wardlaw, 1963; Rowley, 1968; Taylor, 1975). In contrast, smooth serum sensitive GNB have been compared with serum resistant clones derived from these parent strains in the present study. These clones were identical to their parents serologically, biochemically and by the antibiotic sensitivity pattern.

Analysis by PAGE demonstrated that the number of longer polysaccharide O-antigen side chain molecules of LPS is greater in the serum resistant clones than in the serum sensitive parent strains. Overall, hexose:KDO ratios, which provide a measure of the relative amount of core components

(KDO) in relation to O-antigenic components, did not differ significantly between LPS from serum sensitive parent strains and their resistant clones. However, scanning densitometry of PAGE demonstrated that LPS from serum resistant clones contained a higher proportion of long chain molecules than LPS from their serum sensitive parents. Fractionation of LPS after disaggregation demonstrated that sensitive strains contained more short chain molecules than LPS from derived serum resistant clones. The longer (higher molecular weight) polysaccharide O-antigen side chains contained the same proportion of hexose to KDO (20:1) whether from serum sensitive parent or resistant clones, while shorter chains also contained similar ratios averaging 4:1 irrespective of their source. This demonstration of differences in the relative amounts of sub-units of varying length between LPS from serum sensitive and resistant GNB does not establish that these differences, in themselves, are responsible for the variations in serum sensitivity observed. The significance of these differences in the sub-unit composition of LPS as determinants of sensitivity or resistance to serum was established by the use of liposomal membranes, simulating bacterial cell walls, which incorporated LPS and 51Cr, as a "release marker", in the aqueous lumen. Demonstration of equal incorporation of LPS from either serum sensitive or resistant strain into the liposomes has been discussed in a previous publication (Porat et al., 1987). When these liposomes were washed to remove excess LPS and were incubated with fresh NHS, those containing LPS from serum sensitive parents were lysed releasing approximately 50% of the 51Cr they contained. Liposomes with no LPS or containing LPS from serum resistant clones were not affected by incubation with serum. Serum which underwent heat inactivation of complement had no effect on any of the liposomes. Even more striking results were obtained using liposomes incorporating long or short polysaccharide O-antigen side chain molecules of LPS obtained by fractionation on a G 200 column. Liposomes containing equivalent concentrations of the F1 fractions (long polysaccharide O-antigen side chains), from LPS of either serum sensitive or resistant GNB behaved like liposomes containing LPS from serum resistant GNB in that they were not lysed by serum. In contrast, liposomes containing F2 fractions (short polysaccharide O-antigen side chains), from either serum sensitive or resistant GNB, were readily lysed by serum similar to the lysis of liposomes containing LPS from serum sensitive GNB. These findings firmly establish that LPS is a major factor in activating complement mediated killing and is the reactive site for killing of GNB by NHS. In addition, they also demonstrated that resistance to killing by NHS is mediated by the content of longer polysaccharide O-antigen side chains of LPS.

The exact mechanism(s) involved in the blocking of bactericidal activity by these longer polysaccharide O-antigen side chains has yet to be established. Resistance to killing of GNB by serum has been suggested to depend on incorrect deposition of different complement components and on blockage of proper sequencing of the membrane attack complex formation. The results of this study establish that the activity of complement is blocked by a configurational hindrance caused by the length of the polysaccharide O-antigen side chain molecules of LPS. The longer these side chains are the more difficult it becomes for complement and the membrane attack complex to damage the bacterial membrane. The liposome model provides an important method in further needed investigations of bacterial cell wall components and resistance to killing by serum of GNB.

REFERENCES

Fearon DT (1978). Recognition by membrane sialic acid of β1H-dependent decay-dissolution of amplification C3 convertase of the alternative complement pathway. Proc Natl Acad Sci USA 75:1971-1975.
Fierer J, Finley F, Braude AI (1972). A plaque assay on agar for detection of Gram-negative bacilli sensitive to complement. J Immunol 109:1156-1158.
Galanos C, Luderitz O (1976). The role of the physical state of lipopolysaccharides in the interaction with complement. Eur J Biochem 65:403-408.
Goldman RC, Joiner KA, Leive L (1984). Serum resistant mutants of Escherichia coli 0111 contain increased lipopolysaccharide, lack an O-antigen - containing capsule, and cover more of their lipid A core with O-antigen. J Bacteriol 159:877-882.
Grossman N, Leive L (1984). Complement activation via the alternative pathway by purified Salmonella lipopolysaccharide is affected by its structure but not its O-antigen length. J Immunol 132:3766-385.
Grossman N, Schmertz M, Klima EN, Leive L, Joiner KA (1984). Changes in the distribution of length of lipopolysaccharide (LPS) of Salmonella montevideo (0-6,7) alter serum sensitivity. American Association of Immunology, St.

Louis, Missouri, May 1984.
Joiner KA (1985). Studies on the mechanism of bacterial resistance to complement-mediated killing and on the mechanism of action of bactericidal antibody. In Springer-Verlag Berlin, Heidelberg 1985: Current Topics in Microbiology and Immunology, 121:99-133.
Joiner KA, Frank MM (1985). Mechanisms of bacterial resistance to complement-mediated killing. In Springer-Verlag Berlin, Heidelberg 1985: Pathogenesis of Bacterial Infections, pp 122-136.
Joiner KA, Goldman R, Schmertz M, Berger M, Hammer CH, Frank MM, Leive L (1984). A quantitative analysis of C3 binding to O-antigen capsule, lipopolysaccharide, and outer membrane protein of E. coli 0111B4. J Immunol 132:369-375.
Kataoka T, Inoue K, Luderitz O, Kinsky SC (1971). Antibody and complement-dependent damage to liposomes prepared with bacterial lipopolysaccharides. Eur J Bio Chem 21:80-85.
Kazatchkine MD, Fearon DT, Austen KF (1979). Human alternative complement pathway: membrane-associated sialic acid regulates the competition between B and β1H for cell bound C3b. J Immunol 122:75-81.
Kinsky SC, Haxby JA, Zopf DA, Alving CR, Kinsky CB (1969). Complement-dependent damage to liposomes prepared from pure lipides and Forssman hapten. Biochemistry 8:4149-4158
McCabe WR, Kaijser B, Olling S, Uwaydeh M, Hason LA (1978). E. coli in bacteremia: K and O antigens and serum sensitivity of strains from adults and neonates. J Infect Dis 138:33-41.
Moll A, Manning PA, Timmis KN (1980). Plasmid-determined resistance to serum bactericidal activity: a major outer protein, the tra T gene product, is responsible for plasmid-specific serum resistance in Escherichia coli. Infect Immun 28:359-367.
Morrison DC, Kline KF (1977). Activation of the classical and properdin pathways of complement by bacterial lipopolysaccharides (LPS). J Immunol 118:362-369.
Neisser M, Wechsberg F (1901). Uber die Wirkungart bactericider sera. Muench Med Wochenschr 48:697-700.
Ogata RT, Levine RP (1980). Characterization of complement resistance in Escherichia coli conferred by the antibiotic resistance plasmid R 100. J Immunol 125:1494-1498.
Opal S, Cross A, Gemski P (1982). K antigen and serum sensitivity of rough Escherichia coli. Infect Immun 37:956-960.
Osborn MJ (1963). Studies on the Gram-negative cell wall, I. Evidence for the role of 2-keto-3-deoxyoctonate in the lipopolysaccharide of Salmonella typhimurium. Proc Natl

Acad Sci 50:499-506.
Pangburn MK, Morrison DC, Schreiber RD, Muller-Eberhard HJ (1980). Activation of the alternative complement pathway: recognition of surface structures or activators by bound C3b. J Immunol 124:977-987.
Porat R, Johns MA, McCabe WR (1987). Selective pressures and lipopolysaccharide sub-units as determinants of resistance of clinical isolates of Gram-negative bacilli to human serum. Infec Immun 55:320-328.
Roantree RJ, Rantz LA (1960). A study of the relationship of the normal bactericidal activity of human serum to bacterial activity. J Clin Invest 39:72-81.
Root RK, Ellman L, Frank MM (1972). Bactericidal and opsonic properties of C4 deficient guinea pig serum. J Immunol 109:477-486.
Rowley D (1968). Sensitivity of rough Gram-negative bacteria to the bactericidal action of serum. J Bact 95:1647-1650.
Scott TA, Melvin EH (1953). Determination of dextran with anthrone. Anal Chem 25:1656-1661.
Taylor PW (1975). Genetical studies of serum resistance in Escherichia coli. J Gen Microbiol 89:57-66.
Taylor PW (1983). Bactericidal and bacteriolytic activity of serum against Gram-negative bacteria. Microbiol Rev 47:46-83.
Vosti KL, Randall E (1970). Sensitivity of serologically classified strains of Escherichia coli of human origin to the serum bactericidal system. Am J Med Sci 259:114-119.
Vukajlovich SW, Morrison DC (1983). Conversion of lipopolysaccharides to molecular aggregates with reduced subunits heterogeneity: demonstration of LPS-responsiveness in "Endotoxin-Unresponsive" C3H/HeJ splenocytes. J Immunol 130:2804-2808.
Wardlaw AC (1963). The complement-dependent bacteriolytic activity of normal human serum. II Cell wall composition of sensitive and resistant strains. Can J Microbiol 159:41-52.
Westphal O, Luderitz O, Bister F (1952). Uber die Extraktion von Bacterien mit Phenol/Wasser. Z Naturforsch 76:148-152.
Wright BG, Rebers PA (1972). Procedure for determining heptose and hexose in lipopolysaccharides. Modification of the cysteine-sulfuric acid method. Anal Biochem 49:307-319.

Bacterial Endotoxins: Pathophysiological Effects, Clinical Significance, and Pharmacological Control, pages 115–124

QUANTIFICATION OF THE ENDOTOXIN-BINDING CAPACITY OF HUMAN TRANSFERRIN

Dieter Berger and Hans G. Beger

Department of General Surgery, University of Ulm, Steinhoevelstrasse 9, 7900 Ulm, Federal Republic of Germany

INTRODUCTION

Cell wall lipopolysaccharides of gramnegative bacteria are the main pathogenic mediators of gramnegative sepsis in man when reaching the circulation (Beger et al., 1980; Beger et al., 1981; Caridis et al., 1972). The endotoxin carrier system in blood is believed to be the high density lipoprotein fraction (HDL, Ulevitch and Johnston, 1978; Freudenberg et al., 1980). Abdelnoor et al. (1982), however, showed the inability of HDL to neutralize endotoxin in vivo. On the contrary, detoxification activity is found in lipoprotein-free serum. Johnson et al. (1977) were able to isolate an endotoxin inactivating protein from human serum not corresponding to a lipoprotein fraction. Further identification of the protein however was not performed. Recently we demonstrated endotoxin-binding capacity of human transferrin and Gc-globulin in a qualitative manner (Berger and Beger, 1987). In order to follow up the binding reaction, a simple ethanol precipitation procedure for separating protein-bound and free endotoxin was developed.

MATERIALS AND METHODS

Endotoxins of different bacterial strains are obtained from Sigma Co., Munich, FRG. The ^{3}H-labelled lipopolysaccharide of E.coli O 111:B4 was delivered by Amersham/Buchler Co., Braunschweig, FRG. Details of the tritiating procedure and further processing of the crude preparation is described by Berger and Beger (1987). The chromogenic reaction coupled to the limulus amebocyte lysate test of Byk-Mallinckrodt Co., Dietzenbach, FRG used to determine endotoxin concentrations is described in detail by Berger et al. (Eur.Surg.Res., in press). Carbobenzoxy-gly-gly-arg-4-methyl-cumarinyl-7-amid was used as chromogenic substrate. Buffer substances and fine chemicals were purchased from Serva Co., Heidelberg, FRG.

For separating protein-bound and free endotoxin, human transferrin (Behring Co., Marburg, FRG) was solubilized in 0.05 mol/l TRIS/HCl, pH 7.4, containing 2 mmol/l $CaCl_2$ at a final concentration of 1 mg/ml. 20 µl of an ^{3}H-endotoxin-containing aqueous solution were added to 100 µl of the protein solution. The endotoxin-protein complex is precipitated by adding 200 µl ethanol (final concentration 62.5%). Radioactivity is determined in the supernatant and the sediment in a liquid scintillation counter of Beckman Co., Munich, FRG. This system allows the variation of incubation time and temperature as well as the variation of the divalent cation content and the pH-value without further modification of the precipitation procedure. Control experiments were performed using pure human albumin, haptoglobin, and acid-alpha-1-glycoprotein (Behring Co., Marburg, FRG). The efficiency of the precipitation procedure was checked by UV-monitoring of the supernatant at 280 nm. The precipitated protein always exceeded 90%.

RESULTS

Before establishing time and temperature dependence of a receptor-ligand interaction, the saturation concentration of the ligand should be determined. Figure 1 depicts the ^{3}H-endotoxin fraction bound to transferrin at various transferrin concentrations. The incubation took place at 25°C for 20 minutes. The buffer system used is described in the methods section. 500 ng endotoxin per ml (●) could saturate the protein up to a concentration of 1 mg/ml. Higher endotoxin concentrations (▲-1500 ng/ml, ■-2500 ng/ml) were able to saturate higher concentrations of transferrin as would be expected for a receptor-ligand interaction. The interpretation of the results obtained at higher endotoxin and protein concentration ranges seems to be difficult, as complex forming cannot be ruled out. In order to overcome this problem, the respective concentrations used in the following experiments are established as follows: ^{3}H-endotoxin 500 ng/ ml, transferrin 0.83 mg/ml, if not otherwise indicated.

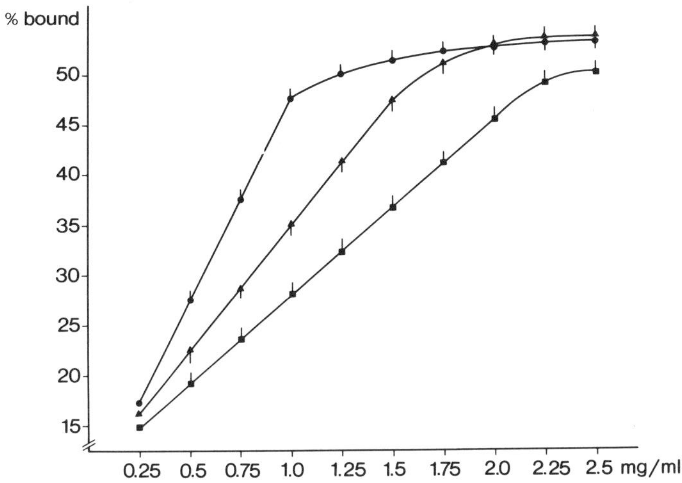

Figure 1. Saturation curve: various transferrin concentrations (abscissa) were incubated with different amounts of endotoxin for 20 min at 25°C. The endotoxin fraction bound is expressed in % of total endotoxin added on the ordinate. Each value is the mean of six experiments ± s.d. (●: 500 ng endotoxin/ml; ▲: 1500 ng endotoxin/ml; ■: 2500 ng endotoxin/ml).

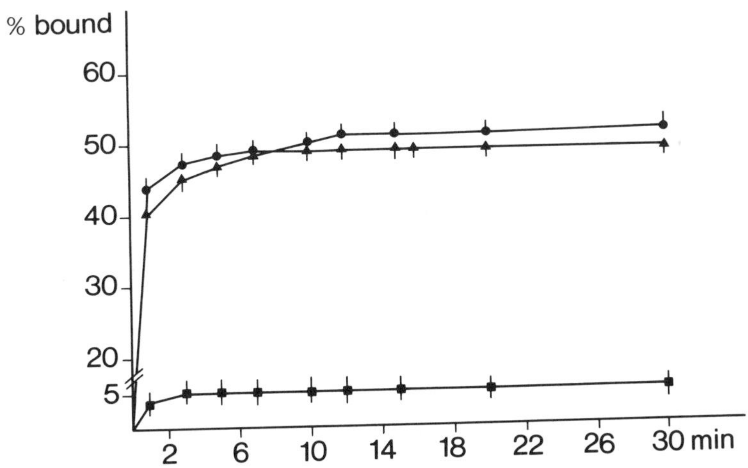

Figure 2. Time and temperature dependence. Statistical analysis according to Fig. 1. (■: 0°C; ●: 25°C; ▲: 37°C).

Figure 2 shows the time and temperature dependence of the binding reaction. At 0°C (■) the maximum endotoxin fraction bound expressed in % of total endotoxin added amounted to 5%. The binding rate increased to 45-50% at 25°C (●) and 37°C (▲). Saturation was achieved after 1-2 min of incubation at each incubation temperature. According to these results the incubation conditions of the following experiments were 20 min at 25°C. The pH-dependence of the binding reaction demonstrated in Fig. 3 exhibited a broad optimum ranging from pH 6 - 7.4 followed by a sharp decrease of the bound endotoxin fraction at higher pH-values. In this experiment TRIS/HCl buffers were used from pH 6 - 8, at pH 5 the incubation was performed in an acetate buffer 0.05 mol/l, each containing $CaCl_2$ at a concentration of 2 mmol/l.

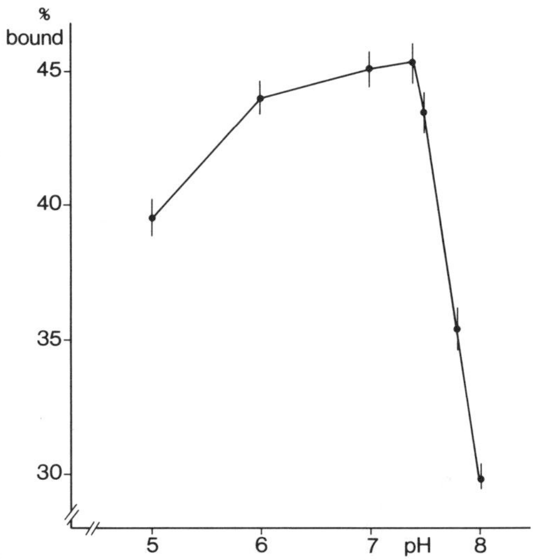

Figure 3. pH-dependence. Endotoxin concentration was 500 ng/ml, the tranferrin concentration amounted to 0.83 mg/ml.

The velocity of the saturation kinetics is identical to those shown in Figure 2 (experiment not shown). An important feature of the binding reaction was represented by the divalent cation requirement. As described in the methods section, $CaCl_2$ can be replaced in the incubation buffer by $MnCl_2$, $MgCl_2$ and even EDTA without modification of the precipitation procedure. The most striking results of these experiments (Table 1) was the almost complete lack of endotoxin binding in the presence of the chelating agent EDTA. Mg^{2+}- and Mn^{2+}-ions were able to replace calcium ions. Saturation kinetics were not changed by different cation or EDTA content of the incubation mixture.

TABLE 1. Divalent cation requirement ($CaCl_2$ was replaced by $MgCl_2$, $MnCl_2$, or EDTA)

ion	Ca^{2+}	Mg^{2+}	Mn^{2+}	EDTA
concentration in mmol/l	2	2	2	2
endotoxin binding capacity in % of total endotoxin added	46	40	42	8

In order to compare the affinity of the transferrin-endotoxin interaction with other proteins, especially therapeutically used ones, an affinity constant should be calculated. This can be done by establishing a scatchard plot (Figure 4).

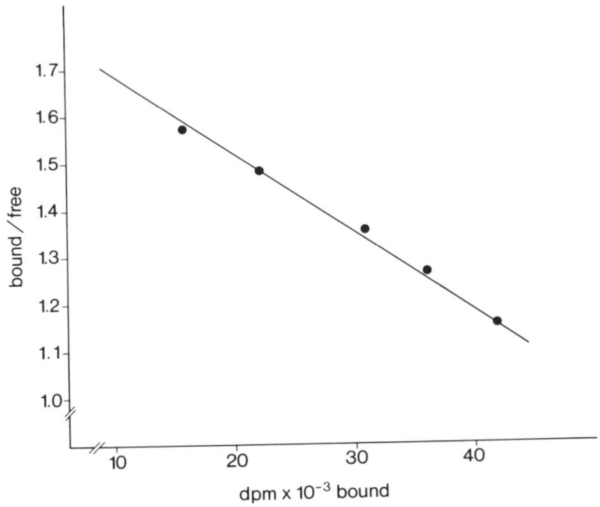

Figure 4. Scatchard plot. The ratio of bound and free endo-

toxin on the ordinate is plotted against the bound fraction in dpm on the abscisse. Transferrin concentration: 3 mg/ml, incubation conditions: 25°C, 20 min).

The endotoxin concentrations tested ranged from 250 ng/ml to 2500 ng/ml. The ratio bound:free was plotted against the bound endotoxin fraction. One experiment out of six is demonstrated. The regression coefficient of 6 determinations varied from 0.85 - 0.87. For calculation of the affinity constant a molecular weight of 20 000 was presumed. Using SDS polyacrylamid gel electrophoresis according to Laemmli (1970), this value was obtained. The affinity constant of the experiment demonstrated amounted to $2.5 \times 10^{-8} \times M^{-1}$. The values obtained from 6 experiments ranged from $1 - 4 \times 10^{-8} \times M^{-1}$. This value, however, is not exactly comparable to the constants of receptor-ligand interaction because the molecular weight of endotoxin in the incubation mixture is not necessarily the same as that obtained by SDS polyacrylamid gel electrophoresis. Furthermore, a correction for unspecific binding was not performed. Up to now this seems impossible because at higher endotoxin concentrations inter-endotoxin-complex forming occurs. Nevertheless, these values provide the possibility to compare the affinity of different endotoxin-binding proteins.

Up to now the evidence of endotoxin-binding capacity of human transferrin is restricted to the tritiated preparation derived from E.coli O 111:B4. If the binding capacity is a general and physiologically important feature of the protein, the tritiated endotoxin should be replaced by non-radioactive preparations of different bacterial strains. In fact this is the case as shown for endotoxins of Salmonella minnesota (✱), Serratia marcescens (●), E.coli O 55:B5 (▲), and E.coli O 111:B4 (■) in Figure 5. The replacement took place in a concentration-dependent manner. The most effective endotoxin is represented by the Salmonella preparation followed by the E.coli strains tested. The endotoxin of Serratia marcescens is much less effective. The varying affinity of different endotoxin preparations is paralleled by changes of the activating potency of the preparations in the limulus amebocyte lysate test (LAL-test).

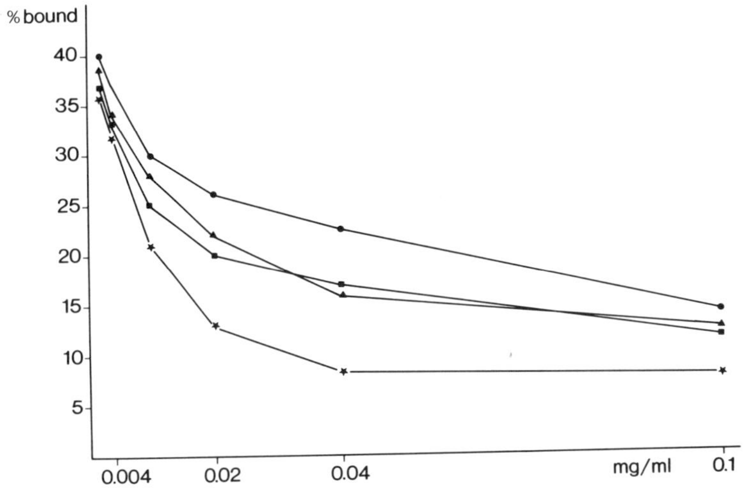

Figure 5. Competitive binding of different endotoxin preparations. The endotoxin fraction bound in % of total endotoxin added is given on the ordinate, the final concentration of non-radioactive endotoxin preparation is shown on the abscissa. (*: Salmonella minnesota; ●: Serratia marcescens; ▲: E-Coli 055:B5; ■: E-coli 0111:B4).

In Table 2 the LAL-activating potency is shown as the ratio between the endotoxin concentration expressed in weight per volume = ng/ml, and the endotoxin concentration in endotoxin units (EU) per ml obtained in the LAL-test, which was standardized according to the EC 5 standard of the US Food and Drug Administration.

TABLE 2. LAL activating potency

	Serratia marcescens	E.coli 0111:B4	E.coli 055:B5	Salmon. minnes.
LAL activating potency expressed as the ratio: $\frac{\text{ET conc. in ng/ml}}{\text{ET conc. in EU/ml}}$	0.22	0.51	0.64	0.96

The value 0.22 for Serratia marcescens means that only 0.22 ng/ml produce 1 EU/ml in the LAL-test. On the other hand, 0.96 ng/ml of the endotoxin preparation of Salmonella minnesota is necessary to produce 1 EU/ml in the LAL-test.

The specificity of the binding can be strongly supported by replacing transferrin by other major serum proteins. Using haptoglobin, acid-alpha-1-glycoprotein, or albumin, up to 2% of the added endotoxin is coprecipitated (Table 3).

TABLE 3. Specificity of the binding reaction. In this experiment transferrin was replaced by other major human serum proteins. The precipitation method can be applied without further modification as checked by the UV absorption of the supernatant at 280 nm. The endotoxin concentration amounted to 500 ng/ml.

	concentration in mg/ml	ET-binding capacity in % of total ET added
haptoglobin	0.83	2
acid α_1-glycoprotein	0.83	1
albumin	0.83	1
albumin	8.3	2
transferrin	0.83	46

DISCUSSION

Using chromatographical, electrophoretical, and immunological methods, endotoxin-binding capacity could be assigned to human transferrin (Berger and Beger, 1987). This study was restricted to an endotoxin preparation of E.coli O 111:B4 and the results are only qualitative. The described ethanol precipitation procedure provides a simple method to follow up the kinetics of the binding reaction and to establish optimal binding conditions. These include a pH-range from 6 - 7.4 and divalent cations such as calcium, magnesium, or manganese. In contrast EDTA strongly inhibits the binding reaction. On the other hand, Munford et al.(1981) showed the concentration dependent inhibition of endotoxin binding to HDL by calcium ions. So the physiological significance of the binding capacity of HDL seems to be doubtful. This view is supported by the lack of endotoxin-neutralizing activity of HDL in vivo (Abdelnoor et al.,1982). Transferrin, which is shown to be an endotoxin binding protein under more physiological conditions than described for the high density lipoproteins, exhibits bacteriostatic action in vitro (Fletcher, 1973). Furthermore, this protein seems to be beneficial in experimental gramnegative sepsis (Martin et al., 1963). Congenital hypo- or atransferrinemia are associated with a higher incidence of infection in man (Weinberg, 1974). These properties are limited to iron-free transferrin and were explained by the competition of transferrin and bacterial siderophores for iron, an essential bacterial growth factor. As shown in Table 4, iron-saturated transferrin also loses its endotoxin binding capacity. So the bacteriostatic action of transferrin can be based on direct binding to the lipopolysaccharide moiety of the bacterial cell wall.

TABLE 4. Endotoxin binding capacity of transferrin at different iron saturation degrees. The essentially iron-free transferrin preparation was partially saturated by adding ferric ammonium citrate and the iron binding capacity as checked by a commercially available test of Boehringer Co., Mannheim, FRG, adapted to 40% and almost 100%, respectively.

iron binding capacity in %	endotoxin bound in % of total endotoxin
99	46
40	25
5	5

Iron-binding capacity of 99% means iron-free tranferrin, the iron-binding capacity of 5% is represented by almost completely iron-saturated transferrin.

A further interesting characteristic of the binding behaviour is the varying affinity of different endotoxin preparations, which is paralleled by changes of the activating potency in the LAL-test. Since pyrogenicity and mitogenicity of different endotoxins are in strong correlation to the LAL-activating potency (Elin et al., 1976), more toxic preparations obviously lose their ability to bind transferrin.

In summary, the study clearly demonstrates the physiological importance of human transferrin as a carrier protein of circulating endotoxin. Hypothetically the substitution of this protein in gramnegative sepsis provides an adjuvant therapy to antibiosis.

REFERENCES

Abdelnoor AM, Harvie NR, Johnson AG (1982). Neutralization of bacteria- and endotoxin induced hypotension by lipoprotein-free human serum. Infect Immunity 38:157-161.
Beger HG, Kraas E, Bittner R (1980). Endotoxinschock: Erkennung und Behandlung. Langenbecks Arch Chir 352:307-310.
Beger HG, Gögler H, Kraas E, Bittner R (1981). Endotoxin bei bakterieller Peritonitis. Chirurg 52:81-88.
Caridis DT, Reinhold RB, Woodruff PWH, Fine J (1972). Endotoxemia in man. The Lancet 24:1381-1386.
Berger D, Beger HG (1987). Evidence for endotoxin binding capacity of human Gc-globulin and transferrin. Clin Chim Acta 163:289-299.
Berger D, Marzinzig E, Marzinzig M, Beger HG (1987) Quantitative endotoxin determination in blood - chromogenic modification of the limulus amebocyte lysate test. Eur Surg Res, in press.
Elin RJ, Sandberg AL, Rosenstreich DL (1976). Comparison of the pyrogenicity, limulus activity, mitogenicity and complement reactivity of several bacterial endotoxins and related compounds. J Immunol 117:1238-1242.
Fletcher J (1971). The Effect of Iron and Transferrin on the Killing of Escherichia coli in Fresh Serum. Immunology 20:493-500.
Freudenberg MA, Bøg-Hansen TC, Back U, Galanos C (1980). Interaction of lipopolysaccharides with plasma high-density lipoprotein in rats. Infect Immun 28:373-380.
Johnson KJ, Ward PA, Goralnick S, Osborn MJ (1977). Isolation from Human Serum of an Inactivator of Bacterial Lipopolysaccharide. Am J Pathol 88:559-574.
Laemmli UK (1970). Cleavage of structural proteins during the assembly of the head of bacteriophage T4. Nature (London) 227:680-685.
Martin CM, Jandl JH, Finland M (1963). Enhancement of acute bacterial infections in rats and mice by iron and their inhibition by human transferrin. J Infect Dis 112: 158-163.
Munford RS, Hall CL, Dietschy JM (1981). Binding of salmonella typhimurium lipopolysaccharides to rat high-density lipoproteins. Infect Immun 34:835-843.
Ulevitch RJ, Johnston AR (1978). The modification of biophysical and endotoxic properties of bacterial lipopolysaccharides by serum. J Clin Invest 62:1313-1324.
Weinberg ED (1974). Iron and Susceptibility to Infectious Disease. Science 184:952-956.

Bacterial Endotoxins: Pathophysiological Effects, Clinical Significance, and Pharmacological Control, pages 125–134

THE ROLE OF LIPOXYGENASES IN THE ACTIVATION OF MACROPHAGES BY LIPOPOLYSACCHARIDE

Ulrich F. Schade, Ingrid Burmeister and Regina Engel

Forschungsinstitut Borstel, Institut für experimentelle Medizin und Biologie, D-2061 Borstel, FRG

INTRODUCTION

Bacterial lipopolysaccharides (LPS, endotoxin) are known to activate various functions and morphological properties of mononuclear phagocytes both *in vivo* and *in vitro* (Rosenstreich and Vogel, 1980; Morland and Kaplan, 1977). Recently, it has been reported that phagocytosis in bone marrow derived macrophage cultures was stimulated with LPS, indicating a direct interaction between endotoxin and the cells (Cooper et al., 1984). The mechanism of the activation of monocytes is unknown. However, some LPS induced phenomena require the presence of arachidonic acid metabolites e. g. the LPS-induced production of collagenase (PGE_2, Bhatnagar et al., 1982) and the release of interleukin 1 (unidentified lipoxygenase product, Dinarello et al., 1984). In order to determine the possible role of cyclooxygenases and lipoxygenases in the activation of macrophages by LPS, the effects of several inhibitors of these enzymes on the uptake of zymosan particles by macrophages primed with LPS was investigated and a lipoxygenase product, 13-hydroxyoctadecadienoic acid (13-HODD) was isolated from activated cells. Our results suggest that 13-HODD is possibly involved in the activation process of macrophages by LPS.

MATERIALS AND METHODS

Lipopolysaccharide

LPS from S. abortus equi and free lipid A from E. coli which had been prepared according to (Galanos and Lüderitz, 1984) were kind gifts of Dr. H. Brade (Forschungsinstitut Borstel).

Macrophages

Resident mouse peritoneal macrophages (NMRI mice, 6-8 weeks) were used throughout the experiments described here (Conrad, 1981).

Assay for Phagocytosis

Quantitative determination of the phagocytic capacity of macrophages was determined as described (Schade, 1986). Adherent macrophages (2×10^5 on glass coverslips) were incubated with LPS (10 ng/ml, 20 h), and fluorescent zymosan was added. The cells were allowed to take up particles (10 min) and the phagocytosis index (PI) was determined using a fluorescent microscope, after washing the cultures and fixation in methanol. (PI = phagocytizing cells x phagocytized particles/ total number of cells counted). Samples were prepared in triplicates and the stimulation was expressed as the multiple of the PI of untreated cell cultures which was set to 1 $\pm$ standard deviation (S.D.).

Determination of 13-Hydroxyoctadecadienoic acid (13-HODD)

Macrophages (2×10^6) were incubated with LPS (100 ng/ ml, 20 h), the supernatants discarded and the remaining cells solubilized (2 ml 0.25% desoxycholate, containing 250 ng prostaglandin B_1 as internal standard. After the addition of ethanol (2 ml), the fluid phase was evaporated in a vacuum centrifuge and the residue (1 N NaOH in methanol, 37° C, 1 h). The hydrolysate was extracted (3 x 1 ml $CHCl_3$), evaporated unter nitrogen and analyzed by reverse phase HPLC (5um RP-18, acetonitril/ H_2O, UV-detection at 280/ 235 nm). Chiralphase HPLC was carried out on a DNBPG-

Chiral phase HPLC-column (Baker) according to (Kühn et al., 1987). 13-HODD, eluting from the RP-HPLC was collected, converted to the methylester and chromatographed on the chiral column with hexane/ isopropanol (99.4/0.6) as eluent (UV-detection at 235 nm). Soybean lipoxygenase (Sigma) was used to produce 13(S)-HODD) as described in (Hamberg and Samuelsson, 1967). Racemic 13-HODD was obtained by photooxidation of linoleic acid in the presence of methylene blue according to (Camp et al., 1983).

RESULTS

Stimulation of Macrophage Phagocytosis

Mouse peritoneal macrophages, after pretreatment with endotoxin <u>in vitro</u> were significantly stimulated in their phagocytic capacity as determined by the uptake of zymosan particles. Optimal stimulation of the uptake was observed at doses between 50-150 ng/ml (Fig. 1).

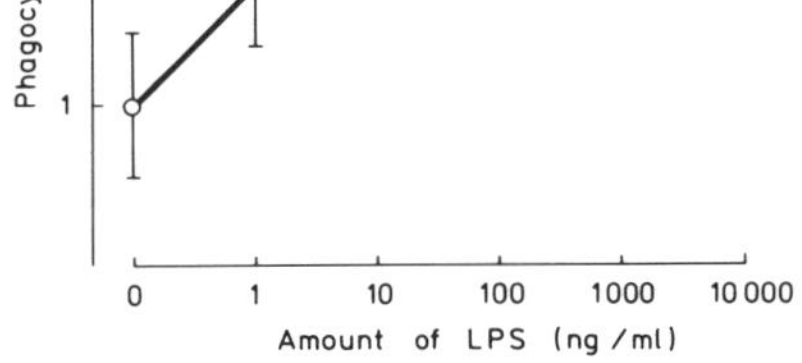

Figure 1. The effect of LPS on the phagocytic capacity of mouse peritoneal macrophages (uptake of zymosan particles)

The activation of phagocytosis was dose dependent (Fig.

1) and could be achieved with bacterial and synthetic free lipid A (data not shown).

Involvement of Lipoxygenases in the activation of macrophages by Endotoxin

In order to test the possible involvement of lipoxygenases or cyclooxygenases in the activation of macrophages, LPS pretreatment of cell-cultures was carried out in the presence of inhibitors of these enzymes. The following compounds were used: diethylcarbamazine (DEC, lipoxygenase-blocker, Hagmann and Keppler 1982), eicosatetraynoic acid (ETYA, lipoxygenase- and cyclooxygenaseblocker, Hamberg and Samuelsson, 1974), BW 755C (lipoxygenase- and cyclooxygenaseblocker, Dinarello et al., 1980), indomethacin (Indo, cyclooxygenaseblocker) and aspirin (cyclooxygenaseblocker). The effects of these substances on phagocytosis are shown in Fig. 2 A-E. The PI of cells treated with LPS was 5.6 ± 1.4. Treatment of macrophages with DEC in addition to LPS resulted in a PI of 3.3 ± 0.6 (2.6 uM) and 2.2 ± 0.3 (26 uM) respectively (Fig. 2 A).

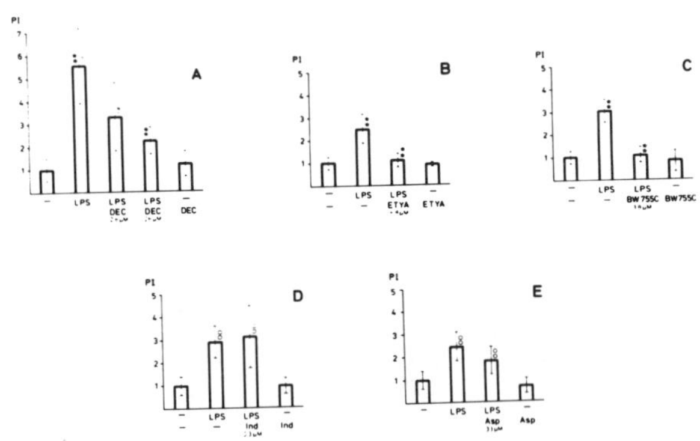

Figure 2. Phagocytosis indices determined in the presence of LPS and inhibitors of lipoxygenases and cyclooxygenases in macrophage cultures.

On incubation with ETYA, 6.8 uM) and LPS a PI of 0.9±0.1 was determined (LPS alone 2.5± 0.2, Fig. 2 B). BW 755C (3.8 uM) inhibited the rise in phagocytosis to about the same degree (PI: 0.9±0.1, LPS alone: 3.1±0.1, Fig. 2 C). In contrast, the cyclooxygenase inhibitors indomethacin (0.3 uM) and aspirin (3.3 uM) did not influence the LPS-induced activation of particle ingestion (Fig. 2 D, E). None of the inhibitors influenced the basal zymosan phagocytosis of untreated macrophages (Fig. 2 A-E).

The Influence of Leukotrienes on Phagocytosis

Synthetic leukotrienes were tested for their ability to stimulate phagocytic properties of resident macrophages. Cell cultures were incubated with LTC_4 and LTB_4 (10^{-9} - 10^{-5} Mol, 20 h) and zymosan phagocytosis was determined. As shown in Fig. 3, pretreatment with LTC_4 did not alter the uptake of particles by macrophages. Also, preincubation with LTB_4 was uneffective (data not shown).

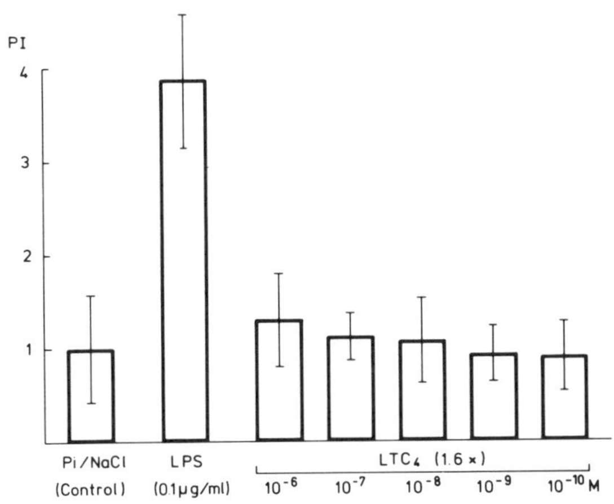

Figure 3. Phagocytosis of zymosan-particles by macrophages pretreated with LPS or LTC_4.

Determination of 13-HODD in LPS-Stimulated Macrophages

Mouse peritoneal macrophages (2×10^7) were incubated with LPS (100 ng /ml, 20 h), the supernatants discarded, the remaining cells lysed and hydrolysates were prepared as described in the materials and methods section. After HPLC, a compound was isolated which eluted with the retention time of synthetic 13-HODD and had the same UV-spectrum. GC-MS analysis of the the methylester, trimethylsilyl ether, confirmed the identity of the material isolated from macrophages as 13-HODD. In order to determine the configuration of the macrophage product, the methylester was applied to chiral phase HPLC. For comparison, a racemic mixture of 13-HODD and 9-HODD methylesters (prepared by photolysis of linoleic acid) and 13-HODD (prepared with the help of soybean lipoxygenase) were also analyzed by chiral phase HPLC. As shown in Fig 4 13-HODD from macrophages eluted with the retention time of 13-HODD from soybean lipoxygenase and, therefore, possesses the (S)-configuration.

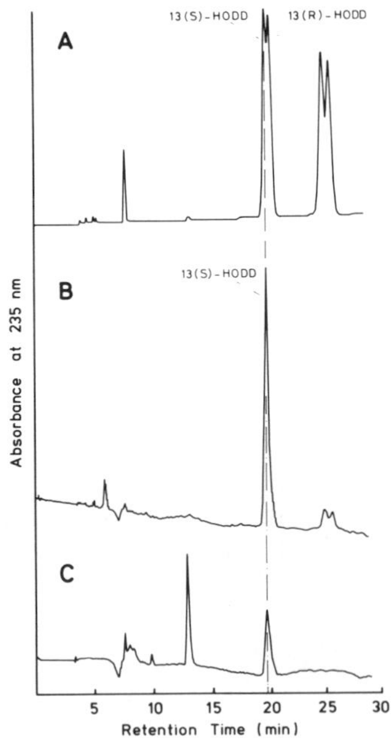

Figure 4. Chiral phase HPLC of a racemic mixture of 13- and 9-HODD (A), 13-HODD prepared by lipoxygenase (b) and 13-HODD from macrophage hydrolysates (C).

Quantitation of 13-HODD in LPS-Stimulated Macrophages

13-HODD from unstimulated and LPS-activated macrophages was isolated and quantitatively determined by HPLC. As shown in Table 1, LPS-treatment increased the content of 13-HODD in hydrolysates of macrophages significantly.

Table 1. Quantitative determination of 13-HODD in quiescent and LPS-stimulated macrophages

Pretreatment of macrophages	13-HODD formed
	(pg/ug protein)
[1]Pi/NaCl	32±5
LPS (100 ng/ml, 20 h)	178±14

[1]Pi/NaCl: phosphate buffered saline

DISCUSSION

The activating potential of lipopolysaccharides for mononuclear phagocytes has been recognized for several experimental parameters like the tumoricidal action (Rosenstreich and Vogel, 1980), release of superoxide anion (Pabst and Johnson, 1980) and the release of lysosomal enzymes (Morland, 1979). We have found that treatment of mouse peritoneal macrophages in vitro with endotoxin increases the association/uptake of zymosan particles by these cells. In order to define the possible involvement of eicosanoids in this LPS effect, inhibitors of the two pathways of arachidonic acid metabolism present in macrophages (lipoxygenase and cyclooxygenase) were tested for their influence on the activation of the cells. When macrophages were incubated with DEC, BW 755C and ETYA, the LPS-induced activation (enhanced phagocytosis) was prevented. Indomethacin and aspirin, however, did not interfere with increased phagocytosis. DEC, BW 755C and ETYA have in common to inhibit lipoxygenase activity, while indomethacin and aspirin are cyclooxygenase blockers. Since only inhibitors of lipoxygenases prevented LPS activity, it is concluded that lipoxygenases are involved in the activation of macrophages with LPS.

It has been shown that LPS induces the synthesis of LTC_4 in macrophages (Lüderitz et al, 1986). In order to find out whether LTC_4 is involved in the activation of macrophages by LPS, it was tested for it's effects on phagocytic properties of macrophages. As shown in Fig. 3, LTC_4 has no influence on this parameter. Since only untreated macrophages have been tested with regard to their responsiveness to LTC_4, it remains, however, to be tested whether stimulated macrophages are able to respond to LTC_4. Moreover, other lipoxygenase derived arachidonic acid products need to be tested (LTD_4, LTE_4, mono- and dihydroxyeicosatetraenoic acids, HETE's).

Recently it was found that lipoxygenase products (mono-HETE's) are integrated into phospholipids of macrophages and polymorphonucleated neutrophils (Pawlowski et al., 1982; Stenson and Parker, 1979). Our findings show, that endogenously synthesized 13-HODD is present in macrophages in a bound form. Since mammalian lipoxygenases catalyze the oxygenation of unsaturated fatty acids with high regio- and stereoselectivity (Hamberg and Samuelsson, 1974), the identification of the product from stimulated macrophages as 13 (S)-HODD (Fig. 4) indicates its enzymatic origine. The amount of 13-HODD, isolated from LPS-treated macrophages was about 6-fold increased compared to unstimulated controls.

In summary, our results show that LPS-primed macrophages produce increased amounts of 13-HODD, suggesting that this linoleic acid metabolite is involved in the activation of macrophages by LPS.

ACKNOWLEDGEMENTS:

The skilled technical assistance of Ms. S. Fuchs and S. Cohrs is kindly appreciated. This study was supported by grant Scha 402/1-1 from the Deutsche Forschungsgemeinschaft. BW 755C was a gift from the Wellcome Research Laboratories, ETYA was provided by Hofmann LaRoche, LTC_4 and LTB_4 were generously donated by Dr. J. Rokach, Merck-Frosst, Canada.

REFERENCES

Bhatnagar R, Schade U, Rietschel ET, Decker K (1982). Involvement of prostaglandin E and adenosine 3',5'-mono phosphate in lipopolysaccharide stimulated collagenase release by rat Kupffer cells. Eur J Biochem 125:125-130.

Camp RDR, Mallet AI, Woollard PM, Brain SD, Kozba Black A, Greaves MW (1983). The identification of hydroxy fatty acids in psoriatic skin. Prostaglandins 26:431-447.

Conrad RE (1981). Induction and collection of peritoneal exudate macrophages In H.B. Herscowitz, H.T. Holden, J.A. Bellanti, A. Ghaffar, (eds.): "Manual of macrophage methodology," New York: Marcel Dekker, pp 5-11.

Cooper PH, Mayer P, Baggiolini M (1984). Stimulation of phagocytosis in bone marrow derived mouse macrophages by bacterial lipopolysaccharide: correlation with biochemical and functional parameters. J Immunol 133:913-922.

Dinarello CA, Bishai J, Rosenwasser LJ, Coceani F (1984). The influence of lipoxygenase inhibitors on the in vitro production of human leucocytic pyrogen and lymphocyte activating factor (interleukin-1). Int J Immunopharmac 6:43-50.

Galanos C, Lüderitz O, (1984). Lipopolysaccharide: properties of an amphipathic molecule In E.Th. Rietschel, (ed.): "Handbook of endotoxin, vol 1: Chemistry of endotoxin," Amsterdam, New York, Oxford: Elsevier Science publishers B.V., pp 46-58.

Hagmann W, Keppler D. (1982). Leukotriene antagonists prevent endotoxin lethality. Naturwissenschaften 69:594-595.

Hamberg M, Samuelsson B. Oxygenation of unsaturated fatty acids by the vesicular gland of sheep (1967). J Biol Chem 242:5344-5354.

Hamberg M, Samuelsson B (1974). Prostaglandin endoperoxides: novel transformation of arachidonic acid in human platelets. Proc Natl Acad Sci USA 71:3400-3404.

Kühn H, Wiesner R, Lankin V Z, Nekrasov A, Alder L, Schewe T (1987). Analysis of the stereochemistry of lipoxygenase-derived hydroxypolyenoic acids by means of chiral phase high-pressure liquid chromatography. Anal Biochem 160:24-

34.
Lüderitz T, Schade UF, Rietschel ET (1986). Formation and metabolism of leukotriene C_4 in macrophages exposed to bacterial lipopolysaccharide. Eur J Biochem 155:377-382.
Morland B, Kaplan G (1977). Macrophage activation in vivo and in vitro. Exp Cell Res 108:279-288.
Morland B (1979). Studies on selective induction of lysosomal enzyme activities of mouse peritoneal macrophages. J Reticuloendothel Soc 26:79-88.
Pabst MJ, Johnston RB (1980). Increased production of superoxide anion by macrophages exposed in vitro to muramyl dipeptide or lipopolysaccharide. J Exp Med 151:101-111.
Pawlowski NA, Scott WA, Andreach M, Cohn ZA (1982). Uptake and metabolism of monohydroxyeicosatetraenoic acids by macrophages. J Exp Med 155:1653-1664.
Rosenstreich DL, Vogel SN (1980). Central role of macrophages in the host response to endotoxin. In Schlessinger D (ed) "Microbiology-1980" Washington: Am Soc Microbiol, pp 11-15.
Schade UF (1986). Involvement of lipoxygenases in the activation of mouse macrophages by endotoxin. Biochem Biophys Res Commun 138:842-849.
Stenson WF, Parker CW (1979). 12-L-Hydroxy-5,8,10,14-eicosatetraenoic acid, a chemotactic fatty acid, is incorporated into neutrophil phospholipids and triglyceride. Prostaglandins 18:285-292.

Addendum:

in the present study, macrophages were incubated for 10 min with zymosan to measure the phagocytic capacity of the cells. During this time, most of the particles are only associated with the cells. The term phagocytosis, therefore, describes the potential of the cells under investigation to bind zymosan, rather than their capacity to ingest the particles.

Bacterial Endotoxins: Pathophysiological Effects, Clinical Significance, and Pharmacological Control, pages 135–143

POSSIBLE INVOLVEMENT OF PLATELET-ACTIVATING FACTOR IN ENDOTOXIN-INDUCED ABORTION.

A. ETIENNE, F. HECQUET and P. BRAQUET.

IHB Research Labs, 17 avenue Descartes,
Le Plessis-Robinson, F-92350 (France).

INTRODUCTION

Endotoxemia induces a high level of miscarriages in pregnant women and similarly, in animal models, sublethal doses of endotoxin injected to mice have been shown to interrupt pregnancy (Rioux-Darrieulat et al., 1972). Nevertheless the exact mechanism underlying the deleterious action of endotoxin in this process is not yet assessed. A strong thrombocytopenia is observed following the i.v. injection of endotoxin in mice and this response is increased during pregnancy ; furthermore, methysergide counteracts the abortive effect of endotoxin confirming the involvement of 5-hydroxytryptamine (5HT) in this model (Darrieulat and Parant, 1971). However the antagonist must be injected several times to be efficient, probably because of the long lasting effect of endotoxin and the involvement of other mediators than 5HT. Recently, platelet-activating factor (PAF), a phospholipid mediator implicated in a variety of pathological states, was proved to be massively released during endotoxic shock (Doebber et al., 1985)). An increase of the amount of this mediator in amniotic fluid during labor has been noted (Nishihira et al., 1984 ; Billah et al., 1985) and its in vitro contractile effect on guinea-pig myometrium has been clearly demonstrated (Montrucchio et al., 1986), suggesting a role of PAF in delivery. As specific PAF-antagonists are now available, (Braquet and Godfroid, 1986) we used two chemically unrelated ones BN 52021 (Braquet, submitted) and PMS 323 (Wichrowski et al., 1986) to investigate the potential role of PAF in the endotoxin-induced abortive process. We also compared the endotoxin-induced abortive effects of i.v. injections of PAF or endotoxin in mice.

MATERIALS AND METHODS

The experiments were carried out on CD1 mice (Charles River) at day 17 of pregnancy, estimated as days after mating. Abortion was induced by i.v. injection in the tail vein of 30µg/kg Salmonella enteritidis toxin (Difco) solubilized in isotonic solution. The abortive activity was controlled, after sacrificing the mice at time 24 hours, by checking the surviving fetuses by their motility and heart beats. The expulsed fetuses were also counted by their scars still perceptible on the uterus. Drugs were given orally just before and 5 hours after endotoxin. 8 to 10 animals per group were used.

The specific action of PAF was studied after one i.v. injection (0.4µg/kg) in the tail of the mouse or two injections (0.5 or 1 µg/kg each) with a 5 hour interval ; the number of newborn mice was checked every 4 hours following PAF injection.

The effect of BN 52021 *per se* on delivery was also sought after oral administration in normal pregnant mice. The number of mice delivered was checked every 4 hours following the first delivery which took place at day 19.

DRUGS

Methysergide was purchased from Sandoz (Basle, Switzerland), dexchlorpheniramine from Shering (USA). PMS 323 was kindly donated by Pr J.J. Godfroid (Paris 7 University, France) ; these drugs were solubilized in water. BN 52021 [3-(1, 1-dimethylethyl) hexahydro-1, 4, 7b-trihydroxy-8-methyl-9H-1, 7-α-(epoxymethano)1H, 6α-H-cyclopenta [C]furo (2,3-b) (3',2':3,4)cyclopenta (1, 2-d) furan-5, 9, 12 (4H)-trione] extracted from Ginkgo biloba leaves was prepared by IHB Labs and was used as a suspension in arabic gum.

RESULTS

- Effects of endotoxin on pregnancy ; activity of PAF antagonists.

The results reported in table 1 confirm that endotoxin administration to pregnant mice, at a low dose (30 µg/kg i.v. ; about 1.5 µg per animal), caused abortion to occur within 24 hours following injection without any toxic effect for the mother ; 32,2 % to 63,8 % of fetuses died depending upon the experiment.

TABLE 1. Effect of drugs on endotoxin-induced abortion in mice.

Drug	unit dose (mg/kg)	% of dead fetuses	rate of normal pregnant mice (NB)
Control	—	39.0	1 / 10
BN 52021	0.5	29.2*	3 / 10
BN 52021	1	27.7*	5 / 10
BN 52021	2	14.0***	6 / 10
Methysergide	0.5	6.4***	7 / 10
Control	—	32.2	8 / 9
BN 52021	2	6.0***	6 / 9
BN 52021	8	33.7	2 / 9
BN 52021	16	46.7*	8 / 9
PMS 323	2	39.5	8 / 9
Control	—	39.4	8 / 10
BN 52021	2	4.1***	7 / 10
PMS 323	0.5	42.1	1 / 10
PMS 323	1	29.4*	6 / 10
Control	—	63.8	1 / 9
Dexchlorpheniramine	10	60.3	3 / 9
Dexamethasone	2.5	45.5***	4 / 10

(NB) Mice without any dead fetuses : Statistical significance calculated by chi-square test ; *p > 95 %, ***p > 99,9 %.

Treatment of the mice with methysergide (0.5 mg/kg) just before endotoxin injection and 5 hours later, protected the animals against the deleterious effects of the toxin : the number of dead fetuses decreased by 83 % and 46 % respectively in two different experiments. In the same protocol, administration of the PAF antagonist BN 52021 (unit dose 0.5 → 16 mg/kg) dose-dependently decreased intra-uterine death and abortion with a maximum effect at 2 mg/kg whereas the higher doses were not effective or even potentiated the action of the toxin (see fig. 1 for graphic representation of the cumulative results from 4 experiments). Similarly, but to a lower extent, the other PAF-antagonist PMS 323 (1 mg/kg) decreased the number of dead fetuses observed after endotoxin injection to the mother.

Dexchlorpheniramine (10 mg/kg) and dexamethasone (2.5 mg/kg) were without any effect against the toxin.

Figure 1. Effect of drugs on endotoxin-induced abortion in mice.

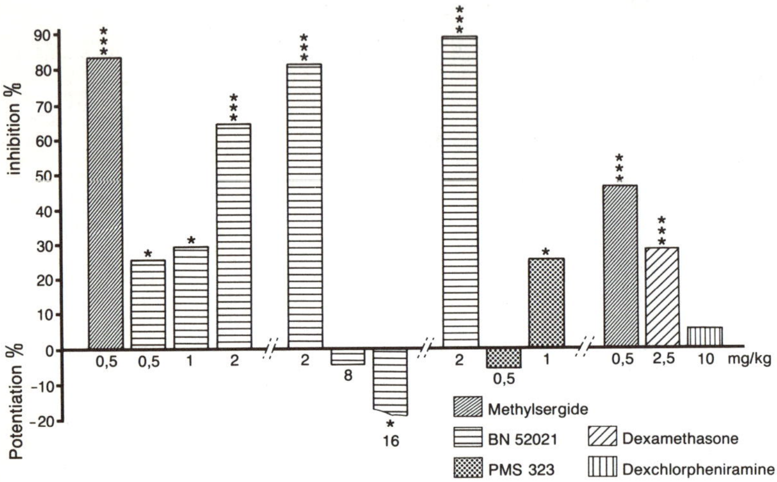

Percentage variation of fetal death induced by endotoxin. Cumulative results obtained from 4 experiments - n = 10 in each group ; *p > 95 %, *** p > 99,9 %.

- Effect of the PAF antagonist BN 52021 <u>per se</u> on termination of pregnancy.

Figure 2. Effect of PAF and BN 52021 on termination of pregnancy. Expulsion patterns of fetuses in control and treated pregnant mice.

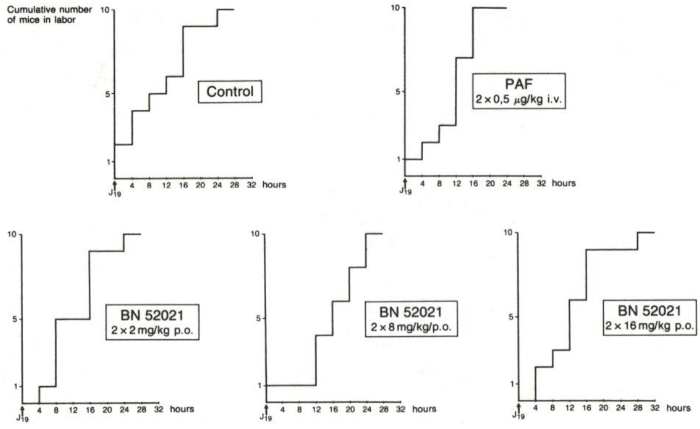

The treatment of normal pregnant mice with BN 52021 at D17 of pregnancy (2, 8 or 16 mg/kg) in the same conditions as those used to inhibit endotoxin effects did not modify significantly the length of gestation or the number of alive fetuses compared to the control group (see fig. 2 and table 2).

- Effects of PAF on pregnancy.

Table 2. Effect of PAF and BN 52021 on termination of pregnancy.

treatment	dose	route	nb of newborns (mean/mice)	% of alive animals
Control	—		12.4	94.5
PAF	2×0.5 µg	i.v.	12.8	84.7
BN 52021	2× 2 mg	p.o.	12.1	91.7
	2× 8 mg	p.o.	12.1	90.1
	2×16 mg	p.o.	11.6	94.0
Control	—	—	9.3	94.1
PAF	2× 1 µg	i.v.	9.9	92.9

The number of newborns was counted at the end of delivery. The percentage of alive animals was calculated with regard to the number of scars on the uterus of the mother.

The i.v. injection of PAF (0.4 µg/kg, 2 x 0.5 µg/kg or 2 x 1 µg/kg) did not induce termination of pregnancy nor shorten the time of expulsion of newborns. The percentage of alive fetuses in the PAF-injected groups was similar to that of the control group (see table 2).

DISCUSSION

The role of PAF in several stages of pregnancy and fetus development has recently been documented : (i) in early pregnancy, this mediator may be involved in the increase of permeability at the site of ovoimplantation (Fim and MacLaren, 1967 ; Psychoyos, 1973 ; O'Neill, 1985) as suggested by the reduction in nidation observed after in utero application of the PAF-antagonist BN 52021 at D 5 of pregnancy (Acker, 1987) or i.p. injection of SRI 63-441 (Spinks and O'Neill, 1987) (ii) in vitro ap-

plication of PAF on guinea-pig myometrium induces contraction followed by rhythmic activity of the tissue ; this PAF contractile response, studied from 0.1 to 100 nM, is dose-dependent (Montrucchio et al., 1986) and is blocked by the specific PAF-antagonist CV 3988 (Tetta et al., 1986) (iii) PAF has been measured in the amniotic fluid of women after labor and amnion cells have been shown to synthetize PAF (Nishihira et al., 1984 ; Billah et al., 1985) In addition, it was well established that endotoxin injection in animals induces abortion which can be prevented by serotonin antagonists (Darrieulat and Parant, 1971). Furthermore, whatever their chemical framework are, PAF-antagonists antagonize lethality and hemodynamic changes observed in endotoxemia (for rev. see Braquet et al., 1987). BN 52021 (10 → 30 mg/kg p.o and 5 → 15 mg/kg sc) provided a significant dose-related protection against the lethality induced in rat by i.v. injection of 16 mg/kg Salmonella enteritidis endotoxin (Etienne et al., 1985). The i.v. injection of BN 52021 also reduced in a dose dependent manner the hypotension due to an injection of Salmonella thyphimurium endotoxin in guinea-pig (Adnot et al., 1986).
Taking account of all this data, this study was performed in order to evaluate whether PAF could trigger, at least in part, the abortifacient effect of endotoxin and if PAF-antagonists could prevent intrauterine fetal death and abortion induced by toxin.

Oral pretreatment of mice with PAF-antagonist before and after endotoxin administration counteracts the deleterious effects of the toxin on pregnancy. A dose-related effect is observed with BN 52021 and at 2 mg/kg of drug the protection is almost total. However i.v. injection of PAF to pregnant mice up to 1 µg/kg in two doses did not induce termination of pregnancy or major changes in the expulsion pattern of fetuses. At the active dose against endotoxin, BN 52021 administration to normal pregnant mice did not modify the pattern of delivery either.

As PAF was shown to induce a down regulation of beta-adrenoceptors in cerebellum (Braquet et al., 1985) and lung tissue (Agrawal and Townley, 1987) and as cyclic AMP altering drugs are known to potentiate the abortifacient effect of endotoxin (Shaw and Cameron, 1979), a long term effect of PAF is perhaps required to alter pregnancy. When PAF is injected directly in blood circulation, its harmful activity on the uterus could not occur since it is rapidly inactivated whereas the long lasting effect of an endotoxin injection would result in a slow release of PAF which would then be efficient against pregnancy.

BN 52021 could thus act directly at the site of action of PAF by inhibiting the binding of the mediator as it acts on platelets (Braquet et al., 1986) or by protecting against the PAF-induced down regulation of beta-adrenoceptors as shown on cerebellum and lung parenchyma.

Nevertheless this hypothesis does not explain the bell-shaped curve of activity observed with BN 52021 treatment and the reverse effect shown by the dose of 16 mg/kg.

Further studies are therefore needed to fully confirm that PAF is a major mediator involved in endotoxin-induced abortion, even if our present results seem to fit in with this hypothesis.

REFERENCES

Acker G, Hecquet F, Etienne A, Braquet P, Mencia-Huerta JM (1987). Rôle des médiateurs de l'inflammation dans l'ovoimplantation chez la ratte. In : immunologie de la grossesse, Ed. G. Chouat. Colloques de l'INSERM. Elsevier, Amsterdam, in press.

Adnot S, Lefort J, Lagente V, Braquet P, Vargaftig BB (1986). Interference of BN 52021, a PAF-acether antagonist, with endotoxin-induced hypotension in the guinea-pig. Pharmacol Res Comm 18:197-200.

Agrawal DK, Townley RC (1987). Effect of platelet-activating factor on beta-adrenoceptors in human lung. Biochem Biophys Res Commun. In press.

Billah MM, Di Renzo GC, Ban C, Truong CT, Hoffman DR, Anceschi MM, Bleasdale JE, Hohnston JM (1985). Platelet-activating factor metabolism in human amnion and the responses of this tissue to extracellular platelet-activating factor. Prostaglandins 30:841-885.

Braquet P, Etienne A and Clostre F (1985). Down-regulation of β_2-adrenergic receptors by PAF-acether and its inhibition by the PAF-acether antagonist BN 52021. Prostaglandins 30:721.

Braquet P, Drieu K et Etienne A (1986). Le Ginkgolide B (BN 52021). Un puissant inhibiteur du PAF-acether iso-

lé du Ginkgo biloba L. Actual Chim Ther 13ème série pp. 237-253.
Braquet P, Touqui L, Vargaftig BB, Shen TY (1987). Perspectives in platelet-activating factor research. Pharmacol Rev. In press.
Braquet P. The Ginkgolides : potent platelet-activating factor antagonists isolated from Ginkgo biloba L. Chemistry, pharmacology and clinical applications. Drugs of the future. Submitted.
Darrieulat F and Parant M (1971). Action d'un antagoniste de la serotonine, le methysergide, sur l'effet abortif ou létal des endotoxines bactériennes chez la souris. Ann Inst Pasteur 121:665-673.
Doebber TW, Wu MS, Robbins JC, MaChoay B, Chary MN and Shen TY (1985). Platelet activating factor (PAF) involvement in endotoxin-induced hypotension in rats. Studies with PAF-receptor antagonist kadsurenone. Biochem Biophys Res Commun 127:799-808.
Etienne A, Hecquet F, Soulard C, Spinnewyn B, Clostre F, Braquet, P (1985). In vivo inhibition of plasma protein leakage and Salmonella enteritidis-induced mortality in the rat by a specific PAF-acether antagonist : BN 52021 Agents and Actions 17:368-370.
Fim CA, Mac Laren A (1967). A study of the early stage of ovoimplantation in mice. J Reprod Fert 13:259-267.
Montrucchio G, Alloatti G, Tetta C, Roffinello C, Emanuelli G, Camussi G (1986). In vitro contractile effect of platelet-activating factor on guinea-pig myometrium. Prostaglandins 32:539-554.
Nishihira J, Ishibashi T, Imai Y, Muramatsu T (1984). Mass spectrometric evidence for the presence of platelet-activating factor (1-0-alkyl-2-acetyl-sn-glycero-3-phosphocholine) in human amniotic fluid during labor. Lipids 19:907-910.
O'Neill C (1985). Examination of the causes of early pregnancy-associated thrombocytopenia in mice. J Reprod Fert 73:567-577.
Psychoyos A (1973). Endocrine Control of egg implantation in : Handbook of physiology R. O Greep EG Astwood SR. Geiger Eds. American physiological Society. Wasghinton. 2:187-215.
Rioux F., Darrieulat F., Parant M, Chedid L (1972). Prevention of endotoxin-induced abortion by treatment of mice with antisera. J. Infec. diseases 137: 7-13.
Shaw R, Cameron JA (1979). Effects of cyclic AMP altering drugs on endotoxin-induced termination of

pregnancy. Research Comm. in Chemical Phathol. and Pharm. 24:43-56.

Spinks NR, O'Neill C (1987). Embryo-derived platelet-activating factor is essential for the establishment of pregnancy in the mouse. The Lancet. January 10:106-107.

Tetta G, Montrucchio G, Alloatti C, Roffinello G, Emanuelli G, Benedetto C, Camussi G, Massobrio M (1986). Platelet-activating factor contracts human myometrium in vitro. Proc Soc Exp Biol and Med. 183:376-381.

Wichrowski B, Jouquey S, Heymans F, Broquet C, Godfroid JJ, Fichelle J and Worcel M (1986). Platelet-activating factor analogs : from agonists to antagonists by synthesis of carboxylate isosters. 6th International Conference on prostaglandins and related compounds. Florence (Italy) June 3-6, 1986.

Endotoxins in Gastrointestinal and Liver Disease

Bacterial Endotoxins: Pathophysiological Effects, Clinical Significance, and Pharmacological Control, pages 147–159

The Role of Intestinal Endotoxins in Gastrointestinal and Liver Diseases

James P. Nolan, M.D.

State University of New York at Buffalo
Department of Medicine
Erie County Medical Center
462 Grider Street, Buffalo, New York 14215, U.S.A.

Whether the intestinal pool of endotoxin is clinically relevant to the pathophysiological events occurring in gastrointestinal and liver diseases remains to be convincingly demonstrated. Much evidence has been presented, however, to suggest that such a relationship exists. At the first International Conference on Endotoxins, the role of LPS in liver injury focussed on mechanisms, mediated and direct by which endotoxins damage hepatocytes (Nolan, 1985). The present chapter will concentrate on clinical studies that bear on the role for intestinal endotoxins in both GI and liver injury.

The pool of endotoxin in the gastrointestinal tract is a large one formed by release of outer cell wall components from Gram-negative aerobic and anaerobic bacteria. This release occurs both by bacterial cell death and by vesicle formation and extrusion in live organisms. In addition, a considerable amount of endotoxins are ingested in food and water.

Concepts that underly a key role for these endotoxins in disease states include the fact that endotoxemia of intestinal origin may occur and be clinically significant without bacteremia being present. Secondly, this endotoxemia from the gut occurs when there is an alteration in endotoxin production, absorption and/or detoxification. Intestinal endotoxin absorption is probably very small normally, but under special conditions can become significant. In animal models, small amounts of labelled LPS do enter the circulation normally (Ravin, 1960), and in everted gut sacs displays Michaelis-Menton kinetics (Nolan et al, 1977). Undoubtedly a major factor in disrupting the mucosal barrier to this toxic macromolecule is hypoperfusion of the gut. For example superior mesenteric artery occlusion in the cat regularly leads to the appearance of LPS in the systemic circulation, and anti-LPS antibodies have been protective in this experimental situation (Gathiram, et al, 1986). In primates undergoing graded hypoxia, endotoxemia occurs which probably reflects both increased absorption and decreased clearance by the reticuloendothelial system (Gaffin et al, 1986). Inflammation of the gut by histamine, serotonin or carageenan have all caused endotoxemia in rodents, and parvovirus infection in dogs can be ameliorated by serum with antibodies to endotoxins (Wessels et al, 1986). General anesthesia alone in rabbits regularly leads to the appearance in the general circulation of a marker endotoxin from the gut (Camara et al, 1983). Gut injury by irradiation also allows escape of LPS into the circulation, and mounting evidence points to gut-derived endotoxemia as the crucial factor in Graft vs. Host Disease (Walker, 1978). While demonstrably important in bone marrow transplants, it may be conjectured that the failure of liver and pancreas transplants may also be related to increased LPS absorption and decreased detoxification.

Demonstrations of portal and systemic vein endotoxemia in humans have been less convincing and both positive and negative studies have been reported. Jacob and his colleagues reported endotoxemia in portal blood of 97% of patients undergoing abdominal surgery (Jacob et al, 1977) while on the other extreme, Fulenwider was unable to detect any endotoxin by the Limulus lysate assay (LAL) in the portal blood of a large group of cirrhotics (Fulenwider, 1980). Recently in a study of 17 patients undergoing elective upper abdominal surgery, no portal endotoxin was demonstrated leading to the conclusion that such endotoxemia was rare in the absence of intestinal disease (Brearley et al, 1985). This conclusion was also supported by another study of portal vein blood in cirrhosis (Holdstock et al, 1982), but LPS has been found in portal blood in some patients with ulcerative colitis at surgery (Palmer et al, 1980). The route by which endotoxin reaches the systemic circulation has been questioned as well. It would now seem that lymphatic transport of absorbed endotoxin from the gut by way of the thoracic duct may be of greater importance than that passing up through the portal vein (Nozickova et al, 1977) and that transmural passage and absorption through the peritoneum also appears to be a major route (Daniele et al, 1970).

Clinically, evidence for the presence of endotoxemia has been sought in a number of gastrointestinal diseases. Kruis and his colleagues measured antibody titers to lipid A, the toxic core component of LPS, in patients with ulcerative colitis and Crohn's disease (Kruis et al, 1984). Despite an inability to regularly demonstrate endotoxin in the systemic circulation by the Limulus lysate assay (LAL), titers to lipid A were found to be significantly higher in the inflammatory bowel disease caused by Crohn's than in either normals or patients with ulcerative colitis. Further the highest titers were found in the Crohn's patients with active disease. A study recently reported (Wellmann et al, 1986), however, was able to demonstrate endotoxemia by the LAL in 17 of 18 Crohn's disease patients hospitalized for acute exacerbations. In an interesting and seemingly effective therapeutic approach, the endotoxemia was rapidly eliminated by intestinal lavage with 5-aminosalicylic acid added to the final saline gavages. Further this form of gut cleansing led to significantly more rapid improvement in the disease parameters and to shortened hospital stays.

The immature gut represents another human condition that leads to endotoxemia. In necrotizing enterocolitis in infants, endotoxemia was regularly found (Scheifele et al, 1981), and it was correlated with thrombocytopenia.

Relationship Between Endotoxinemia and Thrombocytopenia (Platelet Count Less Than 100,000/MM^3 In 40 Infants with Necrotizing Enterocolitis

Patient Group	Number	No. with Thrombocytopenia
Endotoxinemic Infants	19	9 (47%) *,#
No Endotoxinemia In:		
a) Single Test	21	2 (9.5%)*
b) Two Tests	12	0#

* x^2 with Yales' Correction = 5.439, P<0.05
# P<0.005, Fisher's exact test
(Scheifele et al, 1985)

Further in premature infants weighing below 1500 grams with immature guts, endotoxemia was frequently detected in those infants requiring an intensive care setting (Scheifele et al, 1985). Further these sick prematures did not demonstrate endotoxemia until LAL activity could be demonstrated in the stools after the initiation of oral feedings. The significance of endotoxemia in the clinical picture of these immature new borns was not apparent.

Acute Pancreatitis is another gastrointestinal disease where evidence of endotoxemia has been demonstrated. Kivilaakso and his colleagues (Kivilaakso et al, 1984) studied titers to the common enterobacterial antigen (a component of the outer bacterial cell wall) and found that the incidence of a greater than 8 fold increase during hospitalization correlated with the severity of the disease. Suggestive evidence that this endotoxemia was clinically important was found in another study (Foulis et al, 1982) in which the presence of endotoxemia in acute pancreatitis correlated with not only increased severity but with intravascular coagulation, hypoxia and a decrease in alternate pathway CH50.

Lastly Bode and colleagues in a study of bacteria in the jejunal aspirate of chronic alcoholics and hospitalized controls (Bode, et al, 1984) found the alcoholic group had more aerobic and anaerobic organisms in the jejunal fluid, and that the number of organisms correlated closely with the pH of the gastric juice. No correlation was found between the numbers and types of organisms in alcoholics and the presence or absence of liver disease.

It is in both experimental and clinical liver disease, however, where the most impressive evidence exists for a role of enteric endotoxins in liver injury. As with gastrointestinal disease, the problem is whether the associated endotoxemia accompanying liver disease is an epiphenomenon or whether the presence of this toxic material in small quantities is truly injurious. The hypothesis we first proposed in 1975 (Nolan, 1975) states that LPS is constantly being absorbed in small amounts by the gut and passes to the liver where it is removed and detoxified. When metabolic or toxic liver injury occurs, the liver (particularly the fixed macrophages) are unable to perform this function normally, leading in turn to more liver damage, and further impairment of endotoxin detoxification. Increased endotoxin absorption through the gut after alcohol and toxic injury undoubtedly contributes to this process. The resulting increase in hepatic sensitivity to LPS after toxic injury may approach 1,000 fold (Galanos, 1979) and lead to spillover of endotoxin into the systemic circulation resulting in many of the extra-hepatic manifestations of decompensated cirrhosis.

The concept that the integrity of the intestinal flora and its products is essential for toxic liver injury dates back many years. In 1941, sufonamide drugs were shown in animals to protect against carbon tetrachoride (CCl_4) injury and death (Leach, 1941). Animals treated in a germ-free environment showed less liver injury on a necrogenic diet, than those conventionally reared (Luckey, 1954). Further, non-absorbable antibiotics markedly delayed the onset and severity of choline deficiency cirrhosis in rats (Rutenberg, 1957). In an elegant study on this problem, the addition of purified LPS to the drinking water of rats on this diet eliminated the protective effect of neomycin, demonstrating that endotoxin and not the gut bacteria were key to the development of cirrhosis (Broitman, 1964).

While a marked synergism exists between the toxic effects of LPS and other hepatotoxins (Formal, 1960), the mechanism by which endotoxins injure hepatocytes remains to be completely elucidated. LPS causes little evident damage to isolated hepatocytes, and work with the endotoxin resistant C3H/HeJ mouse mutant suggests that most toxic effects are mediated by release of effectors from the Kupffer cells (Morrison, 1978). Possible mediators of liver cell damage by the action of LPS on these monocytic cells include tumor necrosis factor (T.N.F.), platelet activating factor (P.A.F.), toxic oxygen radicals, leukotrienes, and lysosomal enzymes. It is also possible in toxic liver injury that damage to the Kupffer cells may allow LPS to reach altered hepatocytes and cause functional damage. The interaction of LPS and these mediators from the macrophage at the level of the liver cell needs to be elucidated.

A number of animal models of liver injury lend support to the hypothesis that enteric LPS is a critical factor in toxic liver damage. Animals made tolerant to endotoxin or those pretreated with the LPS neutralizing antibiotic polymyxin B are protected against toxic agents as diverse as CCl_4 and the toxic viral protein-frog virus 3 (Nolan, 1973; Nolan, 1978; Gut, 1980). Pre-treatment of rats with D-galactosamine not only markedly increases LPS sensitivity, but this injury can be prevented by colectomizing these animals prior to the administration of the toxin (Grun, 1976).

A great number of clinical studies in patients with liver injury suggest that LPS is present in the systemic circulation in a number of acute and chronic liver diseases, and these studies have been summarized (Nolan, 1981). A major difficulty in comparing the results in different studies is the methodology of the LAL technique used. Indeed one major study (Fulenwider, 1980) failed to detect LPS in the peripheral circulation portal vein, or ascites in a group of decompensated patients with alcoholic cirrhosis. Several recent studies however, suggest that endotoxemia does occur and seems clinically relevant in both liver disease due to alcohol and in states of prolonged extra-hepatic obstruction. A study of

IgA, IgM and IgG antibodies to lipid A were studied in a population of alcoholics with liver disease. Both the IgA and IgG titers were significantly elevated in this group of patients compared to normals and to patients with other gastrointestinal diseases (Nolan, 1986). Figure 1 compares the IgA anti-lipid titers in these groups.

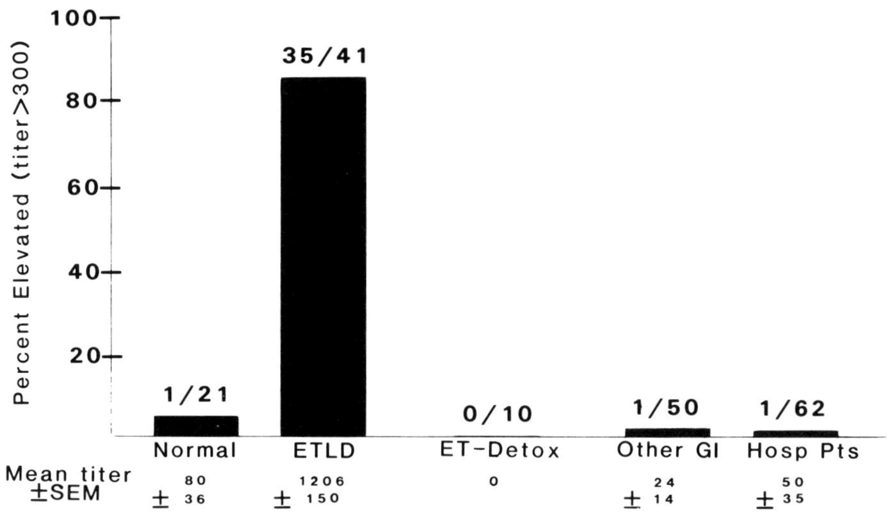

FIGURE 1

Increases in titers of IgA anti-lipid A in serum of patients with alcoholic liver disease.

The percent of patients with serum IgA titers to lipid A >300 are shown for five groups. The mean titers as determined by the ELISA are also depicted.

In another recent study where the LAL was used to detect LPS in the systemic circulation, heavy alcohol intake alone caused transient endotoxemia supporting the hypothesis that gut derived endotoxins may be important in initiating or exacerbating alcoholic liver injury (Bode, 1987). As far as clinical relevance of endotoxemia in this disease, there does seem to be a good correlation with LPS in the systemic circulation and a depressed creatinine clearance. Further the clearance rises significantly when paramocyin is used to decrease the enteric endotoxin burden (Tarao, 1982).

Patients with obstructive jaundice are known to be at risk for the post-operative development of renal failure. A study assaying Limulus positivity in these patients showed that those with endotoxemia preoperatively had significantly lower creatinine clearance that remained low in the immediate post-operative period. In the same paper rats with ligated bile ducts had increased LPS absorption but it could be abolished by the exogenous administration of bile salts which are known to disrupt endotoxins and may be important in reducing the intestinal endotoxin pool (Bailey, 1976). A pre-operative clinical trial on patients with obstructive jaundice revealed that the control group not given sodium deoxydrolate had a 54% incidence of endotoxemia and two-thirds of these patients had impaired creatinine clearance. By contrast, no patients receiving this bile salt every 8 hours for 48 hours prior to surgery had either endotoxemia or renal impairment (Cahill, 1983). In another randomized trial, no such benefit of preoperative bile salt administration was noted in either the incidence of systemic endotoxemia, renal dysfunction or post-operative outcome. However, ursodeoxycholic acid rather than sodium deoxycholate was used (Thompson, 1986). Other evidence that endotoxemia may have deleterious effects in obstructive jaundice is the finding in these patients of an increase in fibrin degradation products associated with LAL positivity (Hunt, 1982).

If LPS absorbed from the intestinal tract is an important cofactor in alcoholic and other toxic liver disease, then attempts to modify the biologic effects of endotoxin might be expected to ameliorate damage from these other hepatotoxins. Surprisingly few such studies have been done in human liver disease although evidence in experimental liver injury suggests that such an approach is feasible. From a theoretical viewpoint, active immunization with LPS from core mutants might protect against liver injury as well as monoclonal antibodies directed against the toxic core. However, as reported in this conference, the protective effect of these antibodies in septic states has been disappointing. Attempts to reduce the endotoxin load in the gastrointestinal tract has been promising in experimental animals, and the use of bile salts and paramomycin in human diseases has been previously cited here as effective in reversing endotoxemia. Testing various absorbants in the intestines for their ability to prevent serotonin induced endotoxemia, bentonite was the most effective, followed by Kaopectate and charcoal particles (Ditter, 1983). Cholestyramine also has been shown to be an effective endotoxin binder (Nolan, 1972). Since macrophage mediators seem to be increasingly important as mediators of LPS toxicity, it is possible in the future that antibodies or receptor antagonists to T.N.F., P.A.F. and other effectors may prove to be therapeutically useful.

This review examines evidence for a pathophysiological role of intestinal endotoxin in a number of gastrointestinal and liver disorders. Accumulating evidence would suggest that such a role may indeed exist for this toxic bacterial product when it is absorbed in abnormally large quantities, and when detoxification mechanisms are impaired. While animal studies support a key role for endotoxins as co-factors for liver injury in particular, an association in human disease remains to be clearly established.

REFERENCES

Chapters in Books:

Nolan JP, Camara DS (1985). The importance of intestinal endotoxins in liver disease. Bacterial Endotoxins Alan R. Liss, Inc. publ. 347-364.

Journal Articles:

Bailey ME, (1976). Endotoxin, bile salts and renal function in obstructive jaundice. Br. J. Surg. 73:774-778.

Bode JC, et al (1984). Jejunal microflora in patients with chronic alcohol abuse. Hepato-gastro. 31:30-34.

Bode C, et al (1987). Endotoxemia in patients with alcoholic and non-alcoholic cirrhosis and in subjects with no evidence of chronic liver disease following acute alcohol excess. J. Hepatology 4:8-14.

Brearley S, et al (1985). Endotoxin levels in portal and systemic blood. Dig. Surg. 2:70-72.

Broitman SA, et al (1964). Influence of neomycin and ingested endotoxin in the pathogenesis of choline deficiency cirrhosis in the adult rat. J Exp. Med. 119:633-641.

Cahill CJ (1983). Prevention of postoperative renal failure in patients with obstructive jaundice-the role of bile salts. Br. J. Surg. 70:590-595.

Camara DS, et al (1983). D-galactosamine liver injury: Absorption of endotoxin and protective effect of small bowel and colon resection in rabbits. Proc. Soc. Exp. Biology Medicine 172:255-259.

Daniele R, et al (1970). Lymphatic absorption of intraperitoneal endotoxin in the dog. Surgery 67:484-487.

Ditter B, et al (1983). Ability of various absorbants to bind endotoxins in vitro and to prevent orally induced endotoxemia in mice. Gastroent. 84:1547-1552.

Formal SB, et al (1960). Experimental shigella infections. III Sensitivity of normal, starved and carbon tetrachloride treated guinea pigs to endotoxin. Proc. Soc. Exp. Biol. Med. 103:415-418.

Foulis AK, et al (1982) Endotoxaemia and complement activation in acute pancreatitis in man. Gut 23:656-661.

Fulenwider JT, et al (1980). Endotoxemia of cirrhosis: An observation not substantiated. Gastoenterol. 78:1001-1004.

Gaffin SL, et al (1986). Hypoxia-induced endotoxemia in primates: Role of reticuloendothelial system function and anti-lipopolysaccharide plasma. Aviat. Space & Environ. Med. 1044-1049.

Galanos C, et al (1979). Galactosamine-induced sensitization to the lethal effects of endotoxin. Proc. Natl. Acad. Sci. 76:939-943.

Gathiram P, et al (1986). Superior mesenteric artery occlusion shock in cats: Modification of the endotoxemia by antilipopolysaccharide antibodies (Anti-LPS). Circulatory Shock 19:231-237.

Grun M, et al (1976). Significance of endotoxemia in experimental "galactosamine hepatitis" in the rat. Acta Hepatogastroenterol (Stuttg) 23:64-81.

Gut JP, et al (1980). Kupffer cell function and frog virus 3 hepatitis in mice and rats. The Reticuloendothelial System and the Pathogenesis of Liver Disease. Amsterdam, Elsevier/North Holland. 211-219.

Holdstock G, et al (1982). Activation of monocytes by portal serum and its relationship to immunoglobulin, immune complex and endotoxin content. Liver 2:222-229.

Hunt, DR, et al (1982). Endotoxemia, disturbance of coagulation, and obstructive jaundice. Amer. J. Surg. 144:325-329.

Jacob AI, et al (1977). Endotoxin and bacteria in the portal blood. Gastroent. 72:1268-1270.

Kivilaakso E, et al (1984). Endotoxaemia and acute pancreatitis: Correlation between the severity of the disease and the antienterobacterial common antigen antibody titre. Gut 25:1065.

Kruis W, et al (1984). Circulating lipid A antibodies despite absence of systemic endotoxemia in patients with Crohn's disease. Dig. Dig. Sci. 29:502-507.

Leach BE, et al (1941). Sulfonamide drugs as protective agents against carbon tetrachloride poisoning. Proc Soc Exp Biol Med 48:361-363.

Luckey TD, et al (1954). Germ free animals and liver necrosis. Ann. NY Acad. Sci. 57:926-932.

Morrison DC, et al (1978). The effects of bacterial endotoxins on host mediation systems, Am. J. Pathol. 93:527.

Nolan JP, et al (1977). In Vitro Studies of Intestinal Endotoxin Absorption. I. Kinetics of absorption in the isolated everted gut sac. Gastroenterology 72:434-439.

Nolan JP (1975) The role of endotoxins in liver injury. Gastroenterol 69:1346-1356.

Nolan JP, et al (1973). Endotoxin and the liver. II Effect of tolerance on carbon tetrachloride induced injury. J. Med. 4:28-38.

Nolan JP, et al (1978). Endotoxin and the liver. III Modification of acute carbon tetrachloride injury by polymyxin B-an antiendotoxin, Gastroent. 75:445-449.

Nolan JP (1981): Endotoxin, reticuloendothelial function and liver injury. Hepatology 1:458-465.

Nolan JP, et al (1986). IgA antibody to lipid A in alcoholic liver disease. Lancet 176-179.

Nolan JP, et al (1972). Effect of cholestyramine on endotoxin toxicity and absorption. Am. J. Dig. Dis. 17:161-166.

Nozickova M, et al (1977). Effect of transient intestinal ischaemia on the thoracic duct lymph absorption of endotoxin. Lymphology 10:161-165.

Palmer KR, et al (1980). Bacteriological and endotoxin studies in cases of ulcerative colitis submitted to surgery. Gut 21:851-854.

Ravin HA, et al (1960). On the absorption of bacterial endotoxin from the gastrointestinal tract of the normal and shocked animal. J. Exp. Med. 112:783-792.

Rutenburg AM, et al (1957). The role of intestinal bacteria in the development of dietary cirrhosis in rats. J. Exp. Med. 106:1-13.

Scheifele, DW, et al (1981). Evaluation of the Limulus test for endotoxemia in in neonates with suspected sepsis. J. Pediat. 98:899-903.

Scheifele, DW, et al (1985). Spontaneous endotoxinemia in premature infants: Correlations with oral feeding and bowel dysfunctions. J. Ped. Gastro 4:67-74.

Scheifele, DW, et al (1985). Endotoxinemia and thrombocytopenia during neonatal necrotizing enterocolitis. Amer. J. Clinical Ped. 83:227-229.

Tarao, et al (1982). Effect of paromomycin sulfate on endottoxemia in patients with cirrhosis. J. Clin. Gastro. 4:263-267.

Thompson JN, et al (1986). A randomized clinical trial of oral ursodeoxycholic acid in obstructive jaundice. Br. J. Surg. 73:634-636.

Walker RI (1978). The contribution of intestinal endotoxin to mortality in hosts with compromised resistance: A review. Exp. Hematol. 6, 172-184.

Wellman W, et al (1986). Endotoxaemia in active Crohn's disease. Treatment with whole gut irrigation and 5-aminosalicylic acid. Gut 27:814-820.

Wessels BC, Gaffin SL (1986). Anti-endotoxin immunotherapy for canine parvovirus endotoxaemia. J. Small Anim. Pract. 27:609-615.

Bacterial Endotoxins: Pathophysiological Effects, Clinical Significance, and Pharmacological Control, pages 161–171

ENDOTOXIN INJECTION AFFECTS THE KUPFFER CELL MORPHOLOGY IN THE RAT LIVER

H. Van Bossuyt, E. Wisse

Laboratory for Cell Biology and Histology, Free University of Brussels (VUB), Laarbeeklaan 103, 1090 Brussels-Jette, Belgium

INTRODUCTION

The liver is the most important organ for clearing endotoxin (lipopolysaccharide, LPS) from the circulation (Freudenberg et al., 1984; Ruiter et al., 1981). Several reports have dealt with the light and electron microscopic alterations in the liver of dog (Boler and Bibighaus, 1967; Rangel et al., 1970), mouse (Levy et al., 1968), rat (DePalma et al., 1967; White et al., 1973; Yoder et al., 1985), baboon (Hoffman et al., 1974) and Rhesus monkey (Balis et al., 1978). In these reports, emphasis was given to the changes in the ultrastructure of the parenchymal cells (Pc), and comparatively little attention has been given to the morphological changes in Kupffer cells.

Kupffer cells (Kc) are resident liver macrophages that play an important role in the clearance of LPS from the blood (Freudenberg et al., 1982; Ruiter et al., 1981). The present study investigates the morphology of rat Kc in situ at different time intervals after intravenous injection of purified LPS.

MATERIALS AND METHODS

Endotoxin was prepared from Salmonella Abortus Equi bacteria (kindly provided by Dr. C. Galanos, Max Planck Insitut fur Immunobiologie, Freiburg, Federal Republic of Germany) according to Galanos et al. (1979). Male <u>Wistar rats</u> (250-280 gm) were kept on normal laboratory diet and water ad libitum. Three rats were used at each time interval (see below). All experiments were carried out under sterile conditions with the rats being under ether anesthesia. The LPS was dissolved in physiological saline by sonication. The injections were given between 08.30 hours and 12.00 hours. One group of animals was injected with 3 mg LPS via the dorsal penile vein; control rats received saline only. A second group of rats was injected (through a 25 gauge needle, Minicath-25, Deseret Pharmaceutical Co., USA) with 0.8 mg LPS into the right small ventral liver lobe via the portal vein. When the time interval for perfusion was more than one hour, the abdomen was closed by suturing the muscle layer and then the skin. All rats were allowed to recover and had free access to water and food. The doses applied in all cases were about one half (for peripherally-injected animals) or one eight (for portally-injected animals) the mean lethal dose.

Before liver fixation, the animals were anesthetized with ether. After 1 min (portal venous administration only), 5 min, 15 min, 30 min, 1 h, 3 h, 6 h 1 day, 3 days and 5 days, liver tissue was fixed by perfusion, via the portal vein, with 1.5 % glutaraldehyde in 0.1 M phosphate buffer (pH 7.4) with 1% sucrose for 5 min at a pressure of 20 cm of water. Twenty 50 μm vibratome sections were prepared from the right ventral liver lobe. Ten sections were postfixed in 1 % OsO_4 in phosphate buffer, dehydrated in ethanol and embedded in Epon. The remaining sections were incubated for 1 h at room temperature in a medium to demonstrate peroxidase activity : 0.14 M sodium cacodylate buffer (pH 6.9) containing 0.1 % diaminobenzidine and 0.01 % H_2O_2 (modified procedure of Wisse et al., 1974); then washed and postfixed in OsO_4. Ultra-thin sections, stained with uranyl acetate (10 min) and lead citrate (20 min), were examined with a Philips EM 400.

RESULTS

A) Control Animals

Only those ultrastructural details that are necessary to elucidate the changes in animals given an LPS injection will be described. For description of the appearance of other organelles we refer to previous reports (Wisse, 1977; Wisse et al., 1974). Several types of vacuoles occurred in the Kc, differing in size, shape and density. Some dense bodies, presumably representing primary lysosomes, consisted of rounded, dense, membrane limited organelles with a diameter of about 150 nm (Fig. 1). They were composed of a membrane, a clear zone, a "second" membranous layer and a uniformly dense content. These dense bodies were often found close to the cytocenter. A second type of dense bodies was interpreted as phagosomes and secondary lysosomes. They were medium sized with a diameter up to about 800 nm, with a content that was never uniformly dense or lucent. They sometimes had a clear zone as did primary lysosomes. Large vacuoles (Fig. 1) containing whole cells (e. g. erythrocytes) or membranous structures were considered to represent phagosomes. They were found occasionally and seldom had a round shape. A third type of dense body, which occurred less frequently, comprised large lucent vacuoles with a diameter up to 1.1 μm, often round in shape, and containing only some granular or fibrillar material. These vacuoles may have been of pinocytotic origin, as judged from a fuzzy coat-like rim on the inner side of the membrane, which suggests recent internalization of the plasma membrane.

Two types of rough endoplasmic reticulum (RER) were present (Fig. 1). The first, more frequent type had a matrix denser than the cytoplasm and with membranes which were difficult to distinguish. The other type of RER had a lucent, more narrow matrix, and the membranes were clearly seen. The ribosomal layer was discontinuous. The two types were found close to each other, even within one cisterna. Some short cisternae were ribosome-free and presumably represent SER.

B) Animals Injected With LPS

After administration of endotoxin, changes in mitochondria and RER were restricted to about 20 % of the Kc, whereas vacuolar changes were found in almost all Kc. The alterations were identical in periportal and centrolobular Kc. Some Kc were located in the space of Disse, indicating that they had crossed the endothelial lining. Between 1 h and 1 day necrotic Kc were observed.

Changes in the vacuolar apparatus. One and 5 min after injection via the portal vein, Kc showed vacuoles of variable size. They were filled with dense material, often containing membranous structures, which resembled the

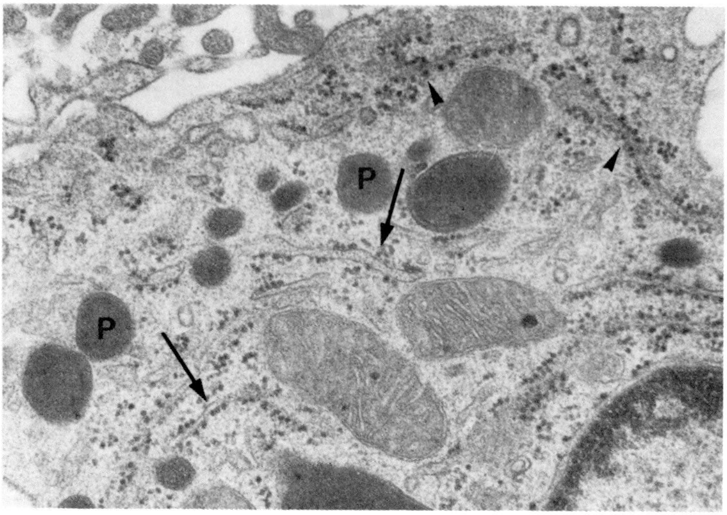

Fig. 1: Kupffer cell from a control rat showing mitochondria, the two types of RER (dense matrix arrowheads; lucent narrow arrows), primary lysosomes P. x41 800

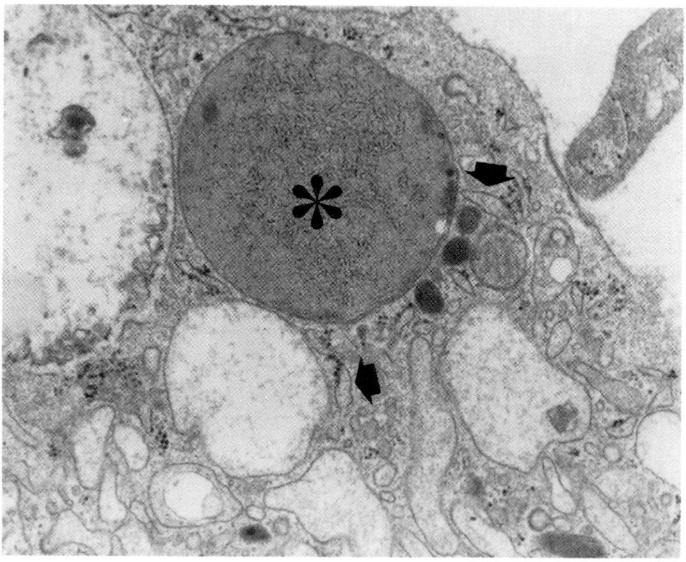

Fig. 2: Thirty minutes after portal venous injection of LPS many vacuoles were present. They contained granular material and in some, LPS could be recognized (asterisk). There were short profiles of enlarged lucent RER (arrows) without ribosomes. x26 200

structure of pure preparations of LPS. There were also many lucent vacuoles with some fibrillar and granular material. At 5 min, the Kc of animals injected via the peripheral vein displayed the same changes in the vacuolar apparatus but membranous LPS structures were not present. The number of medium and large vacuoles, containing granular and/or fibrillar material, increased between 15 min and 30 min (Fig. 2). For animals injected via the portal vein, this effect continued up to 1 h after administration of LPS. By 15 min, platelets were often associated with Kc, and some phagocytosis of platelets occurred. Phagocytosis of neutrophilic granulocytes and erythrocytes was observed at 30 min and increased until 6 h (Fig. 3). Between 3 h and 1 day, a decreasing number of Kc showed the small dense bodies (Fig. 5). By day 1, the cells mostly contained medium-sized vacuoles filled with different material. At this time, recent phagocytosis of whole cells was not frequent in portally-injected animals. The increase in vacuolar inclusions resulted in enlarged Kc. The cells reached a maximum volume between 3 h and 1 day, returning to almost normal on day five.

Changes in the RER. Five min after injection via the portal vein, the RER in some Kc was mainly of the lucent type, and focally showed cisternal swelling. Short, irregular profiles of RER were frequent at 15 min - 1 h (Fig. 2). They had an enlarged lucent matrix and occurred especially around mitochondria. This effect was less frequent in peripherally-injected animals. By 3 and 6 h, enlarged lucent cisternae with and without ribosomes were found. One day after injection of LPS the lucent type of RER was still predominant, often irregular and swollen (Fig. 4). At later time intervals the Kc showed RER with normal morphology.

Changes in the mitochondria. From 5 min to 1 h after portal injection of LPS, mitochondria occasionally showed an irregular, crenated outline and focal narrowing of the intermembranous space. The cristae were slightly swollen. This effect was seen from 15 min in some Kc mitochondria of animals injected via the peripheral vein. Mitochondria showed a rough outline and contained less cristae at 3 h - 6 h. The outer membrane and inner membrane were in intimate contact. The matrix was cloudy and the intracristal space was irregular. Some cristae had disappeared and the matrix granules were small or absent. By 6 h few Kc showed a mixed mitochondrial population within a single cell. Mitochondria occasionally had a denser matrix and dilated

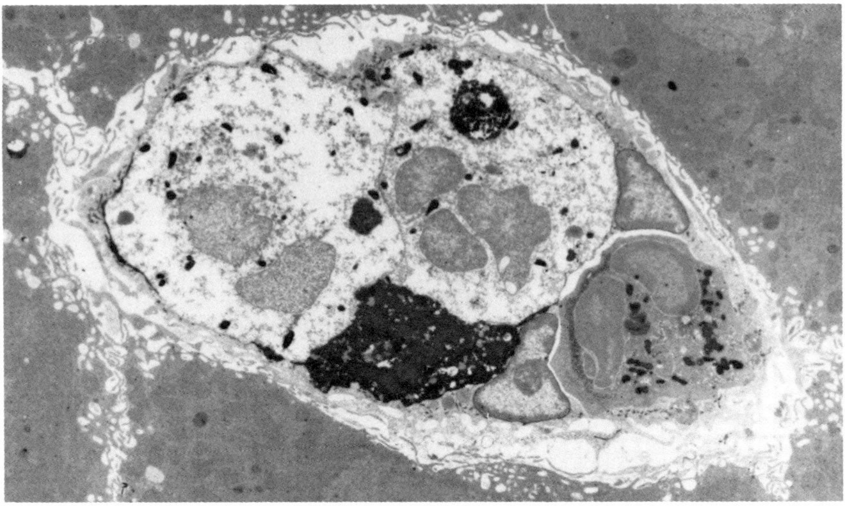

Fig. 3: Three hours after injection of LPS. Three granulocytes phagocytozed by a Kupffer cell, apparently blocking the sinusoid. Stained for peroxidase. x4 400

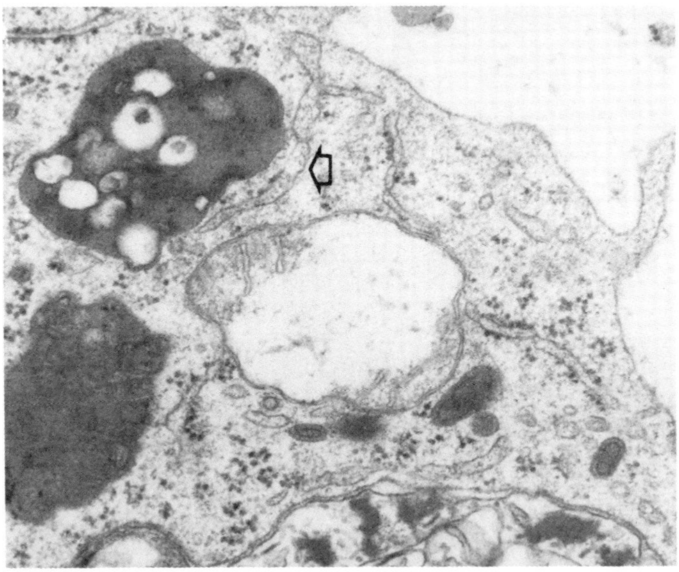

Fig. 4: Twenty-four hours after injection of LPS, impressive alterations were seen in mitochondria, e.g. bleb formation and disappearance of matrix and cristae. The RER was predominantly of the lucent type, often swollen (arrow). x42 900

intracristal space. One day after injection of LPS, the inner membranes of mitochondria were irregular; the cristae were sometimes curled, and there was a partial disappearance of both cristae and matrix (Fig. 4). Matrix granules were seldomly present in such mitochondria, and bleb formation due to focal swelling occurred. These mitochondria were frequently present in cells with an empty or cloudy cytoplasmic matrix, whereas the neighbouring cells were well preserved. Lysosomes with an interrupted membrane were seen in those cells. Concomitantly, the plasma membrane was discontinuous, suggesting increased fragility and rupture, possibly during tissue processing. Fusion between lysosomes and other organelles was never seen. By the third day after injection via the portal vein, some Kc showed mitochondria of similar structure as seen at 3 and 6 h. Mitochondria with a dense matrix and enlarged cristae were observed at 3 and 5 days after peripheral administration of LPS. The animals injected via the portal vein showed these same changes at 5 days.

Changes in the Golgi apparatus. Between 1 h and 6 h, the Golgi apparatus was smaller, with less associated vesicles. One day after injection of LPS, one or more lamellae of the Golgi apparatus were swollen in Kc with altered mitochondria (Fig. 5). The organelle showed again normal morphology by day 3 and 5.

Free, cytoplasmic fat droplets in Kc were occasionally found at 3 h. The number of Kc containing free fat droplets increased until 3 days after injection of LPS. No Kc with free fat droplets were found at 5 days.

Fibrin was found in the sinusoids at 3 h and 6 h after peripheral administration of LPS. It was associated with the Kc plasma membrane, but not phagocytosed.

Small Kupffer cells. Kc with a lower cytoplasm to nucleus ratio, which we call small Kc, were observed at 3 days (Fig. 6). Such cells had fewer phagocytic vacuoles and many cytoplasmic protrusions. Long cisternae of the RER were abundant and were connected to annulate lamellae. The cells showed peroxidase staining in the RER and the nuclear envelope. These Kc became more frequent at 5 days after injection of LPS.

DISCUSSION

After administration of LPS via portal or peripheral veins, Kc showed changes in their mitochondria, dense

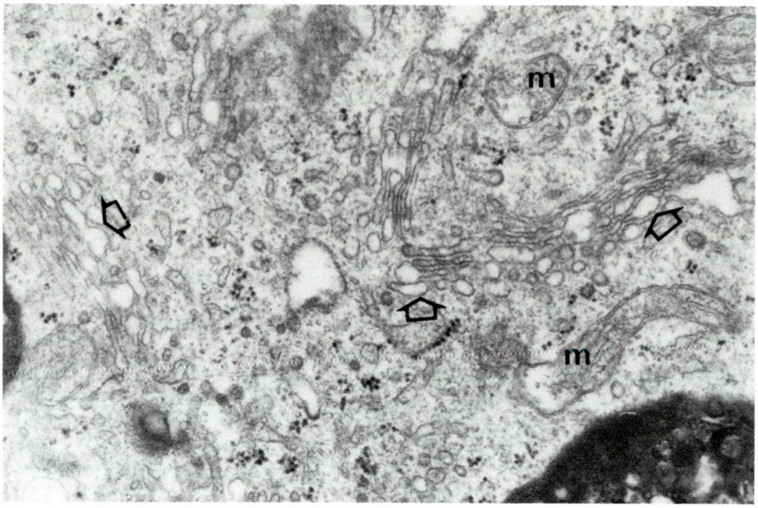

Fig. 5: One day after administration of LPS, Golgi complexes with swollen cisternae (arrows) were present. Mitochondria with dissolved matrix (m). There were no small dark vacuoles present around the cytocenter. x34 900

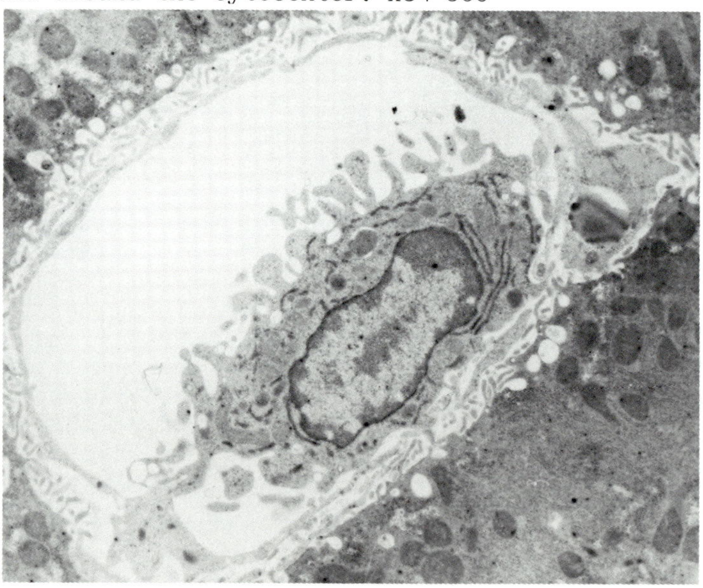

Fig. 6: Small Kupffer cells were regularly seen at three and five days after injection of LPS. They were peroxidase positive and showed a low cytoplasm/nucleus ratio and many cytoplasmic protrusions. x4 700

bodies and vacuoles, RER, Golgi apparatuses and the presence of free fat droplets. The presence of free lipid droplets in Kc has been observed by others (Boler and Bibighaus, 1967; Hoffman et al., 1974). Striking alterations occurred in the mitochondria, the vacuolar apparatus and the lucent type of RER. Dilatation of the smooth ER and disorganization of the rough ER may indicate altered protein synthesis and have also been described for Pc (Boler and Bibighaus, 1967; DePalma et al., 1967; Levy et al., 1968; Rangel et al., 1970).

Shortly after injection of LPS, the number of endocytic vacuoles increased. Most of them were of the pinocytotic type. They occurred more in portally-injected than in peripherally-injected animals. Typical membranous LPS structures, differing in size, found in vacuoles after portal venous injection, were comparable to structures seen after positive staining of pure LPS (Gaeta and Wisse, 1983; Balis et al., 1978; Ruiter et al., 1981). At one hour, the phagocytic vacuoles increased, and at three hours the primary lysosomes had disappeared in many Kc. Three and six hours after injection of LPS, many Kc had ingested polymorphonuclear leucocytes, as was also reported earlier (Freudenberg et al., 1982). This increase in vacuolar volume and blood cell phagocytosis resulted in swelling of Kc (Hoffman et al., 1974; Levy et al., 1968; McCuskey et al., 1983; Ruiter et al., 1981). The swollen Kc obviously narrowed the sinusoids and hampered blood flow. Other mechanisms also retard the blood flow, such as early platelet adhesion to sinusoidal cells, platelet aggregation, fibrin deposits and marginating neutrophilic granulocytes. Similar observations have been reported (McCuskey, 1983; Boler and Bibighaus, 1967; Balis et al., 1978; Freudenberg et al., 1982; Hoffman et al., 1974).

Several minutes after injection of LPS, platelets adhered to the Kc and endothelial cells. Between one and six hours, neutrophilic granulocytes accumulated in the liver sinusoids up to twenty-fold (data not shown), resulting in at least partial obstruction of the sinusoids. The sinusoidal blood flow might become retarded in those sinusoids where plugging occurs. In vivo microscopy revealed a reduced blood flow in the liver sinusoids two hours after endotoxin injection (McCuskey, 1983). This will result in local ischaemia causing ultrastructural changes in the surrounding cells. Indeed, some of the mitochondrial alterations have also been described as a result of hypoxia (White et al., 1973). However, mitochondrial swelling and internal disintegration occurred mainly at day one, at which time the number of granulocytes was already lower. Lysosomes with a broken membrane were observed in those cells. This leakage probably caused the cloudy cytoplasm and the mitochondrial damage.

Analogous changes have been reported for parenchymal cell mitochondria after a single dose of endotoxin (Boler and Bibighaus, 1967; Hoffman et al., 1974; Levy et al., 1968; White et al., 1973). The changes displayed at various time intervals were focal, and some Pc showed a mixed mitochondrial population within a single cell. This is in keeping with our results for Kc.

The ultrastructural changes of Kc did not differ substantially as a result of the two routes of administration.

These observations indicate that the ultrastructural alterations in Kc are mediated eighter through ischaemia as a result of sinusoidal obstruction, or through lysosomal enzyme release, or through other mediators, (e.g. secreted by neutrophils). A direct LPS effect on Kc is possible since the subcellular localization of LPS at the different time intervals is not yet elucidated. The ultrastructural alterations may reflect changes in the glucose and protein metabolism of Kc. Biochemical studies are underway to investigate this.

REFERENCES

Balis JU, Rappaport ES, Gerber L, Fareed J, Buddingh F, Messmore HL (1978). A primate model for prolonged endotoxin shock. Blood-vascular reactions and effects of glucocorticoid treatment. Lab Invest 38:511-523.

Boler RK, Bibighaus AJ (1967). Ultrastructural alterations of dog livers during endotoxin shock. Lab Invest 17:537-561.

DePalma R, Coil J, Davis J, Holden WD (1967). Cellular and ultrastructural changes in endotoxemia : a light and electron microscopic study. Surgery 62:505-515.

Freudenberg MA, Freudenberg N, Galanos C (1982). Time course of cellular distribution of endotoxin in liver,

lungs and kidneys of rats. Br J Exp Path 63:56-65.
Freudenberg MA, Kleine B, Galanos C (1984). The fate of lipopolysaccharide in rats: evidence for chemical alteration in the molecule. Rev Infect Dis 6:483-487.
Gaeta GB, Wisse E (1983). Endotoxin effect on the isolated perfused rat liver : functional and ultrastructural observations. J Submicrosc Cytol 15:705-712.
Galanos C, Lüderitz O, Westphal O (1979). Preparation and properties of a standardized lipopolysaccharide from Salmonella Abortus Equi (Novo-Pyrexal). Zentralbl Bakteriol Mikrobiol Hyg(A) 243:226-244.
Hoffmann E, DiLuzio N, Holper K, Brettschneider L, Coover J (1974). Ultrastructural changes in the liver of baboons following lead and endotoxin administration. Lab Invest 30: 311-319.
Levy E, Path FC, Slusser RJ, Ruebner BH (1968). Hepatic changes produced by a single dose of endotoxin in the mouse. Am J Pathol 52: 477-502.
McCuskey RS (1983). Endotoxins and the pathophysiology of the hepatic microvascular system. In Koo A (ed): "Microcirculation of the alimentary tract", Singapore: World Sci Publ Co, pp 345-350.
Rangel DM, Byfield JE, Adomian GE, Stevens GH, Fonkalsrud EW (1970). Hepatic ultrastructural response to endotoxin shock. Surgery 68:503-511.
Ruiter DJ, Van der Meulen J, Brouwer A, Hummel MJR, Mauw BJ, Van der Ploeg JCM and Wisse E (1981). Uptake by liver cells of endotoxin following its intravenous injection. Lab Invest 45: 38-45.
White RR, Mela L, Bacalzo LV, Olofsson K, Miller LD (1973). Hepatic ultrastructure in endotoxemia, hemorrhage and hypoxia : emphasis on mitochondrial changes. Surgery 73:525-534.
Wisse E (1977). Ultrastructure and function of Kupffer cells and other sinusoidal cells in the liver. In Wisse E, Knook DL (eds): "Kupffer cells and other liver sinusoidal cells", Amsterdam: Elsevier/North-Holland Biomedical Press, pp 33-60.
Wisse E, Van der Meulen J, Van't Noordende, JM (1974). Observations on the fine structure and peroxidase cytochemistry of normal rat liver Kupffer cells. J Ultrastruct Res 46:393-426.

Bacterial Endotoxins: Pathophysiological Effects, Clinical Significance, and Pharmacological Control, pages 173–184

IMMUNE RESPONSES IN LIVER AND BILE TO *ESCHERICHIA COLI* 0111:B4 LIPOPOLYSACCHARIDE

Sharon A. Paull, Lisa Carter and Graham D.F. Jackson.
School of Microbiology, University of N.S.W.
P.O. Box 1, Kensington, 2033, N.S.W.
AUSTRALIA

INTRODUCTION

Due to their wide range of biological effects, bacterial endotoxins are among the most widely studied of the naturally occurring antigens. Interaction with the host's immune apparatus has been intensively studied (reviewed by Morrison and Ryan, 1979) with particular emphasis on initiation and modulation (e.g. Jacobs, 1982; McGhee *et al.*, 1982) of humoral immune responses. Our current interest in antibody responses to lipopolysaccharide (LPS) itself, arises from studies where anti-LPS activity associated with IgM and IgG rather than IgA was detected in the bile of rats which had received either killed *Salmonella enteritidis* intravenously (Jackson and Walker, 1983) or live *Vibrio cholerae* intestinally (Cooper *et al.*, 1984). The finding of specific IgA antibodies also in bile in the second instance was not surprising as several laboratories have reported such a response to a variety of antigens (e.g. Lemaitre-Coelho *et al.*, 1978; Andrew and Hall, 1982; Dahlgren *et al.*, 1983) and has been considered to stem from the well documented hepatobiliary transport process (Jackson *et al.*, 1978; Orlans *et al.*, 1978; Renston *et al.*, 1980).

It has now been determined that the secretion of high levels of specific IgM antibody in bile following IV immunisation is not restricted to bacterial antigens (Jackson *et al.*, 1985) and that there is a correlation between the presence of the IgM in bile with the occurrence of antibody secreting cells in the liver (Carter *et al.*, 1987). Since the liver is a target organ for endotoxic activity and is involved in its elimination, immunological activity in this site could well bear on the resolution of these events. In this study we have investigated immune responses in liver and bile to a LPS injection given intra Peyer's patches (IPP) or IV, which reflect the sites of entry and dissemination in the host of materials from the gastrointestinal tract.

MATERIALS AND METHODS

Animals: Male Wistar rats (10-20 weeks of age) were obtained from the University of N.S.W. animal breeding unit.
Lipopolysaccharide: *Escherichia coli* 0111:B4 LPS (Sigma Chemical Co., St. Louis, Mo.), prepared by trichloroacetic acid extraction, was dissolved in phosphate buffered saline (PBS) pH 7.2.
Injection of animals: IV injections were made into a tail vein of anaesthetised rats. For IPP injections, the small bowel was drawn through a small abdominal incision. The total dose of antigen was injected into 6-8 patches. A suspension of bentonite particles (Sigma Chemical Co.) was prepared according to the method of Gallily and Garvey (1968).
Preparation of serum and bile: Serum samples were prepared from blood taken from a tail vein and bile was collected according to the method of Lambert (1965).
Serology: Anti-LPS activity in serum and bile was measured in a standard passive haemagglutination assay (HA) and total immunoglobulin (Ig) levels were determined by radial immunodiffusion (Mancini *et al.*, 1965). Ig standards were purchased from Miles Laboratories, Elkhart, In.
Gel filtration: Pools of bile freshly collected from immunised rats were chromatographed on a column of Ultrogel AcA22 and the fractions assayed for Ig classes and anti-LPS activity as previously described (Carter and Jackson, 1987).
Antisera: Procedures have been described for preparation of heavy chain specific antisera against rat IgA and IgM (Jackson and Walker, 1983) and IgG (Carter *et al.*, 1987). Goat anti-rat IgG2b was purchased from Pel-Freez Biologicals, Rogers, Ar.
Haemolytic plaque assay: The preparation of mononuclear cell (MNC) suspensions from perfused livers and spleens has been described (Manning *et al.*, 1984). Direct (IgM) and

indirect (IgG and IgA) antibody forming cells (AFC) were assayed using the method of Henry (1980) modified as described by Carter et al. (1987) using horse erythrocytes (HRBC) (CSL, Melbourne, Australia) sensitised with LPS according to the method of Möller et al. (1965). The total number of cells secreting Ig (ISC) were detected by the staphylococcal protein A plaque assay of Gronowicz et al. (1976).

Histology: Formalin-fixed tissues from test rats which had received an IV injection of 1.0ml of a suspension of colloidal carbon were embedded in paraffin, sections were cut at 5μm and stained with haematoxylin and eosin (H & E).

RESULTS

Responses in serum and bile to IV injection of LPS: Rats were injected IV with 75μg E.coli 0111:B4 LPS, a dose predetermined to result in a high serum antibody response. Serum and bile were collected up to 10 days after injection and assayed for anti-LPS activity and total Ig concentration. HA activity appeared in both fluids on day 3 and reached maximum levels on day 5 (Figure 1A). The activity in bile declined thereafter whilst that in serum persisted. Sensitivity to mercaptoethanol and association with the IgM region after gel filtration indicated that this class of Ig was predominant in the response (results not shown).

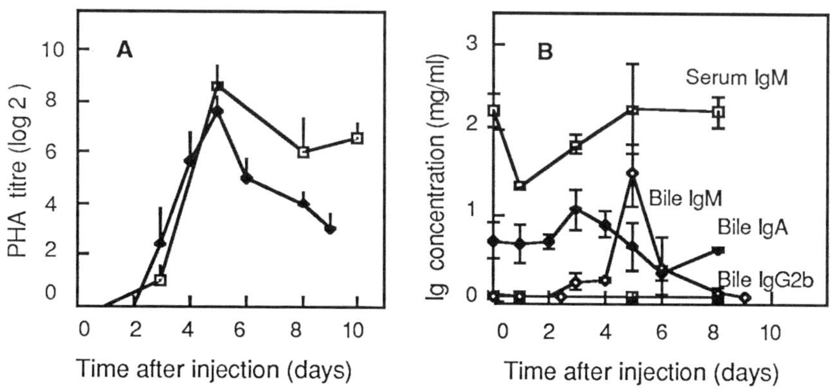

Figure 1. Anti-LPS activity in serum (□) and bile (◆) (A) and immunoglobulin levels in serum and bile (B) of rats injected IV with E.coli 0111:B4 LPS.

During the biliary antibody response there was a transient surge in total IgM concentration with a 35-fold increase ($p < 0.01$) at day 5 (Figure 1B). There was an initial decrease in the level of serum IgM but the pre-immunisation level was restored by day 3. Little or no change was observed in the concentrations of biliary IgA or IgG2b, chosen as a representative sub-class of IgG. It is noted that the immunologically inert material bentonite did not result in any change in biliary IgM levels when given IV.

The role of the spleen in the biliary IgM response was demonstrated by determining the responses to LPS in splenectomised rats. The results presented in Table 1 indicate reductions of 97% and 75% in the HA activity in serum and bile respectively, and only a four-fold increase in the total level of IgM in bile. The lower concentration of IgM and IgA in the sera of splenectomised animals is noted.

TABLE 1. Antibody responses to LPS and immunoglobulin levels in serum and bile of splenectomised rats.

Treatment[a]	Sample[b]	Anti-LPS titre[c] (range)	Immunoglobulin concentration[c] (mg/ml) ± SEM IgM[d]	IgA	IgG2b
sham operated	serum	384 (256-512)	2.262± 0.549	0.227± 0.037	8.500± 3.900
	bile	192 (128-256)	1.490± 0.360	0.644± 0.286	0.074± 0.010
splenectomy	serum	12 (8-16)	1.115± 0.180	0.106± 0.005	6.115± 2.260
	bile	48 (32-64)	0.158± 0.110	0.856± 0.123	0.124± 0.050

a. Rats injected IV with 75µg E.coli 0111:B4 LPS at time of treatment.
b. Serum and bile samples taken 5 days after treatment.
c. Represents mean value from 3 - 6 rats.
d. Concentration of IgM in bile of non-immunised rats is <0.040 mg/ml.

Antibody secreting cells in spleen and liver after IV injection of LPS: Previous studies in this laboratory as quoted above have shown that the most likely source of biliary IgM is from synthesis within the liver. Figure 2 depicts the numbers of AFC and ISC in the spleens and livers of rats at the time of and 5 days after IV injection with LPS.

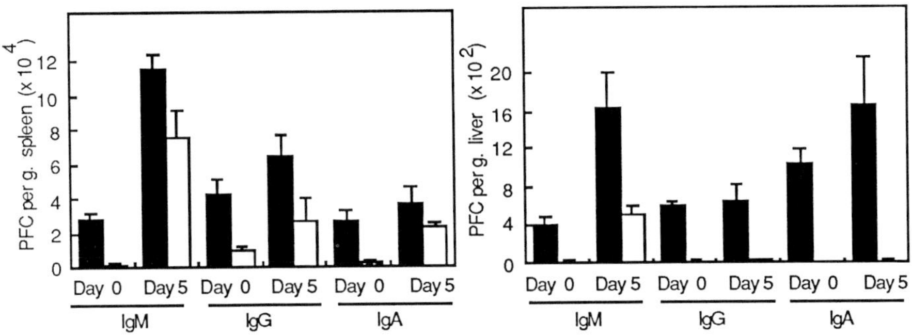

Figure 2. Numbers of IgM, IgG and IgA AFC (anti LPS secreting) (□) and ISC (Ig secreting) (■) in liver and spleen of rats given an intravenous injection of E.coli 0111:B4 LPS.

The response in liver was associated with IgM where along with the occurrence of AFC against LPS there was a significant increase ($p < 0.001$) in ISC IgM numbers above those contributed by the anti-LPS AFC. No significant increases in IgA and IgG AFC or ISC were detected. The major response in the spleen was also associated with IgM although anti-LPS IgG and IgA AFC were present. Small numbers of IgM AFC were detected in blood (results not shown).

Antibody responses after IPP injection of LPS: Since LPS normally present in the gastrointestinal tract is likely to stimulate the gut associated lymphoid tissues, a group of rats was injected IPP with LPS. Five days later the biliary and serum antibody responses were determined along with enumeration of anti-LPS AFC and ISC in the spleen and liver. The results in Table 2 show a greater than four-fold increase in the concentration of IgM in bile together with the presence of anti-LPS activity. Gel filtration of this bile and analysis of the fractions revealed that specific activity was associated with IgM, IgG and IgA (results not shown). Table 3 reports the presence of anti-

LPS AFC expressing all 3 Ig classes in liver and spleen. It is noted that the numbers of IgA AFC in the liver are higher than the numbers of IgG AFC.

TABLE 2. Antibody responses and immunoglobulin levels in serum and bile of rats injected intra Peyer's patches with LPS.[a]

	Anti-LPS titre[b] (range)	Immunoglobulin concentration[b] (mg/ml) ± SEM		
		IgM[c]	IgA	IgG2b
serum[d]	416 (128-512)	2.231± 0.077	0.226± 0.025	6.317± 1.227
bile	208 (64-256)	0.154± 0.032	0.652± 0.108	0.115± 0.063

a. Rats injected IPP with approximately 0.05ml of a 75μg/ml solution of E. coli 0111:B4 LPS in saline.
b. Represents mean value from 4 rats.
c. Concentration of IgM in bile of non-immunised rats is <0.040 mg/ml.
d. Samples taken 5 days after injection.

TABLE 3. Numbers of antibody forming cells in liver and spleen of rats injected intra Peyer's patches with LPS.[a]

	Anti-LPS AFC/g tissue[b]		Ig secreting cells (ISC)/g tissue[b]	
	Spleen	Liver	Spleen	Liver
IgM	19,000± 1,070	733± 206	71,866± 30,233	1,210± 66
IgG	19,602± 3,151	326± 66	41,477± 10,584	1,210± 460
IgA	23,864± 13,855	995± 640	56,818± 7,000	1,862± 263

a. Livers and spleen cell suspension from same rats used to determine results in Table 2.
b. Results expressed as the arithmetic mean ± standard deviation.

Cellular infiltration in the liver: In preparing MNC suspensions from the perfused livers it was noted that a significantly higher number of cells ($p < 0.001$) was obtained from the immunised rats ($1.88 \pm 0.25 \times 10^7$ cells/liver from IV immunised animals versus $3.25 \pm 0.56 \times 10^6$ cells/liver from unimmunised rats). It is calculated that less than 0.1% of the infiltrate was associated with Ig secretion. Histology reveals (Figure 3) an infiltration of cells with lymphocytic morphology principally in the portal regions with occasional cells located in sinusoids. Immunohistochemical staining with monoclonal antibodies directed against rat lymphocyte membrane markers revealed that the majority of infiltrating cells express T-cell markers (results not shown).

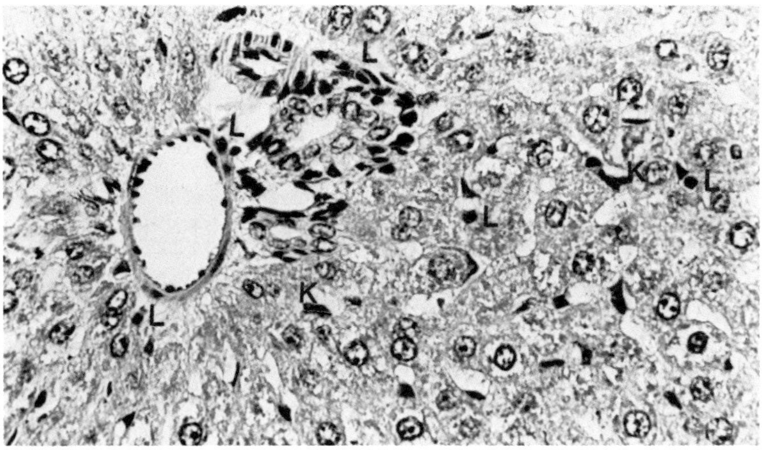

Figure 3: Perfused liver from rats 6 days after IV injection of E.coli LPS. Note lymphocyte (L) infiltration in sinusoidal and portal regions. Kupffer cells (K) are associated with colloidal carbon (H & E x 960).

DISCUSSION

The presence of endotoxin within the gastrointestinal tract either as free molecules or in association with bacteria is a constant threat to the integrity of host tissues. Inadvertant entry into the bloodstream leads to contact with the elements of non-specific and specific resistance; in particular events in the liver associated with reticuloendothelial system activity are generally considered to be of paramount importance in protection. While the role of antibody in enhancing protection is somewhat controversial, anticore region or lipid A antibodies have been shown to inhibit pathophysiologic activities (Rioux-Darrieulat et al., 1978) and monoclonal anti-O-polysaccharide antibodies can be protective in an animal model of Salmonellosis (Colwell et al., 1984).

In this paper we have demonstrated the presence of Ig secreting cells in the liver following IV and IPP injection of an E.coli LPS and suggest they are the source of much of the specific antibody activity appearing in bile. It is further suggested that these events may augment defence processes against this and other agents which lodge in the liver to be removed or inactivated. Also these observations support our recent studies involving antibody synthesis in the livers of rats challenged with horse erythrocytes (Carter et al., 1987; Carter and Jackson, 1987).

The surge of IgM in bile is most marked following IV immunisation and reflects a typical primary response to this antigen. The absence of any significant change in the concentration of IgA and IgG2b in bile and the presence of normal levels of all 3 classes in serum after 3 days indicates that a highly individual event in the liver causes the delivery of IgM into bile. The presence of anti-LPS AFC and the large increase in total numbers of IgM secreting cells, presumably as a result of polyclonal activation, offer the most likely explanation for the phenomenon. The decline in serum IgM concentration at 24 hours is likely to result from the early suppressive effect of LPS on IgM secreting cells in the spleen as reported by Michael (1966). The reduced response to LPS in serum and bile following splenectomy strongly suggests this organ as the major source of the IgM AFC in the liver. This is in agreement with our previous studies (Carter et al., 1987). The lower but significant response in the bile of splenectomised rats could be from cells which have emigrated from other sites such as the bone marrow (Benner and van Oudenaren, 1975).

The occurrence of specific anti-LPS IgA, IgG and IgM in bile following IPP challenge reflects the site where the immunocompetent cells are challenged and is similar to the response noted following IPP challenge with HRBC (Carter and Jackson, 1987). The presence of the respective AFC in the liver supports the notion of their being the source of this biliary Ig. With regard to IgA in bile it seems that up to 25% of the natural IgA in bile is not derived from intestinal synthesis (Manning et al., 1984). Also Dahlgren et al.(1983) have shown that the IgA resulting from IPP immunisation with E.coli has sources other than from the intestine and Jones et al. (1987) have detected IgA cells in the livers of rats immunised intraduodenally with cholera toxin. These observations along with our data, suggest that early in a response, the bulk of specific IgA appearing in bile is from synthesis within the liver.

The argument that in situ synthesis provides the source of biliary IgM and IgG is much easier to sustain since neither class is actively transported from blood (Fisher et al., 1979). The mechanism of secretion of these Ig into bile is unknown but is suggested to be independent of SC (Jackson et al., 1985). Insight into this matter is dependent on determining the exact location of the cells and the receptor-ligand interactions in the liver which are responsible.

As well as secretion into bile it is possible that a portion of the Ig could be retained in the liver, at least at focal points, to participate in effector processes. Classically, antibody can act as an opsonin, and with the recent demonstration of the uptake of IgA-antigen complexes by Kupffer cells (Rifai and Mannik, 1984) all classes of Ig may operate in this fashion. It has been proposed (e.g. Russell et al., 1981) that IgA-antigen complexes may be cleared from the circulation in the liver via the SC mediated active transport process and that bacterial antigens are good candidates for such removal (Russell et al., 1984). Because of limitations to the sizes of the antigen and antibody molecules (reviewed by Underdown and Schiff, 1986) its relevance has not been clearly established. However, as we have described here the ability of copious quantities of IgM and the other classes to be secreted, it is suggested that a reappraisal of this process might be worthwhile.

The cellular infiltration noted in these studies involves predominantly T-cells and can be considered to aid in clearance of endotoxin by enhancing phagocytic processes, although their involvement may be more significant in dealing with intact microorganisms. LPS specific T cells have not been characterised in this study but an earlier report from our laboratory described their induction following infection with Salmonella enteritidis (Marks et al., 1975).

Mononuclear cell infiltration in the liver is commonly equated with the pathology of many diseases yet it is also known that lymphocytes accumulate in or are recruited to the liver during responses to endotoxin (Likovsky and Konickova, 1977) allogeneic lymphocytes (Emerson, 1977) and foreign erythrocytes (Carter et al., 1987). This event therefore may be a normal feature of immune responses and we would postulate that underlying pathological infiltrates in many diseases a normal accumulation of lymphocytes occurs. Their functions in inductive and expressive phases of immune activity are yet to be determined.

ACKNOWLEDGEMENT

This work was supported by a grant from the National Health and Medical Research Council of Australia.

REFERENCES

Andrew E, Hall JG (1982). IgA antibodies in the bile of rats. I. Some characteristics of the primary response. Immunology 45: 169-175.

Benner, R, Van Oudenaren A (1975). Antibody formation in mouse bone marrow. IV. The influence of splenectomy on the bone marrow plaque-forming cell response to sheep red blood cells. Cell Immunol 19: 167-182.

Carter, L, Barrington PJ, Cooper GN, Jackson GDF (1987). Antibody synthesis in the rat liver: An association between antibody-forming cells in the liver and biliary antibodies following intravenous injection of horse erythrocytes. Int Archs Allergy appl Immun 82: 153-158.

Carter L, Jackson GDF (1987). Antibody synthesis in the rat liver: A source of biliary IgM, IgA and IgG following injection of horse erythrocytes into the intestine or Peyer's patches. Recent Developments in Mucosal Immunology, in press.

Colwell DE, Michalek SM, Briles DE, Jirillo E, McGhee JR (1984). Monoclonal antibodies to Salmonella lipopolysaccharide: Anti-O-polysaccharide antibodies protect C3H mice against challenge with virulent Salmonella typhimurium. J Immunol 133: 950-957.

Cooper GN, McNab CE, Walker PG, Jackson GDF (1984). Intestinal antibodies in rats following exposure to live Vibrio cholerae. Aust J Exp Biol Med Sci 62: 465-477.

Dahlgren U, Ahlstedt S, Anderson T, Hedman L, Hanson LA (1983). IgA antibodies in rat bile are not solely derived from thoracic duct lymph. Scand J Immunol 17: 569-574.

Emerson EE (1977). Migratory behaviour of lymphocytes with specific reactivity to alloantigens I. J Immunol 119: 1204-1212.

Fisher MM, Nagy B, Bazin H, Underdown BJ (1979). Biliary transport of IgA: role of secretory component. Proc Natl Acad Sci USA 76: 2008-2012.

Gallily R, Garvey JS (1968). Primary stimulation of rats and mice with hemocyanin in solution and adsorbed on bentonite. J Immunol 101: 924-929.

Gronowicz E, Coutinho A, Melchers F (1976). A plaque assay for all cells secreting Ig of a given type or class. Eur J Immunol 6: 588-590.

Jackson GDF, Lemaitre-Coelho I, Vaerman JP, Bazin H, Beckers A (1978). Rapid disappearance from serum of intravenously injected rat myeloma IgA and its secretion in bile. Eur J Immunol 8: 123-130.

Jackson GDF, Walker PG (1983). The transient appearance of IgM antibodies in the bile of rats injected with Salmonella enteritidis. Immunol Lett 7: 41-45.

Jackson GDF, Walker PG, Schiff JM, Barrington PJ, Fisher MM, Underdown BJ (1985). A role for the spleen in the appearance of IgM in the bile of rats injected intravenously with horse erythrocytes. J Immunol 135: 152-157.

Jacobs DM (1982). Lipopolysaccharide and the immune response. In Strober W, Hanson LA, Sell KW (eds): "Recent Advances in Mucosal Immunity", New York: Raven Press, pp 47-55.

Jones AL, Altorfer J, Hardesty SJ (1987). Specific IgA secretion by mononuclear cells within the rat liver. Recent Developments in Mucosal Immunology, in press.

Lambert R (1965). "Surgery of the digestive system in the rat". Springfield: Thomas.

Lemaitre-Coelho I, Jackson GDF, Vaerman JP (1978). Relevance of biliary IgA antibodies in rat intestinal immunity. Scand J Immunol 8: 459-463.

Likovsky Z, Konickova Z (1977). Accumulation of activated lymphocytes in liver of endotoxin treated rabbits. Life Sciences 21: 1425-1428.

Mancini G, Carbonara AD, Heremans JF (1965). Immunochemical quantitation of antigens by single radial immunodiffusion. Immunochem 2: 235-254.

Manning RJ, Walker PG, Carter L, Barrington PJ, Jackson GDF (1984). Studies on the origins of biliary immunoglobulins in rats. Gastroenterol 87: 173-179.

Marks RM, Jackson GDF, Cooper GN (1975). Salmonella enteritidis infection in rats: Antigens involved in cell-mediated immunity. Aust J Exp Biol Med Sci 53: 315-324.

McGhee JR, Michalek SM, Kiyono H, Babb JL, Clark MP, Mosteller LM (1982). Lipopolysaccharide regulation of the IgA immune response. In Strober W, Hanson LA, Sell KW (eds): "Recent Advances in Mucosal Immunity", New York: Raven Press, pp. 57-70.

Michael,JG (1966). The release of specific bactericidal antibodies by endotoxin. J Exp Med 123: 205-212.

Möller G (1965). 19S Antibody production against soluble lipopolysaccharide antigens by individual lymphoid cells in vitro. Nature 207: 1166-1168.

Morrison DC, Ryan JL (1979). Bacterial endotoxins and host immune responses. Adv Immunol 28: 293-450.

Orlans E, Peppard J, Reynolds J, Hall J (1978). Rapid active transport of IgA from blood to bile. J Exp Med 147: 588-592.

Renston RH, Jones AL, Christiansen WD, Hradek GT, Underdown, BJ (1980). Evidence for a vesicular transport mechanism in hepatocytes for biliary secretion of IgA. Science 208: 1278-1278.

Rifai A, Mannik M (1984). Clearance of circulating IgA immune complexes is mediated by a specific receptor on Kupffer cells in mice. J Exp Med 160: 125-137.

Rioux-Darrieulat F, Parant M, Chedid L (1978). Prevention of endotoxin-induced abortion by treatment of mice with antisera. J Inf Dis 137: 7-13.

Russell MW, Brown TA, Mestecky J (1981). Role of serum IgA. Hepatobiliary transport of circulating antigen. J Exp Med 153: 968-976.

Russell MW, Brown TA, Claflin JL, Schroer K, Mestecky J (1983). IgA-mediated hepatobiliary transport constitutes a natural pathway for disposing of bacterial antigens. Inf & Immun 42: 1041-1048.

Underdown BJ, Schiff MJ (1986). IgA: Strategic defense initiative at the mucosal surface. Ann Rev Immunol 4: 389-417.

Bacterial Endotoxins: Pathophysiological Effects, Clinical Significance, and Pharmacological Control, pages 185–194

UPTAKE AND DEACYLATION OF BACTERIAL LIPOPOLYSACCHARIDES IN CULTURED RAT HEPATOCYTES

Ichiro. Fukuda*, Ken-ichi. Tanamoto and Shiro. Kanegaseki

Institute of Medical Science, University of Tokyo, Minatoku, Tokyo, 108 Japan
*Department of Internal Medicine, Tohoku University School of Medicine, Sendai, 980 Japan

INTRODUCTION

Endotoxin is a constituent of the gram-negative bacterial cell wall and exhibits numerous biological effects both in vivo and in vitro. Endotoxin is chemically a lipopolysaccharide (LPS) composed of a polysaccharide part, O-antigen and core, and a lipid part called lipid A. Lipid A consists of a glucosamine disaccharide backbone which is substituted with fatty acid residues in ester or amide linkage (Fig. 1).It is the active center, responsible for almost all biological effects of LPS. Fatty acid residues are essential to fully express the biological activities of LPS (Tanamoto et al., 1979; Galanos et al., 1983; Kanegaseki et al., 1986). LPS is trapped mainly in the liver when injected in vivo. Furthermore, non-septic endotoxemia was found to occur in hepatic failure (Wilkinson et al., 1974; Yajima et al., 1985). Hence, it is thought that the liver may play an important role for metabolism, especially for the detoxification of LPS. However, the fate of LPS in the liver has not been well documented. The purpose of the present study was to elucidate the role of this organ for the metabolism of LPS and to clarify whether or not cultured hepatocytes incorporate and modify the LPS in vitro. Here, we found that the cells incorporate R-form LPS but not S-form LPS, in accordance with the results of Freudenberg et al. (Freudenberg et al., 1982; Kleine et al., 1985), and also that the cells partly degraded the molecule.

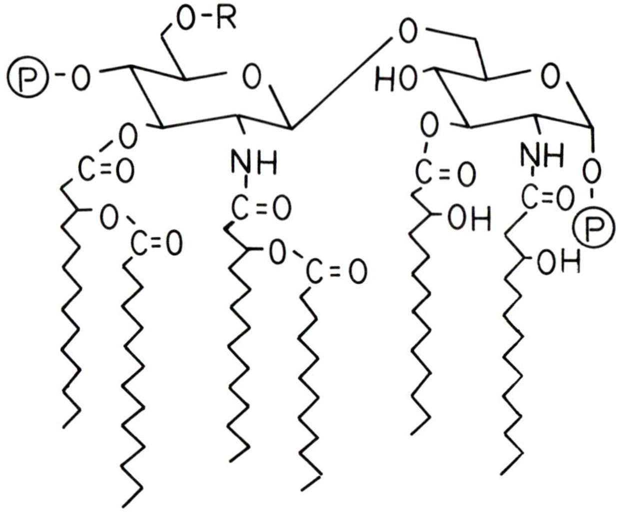

Figure 1. Structure of E. coli lipid A. Shown is the lipid A backbone with substituents. Symbol P designates phosporyl groups and R 3-deoxy-D-mannocutulosonic acid in undegraded lipopolysaccharide and H in free lipid A.

MATERIALS AND METHODS

LPS was obtained from E. coli E112 and Salmonella anatum A1 epi by hot phenol-water extraction (Westphal et al., 1952). Both strains were deficient in UDP-glucose-4-epimerase and E. coli additionally lacks glucosamine deaminase. LPS from E. coli was biosynthetically radiolabeled in the glucosamine backbone (^{3}H) or fatty residues (^{14}C) by culturing them in PPBE broth containing N-acetyl-D-(1-^{3}H)-glucosamine or (1-^{14}C)-acetate. LPS from Salmonella was radiolabeled in the polysaccharide portion with (1-^{3}H)-galactose. S- and R-form LPS from E. coli were obtained after cultivation of the bacteria in the presence or absence of galactose. All radioisotopes were purchased from New England Nuclear (NEN). The specific activities of LPS from E. coli were as follows: 800 cpm/μg and 1750 cpm/μg ^{3}H-glucosamine in S- and R-form LPS, respectively, 650 cpm/μg ^{14}C-fatty acids in R-form LPS. LPS from Salmonella contained 1200 cpm/μg.

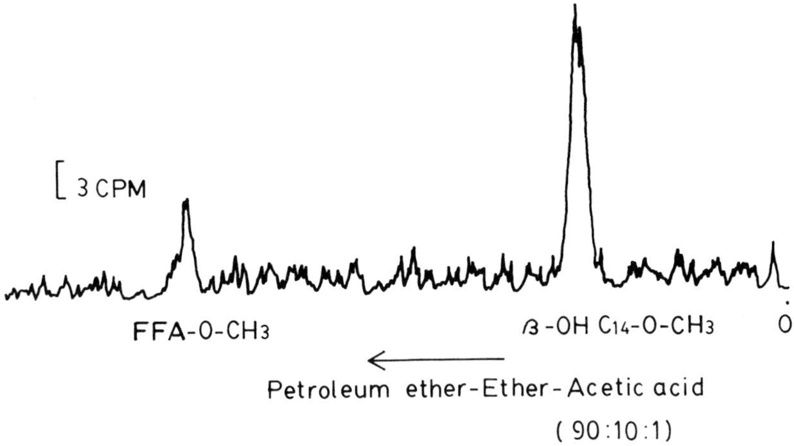

Figure 2. Chromatoscan of a thin layer chromatogram of fatty acid methyl ester from E. coli LPS after methanolysis using petroleum ether:ether:acetic = 90:10:1 as a solvent.

Almost all radioactivity (97%) was extractable in the chloroform:methanol layer after hydrolysis, showing that the lipid A part was labeled. These results confirmed that E coli LPS was radiolabeled in the fatty acid residues (^{14}C) and in D-glucosamine (^{3}H). Salmonella LPS labeled in the polysaccharide was treated with 1% acetic acid at 100°C, 15 h, and was applied to Sephadex G150 chromatography. All radioactivity was recovered from the fractions corresponding to the heterogeneous polysaccharide but not to those of lipid A. After hydrolysis of the LPS with 4N HCl (100°C, 15 h), the LPS was analyzed by paper chromatography using the solvent system of butanol:pyridine:water (6:4:3). Almost all radioactivity was detected on the site corresponding to galactose.

The binding of LPS to hepatocytes was studied using both radiolabeled S- and R-forms of LPS. As the S-type of LPS, both E coli and Salmonella LPS were tested. The association of LPS with the cells was detected by counting the radioactivity of the cells and the medium after their separation. In the case of S-form LPS, no uptake was detected up to 72 h of cultivation and all the radioactivity added was recovered from the medium when 35 μg of each LPS were added.

Hepatocytes were isolated using a method described in detail elsewhere (Obrink, 1982). Briefly, male Wistar rats were anesthetized with ether. After the completion of a median abdominal incision, the portal veins were perfused at 37°C with Hank's balanced salt solution (HBSS) containing 0.05% collagenase (Sigma, Type VI). Hepatocytes were then separated from other cells by centrifugation at 50 x g and placed in fibronectin-coated wells. We also used hepatocytes from Otsuka Assay Institute (Tokushima), commercially called HEPAPLATE. These cells were cultured in an L15 medium at 37°C, 5% CO_2. 1 x 10^5 cells/wells were used for the experiments.

After cultivation with radiolabeled LPS, hepatocytes were washed with PBS twice and released from the wells by incubation with 0.05% trypsin and 0.02% EDTA solution at 37°C, 15 min.

Lipophilic substances from the culture medium and hepatocytes were extracted with chloroform:methanol (2:1) 24, 48 and 72 h after the addition of radiolabeled LPS. Two dimensional thin layer chromatography of the extraction after 48 h was performed in chloroform:methanol:water (65:25:4) for the first dimension and petroleum ether: ether:acetic acid (70:30:2) for the second dimension. Radioactivity was measured by a scintilation counter (Beckman 2000). GPT and GOT activities of the culture medium were measured by Karmen's method.

RESULTS

To confirm the labeling sites of these LPS, the following treatments were performed. E. coli LPS labeled in fatty acid residues were methanolized with 5% HCl methanol at 100°C, 18 h and liberated fatty acid methyl esters were extracted 5 times with petroleum ether. 92% of radioactivity of E. coli LPS labeled on fatty acids was recovered in the extract. The extracted fatty acids were analyzed by thin layer chromatography using a solvent system of petroleum ether:ether:water (90:10:1). ß-hydroxytetradecanoic acid and non-hydroxylated fatty acids were found at a ratio of about 4:1 (Fig. 2). E coli LPS radiolabeled in diglucosamine backbone were treated with 1% acetic acid at 100°C, 90 min and free lipid A was extracted with chloroform:methanol (5:1).

On the other hand, R-form LPS bound to the cells. When 15 ug were incubated with 1 x 10^5 cels, 4175 $\pm$ 250, 6745 $\pm$ 210 and 8610 $\pm$ 365 cpm (N=4, mean $\pm$ S.D.) of radioactivity were incorporated into the cells after 24, 48 and 72 h incubation, respectively (Fig. 3). The dose dependency for the uptake of LPS by hepatocytes using the R-form LPS is shown in Fig. 4. The uptake of LPS increased in a dose-dependent manner in the range tested (up to 35 μg). To exclude the possibility that the difference between S- and R-form LPS in the uptake by hepatocytes was due to the difference of their cytotoxicity, we measured GPT and GOT levels in the medium. When 40 μg of R-form E. coli LPS or S-form Salmonella LPS were added, no increase of GPT and GOT was observed (Table 1) indicating that LPS at this concentration exhibits little cytotoxicity.

The degradation of LPS by hepatocytes was studied using R-form LPS whose fatty acids were radiolabeled. After various lengths of incubation with 40 μg of LPS, the medium and the cells were separated by centrifugation. The lipophilic substances of each fraction were then extracted with chloroform:methanol (2:1). Radioactivity was found in the extract from cultured medium with chloroform:methanol (2:1). The radioactivity increased depending on the culture time, reaching a plateau after 48 h (Fig. 5). Two dimensional thin layer chromatography of the extract after 48 h of cultivation was performed (Fig. 6). The radioactivities were found at the position corresponding to that of ß-hydroxy fatty acids and triglyceride. The molar ratio of these lipids was 7:1. These lipids found in the medium were not detected in the cell extracts. Also no radioactivity was found in the extract from the medium containing the same amount of LPS but no hepatocytes.

SUMMARY AND DISCUSSION

When liver LPS is injected into animals, it accumulates mainly in the liver (Nolan et al., 1982). Under the clinical picture of hepatic failure, non-septic endotoxemia appears. Accordingly, we focussed our interest on liver cells, especially on hepatocytes, whose role in the metabolism and detoxification of endotoxin is unknown. Freudenberg et al. found that in experimental rats, S- and R-form LPS were incorporated into Kupffer

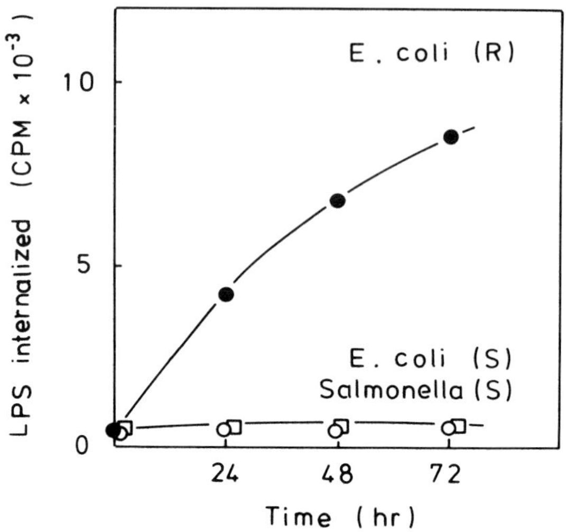

Figure 3. Time dependency of the uptake of S- and R-form LPS by hepatocytes. S-form Salmonella LPS (36,000 cpm) and S- and R-form E. coli LPS (S:24,000 cpm, R:26,250 cpm) were incubated with 1 x 10^5 hepatocytes. The radioactivity incorporated was measured after washing the cells at indicated times.

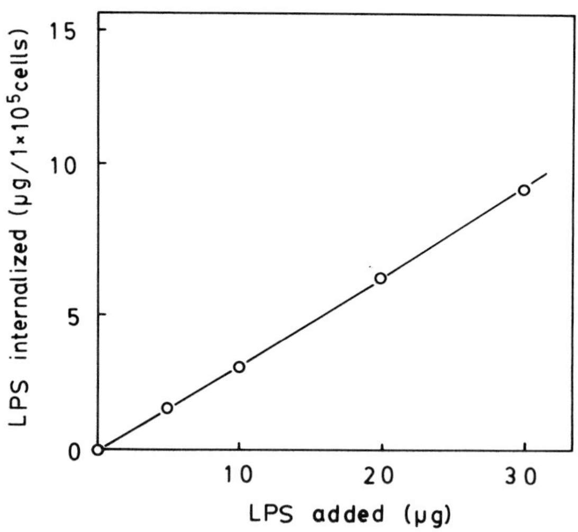

Figure 4. Dose-dependent uptake of R-form E. coli LPS by hepatocytes. The uptake of LPS increase in a dose-dependent manner up to 35 µg. 1 x 10^5 hepatocytes/well were used. The means of two wells were plotted.

GOT

	24	48 hr	
Control	45.8 ± 3.2	76.5 ± 4.9	(N=4)
E. coli (R)	42.5	67.5	(N=2)
Salmonella (S)	40.5	68.0	

GPT

	24	48 hr	
Control	11.5 ± 3.6	13.3 ± 0.8	(N=4)
E. coli (R)	6	10	(N=2)
Salmonella (S)	8.5	11	

Table 1. GPT and GOT values in cultured medium. LPS (40μg/well) were incubated with 1 x 10^5 hepatocytes. GPT and GOT of cultured medium without LPS were measured as a control. The values are means or means ± SD.

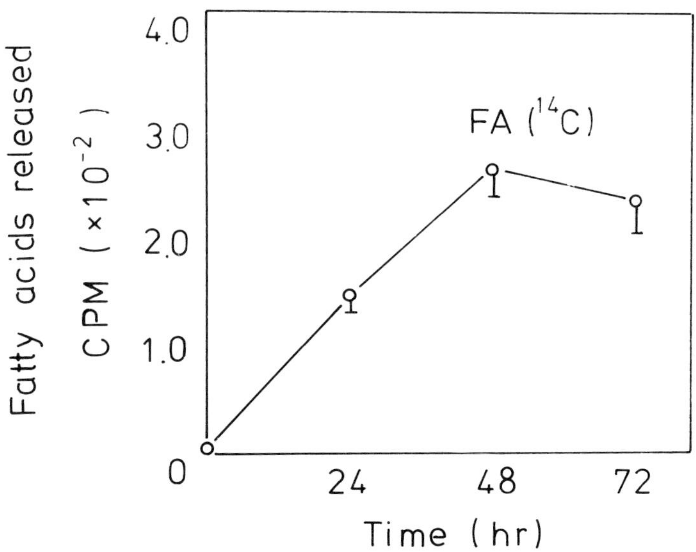

Figure 5. The radioactivity of lipophilic substances which were extracted from cultured medium with chloroform: methanol (2:1) at different times. The radioactivity increased depending on the time and reached a plateau value after 48 h.

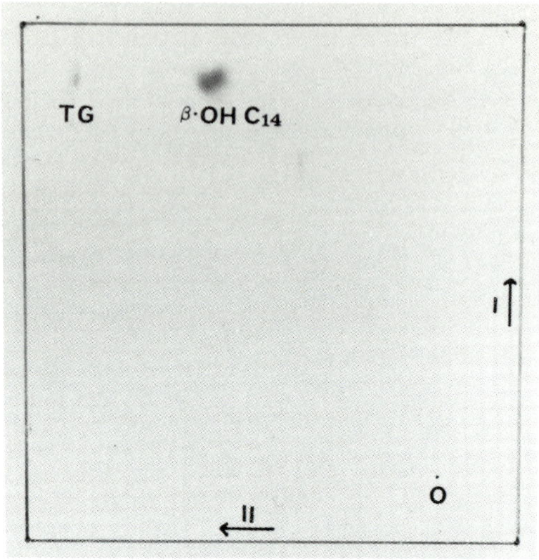

Figure 6. Two-dimensional TLC analysis of the extract with chloroform:methanol from culture medium after 48 h. 1 x 10^6 cells and 400 μg of LPS labeled in the fatty acids were used. TG=triglyceride, ß-OH C_{14}-3-hydroxytetradecanoic acid.

cells whereas R-form LPS was only detected in hepatocytes. After 3 days, almost all LPS accumulated in hepatocytes (1982).

We also demonstrated that cultured hepatocytes incorporated R-form LPS in vitro but not S-form LPS in accordance with the result of Freudenberg et al.. The difference of the incorporation between S- and R-form was not due to the difference of the cytotoxicity as shown in the results of GPT and GOT change.

Hall et al. first found that LPS was deacylated in human neutrophils (1983). The importance of the fatty acids for manifestation of biological activities has been emphasized using both natural endotoxin and chemical synthesized lipid A analogues (Tanamoto et al., 1979; Galanos et al., 1983; Kanegaseki et al., 1986). Kleine et al. found the LPS after the injection were deacylated in the liver and excreted into the bile (1985). We have shown that LPS was deacylated also in vitro after the cultivation with hepatocytes and that fatty acids released were detected in the cultured medium in the form of free ß-hydroxytetradecanoic acids and triglycerides. These results indicate that the hepatocytes also in vitro incorporated and deacylated LPS. This means that the cells play an important role for the detoxification of endotoxin.

REFERENCES

Freudenberg, MA, Freudenberg N, Galanos C (1982). Time course of cellular distribution of endotoxin in liver, lungs and kidneys of rat. Br. J. Exp. Path 63:56-65.

Galanos C, Lüderitz O, Rietschel ET, Westphal O, Brade H, Brade L, Freudenberg M, Schade U, Imoto M, Yoshimura H, Kusumoto S, Shiba T (1985). Synthetic and natural Esheriichia coli free lipid A express identical endotoxic activities. Eur. J. Biochem 148:1 5.

Hall CL, Munford RS (1983). Enzymatic deacylation of the lipid A moeity of Salmonella typhimurium lipopolysaccharides by human neutrophils. Proc. Natl. Acad. Sci. U.S.A. 80: 6671-6675.

Kanegasaki S, Tanamoto K, Yasuda T, Homma JY, Matuura M, Matatsuka M, Kumazawa Y, Yamamoto A, Shiba T, Kusumoto S, Imoto M, Yoshimura H, Shimamoto T (1986). Structure-activity relationship of lipid A: Comparison of biological activities of natural and synthetic lipid A's with different fatty acid compositions. J. Biochem 99: 1205-1210.

Kleine B, Freudenberg MA, Galanos C (1985). Excretion of radioactivity in faeces and urine of rats with ^{3}H, ^{14}C-lipopolysaccharide. Br. J. Exp. Path. 66: 303-308.

Nolan JP (1982). Endotoxin, sinusoidal cell and liver injury in Progress in Liver Disease (Grude & Stratton, New York) 361.

Obrink B (1982). Hepatocytes-collagen adhesion in Methods in Enzymology (Academic Press, Inc.) 82: 513-530.

Tanamoto K, Abe C, Hamma JY, Kojima Y (1979). Region of the lipopolysaccharide of Pseudomonas aeruginosa essential for antitumor and interferon inducing activities. Eur. J. Biochem. 97: 623-629.

Yajima Y, Fukuda I, Otsuki M, Suzuki H, Ota S, Ishii M, Mori K, Goto Y (1985). Endotoxemia in liver diseases: Detection by Quantitative assay using chromogenic substrate with perchloric acid pretreatment. Tohoku. J. exp. Med. 147: 411-419.

Westphal O, Lüderitz O, Bister F (1952). Uber die Extraktion von Bakterien mit Phenol/wasser. Z. Naturforsch. 7b 148-150.

Wilkinson SP, Arroyo V, Gassard BG, Moodie H, William R, (1974). Relation of renal impairment and hemorrhagic diathesis to endotoxemia in fulminant hepatic failure. Lancet 1: 521-524.

Bacterial Endotoxins: Pathophysiological Effects, Clinical Significance, and Pharmacological Control, pages 195–206

THE INFLUENCE OF THE GUT FLORA ON GRAFT VERSUS HOST DISEASE (GvHD) FOLLOWING ALLOGENEIC BONE MARROW TRANSPLANTATION - EXPERIMENTAL OBSERVATIONS AND POSSIBLE MECHANISMS.

L Faulkner[1], N Rapson[2], R Moore[1], J Cohen[1]

[1]Infectious Diseases Unit, Dept of Medicine and Bacteriology, RPMS, London and [2]Dept of Experimental Immunobiology, Wellcome Research Laboratories, Beckenham, Kent.

INTRODUCTION

Bone marrow transplantation (BMT) has emerged as a valuable therapeutic option in a variety of conditions including several haematological malignancies, aplastic anaemia, and some inherited metabolic abnormalities. The two principal barriers to success in BMT are intercurrent infection and graft versus host disease (GvHD).

Acute GvHD usually develops within two to ten weeks following BMT. The main target organs are the skin, mucous membranes, gastrointestinal tract, liver, and lymphoid tissues. The clinical syndrome may be confined to a maculopapular rash involving the palms and soles accompanied by minor abnormalities of hepatic transaminases, or may develop into a serious systemic disease accompanied by abdominal pain, intractable diarrhoea and severe hepatic dysfunction (Deeg and Storb, 1984). Up to one half of the patients given HLA identical marrow grafts may develop some form of graft versus host disease; for those with severe (grade III-IV) involvement mortality increases steeply, usually as a consequence of infectious complications secondary to the immunosuppressive therapy given for the GvHD (Wick et al, 1983).

Extensive experimental and clinical observations have established that for GvHD to occur three basic requirements must be fulfilled. These are firstly, genetically determined histoincompatability between graft donor and recipient; secondly, the presence of immunocompetent cells in the graft capable of recognising and reacting against host antigens; and thirdly, inability of the recipient to reject the donor graft. It seems clear that the effector arm of this response are donor cytotoxic T lymphocytes which are targeted on recipient cells which bear class II receptor antigens. This summary, true as it is, is only part of the story and it has become apparent that a variety of the other factors are operating which determine whether or not clinical GvHD is apparent, and if so to what degree (Reviewed in Deeg and Storb, 1984; Wick et al, 1983; Sullivan and Parkman, 1983). One of these factors which has been known about for many years is the microflora of the gut.

A number of early experimental observations clearly implicated the gut microflora in the pathogenesis of GvHD. Conventional mice which received allogeneic BMT all developed GvHD and died. In contrast, germ-free recipients remained well and survived indefinitely (Jones et al, 1971). Total gastrointestinal decontamination of conventional mice with oral non-absorbable antibiotics was equally effective at preventing GvHD (van Bekkum et al, 1974), as was selective decontamination which eliminated only the aerobic gram negative flora (Heit et al, 1973). Conversely, GvHD could be provoked in germ-free mice by colonisation with gram negative organisms such as Salmonella species, but not by gram positive organisms such as Staphylococcus epidermidis (Pollard et al, 1976). It was of interest that although germ-free and decontaminated mice exhibited no macroscopic or 'clinical' features of GvHD, microscopic lesions were present, suggesting that a component of the bowel flora (and specifically the aerobic coliform population) was able to convert subclinical to clinically apparent GvHD; bacterial endotoxin is a strong candidate as this component.

Large amounts of endotoxin are present in the gut in association with the faecal flora. Oral non-absorbable antibiotics suppress the gram negative flora and faecal endotoxin levels fall (Rogers et al, 1985). Endotoxin can be absorbed from the gut under various circumstances, including after radiation (Maxwell et al, 1986; Gans and Matsumoto, 1974; Walker, 1978; Smith et al, 1963). Endotoxin is a potent immunostimulant and several observations suggest that it may be playing a critical role in the expression of GvHD. In vitro, endotoxin enhances the proliferation of cells stimulated by alloantigens (Forbes et al, 1975) and augments the generation of antigen-specific cytotoxic lymphocytes in mixed lymphocyte culture (Narayanan et al, 1978). In animal experiments, neonatal mice injected repeatedly with endotoxin develop a 'runting syndrome' with macroscopic and microscopic features very similar to those of GvHD (Keast and Walters, 1968). In models of GvHD based upon the intraperitoneal injection of allogeneic spleen cells, cells harvested from endotoxin tolerant donor mice are ineffective (Rose et al, 1976). Finally, in an Fl murine model of GvHD, endotoxin injection four days after parental strain spleen cell grafts enhances the splenomegaly and mortality of GvHD (Skopinska, 1972).

In addition to these experimental data, a number of clinical observations have pointed to the gut flora as one of the environmental factors which may influence GvHD. A number of small clinical studies have suggested that patients receiving total gastrointestinal decontamination in a laminar flow environment are less at risk from GvHD (Storb et al, 1983; Mahmoud et al, 1984; Gratama et al, 1987). Successful suppression of the gram negative intestinal flora, low faecal endotoxin levels and the absence of GvHD are interrelated (Vossen et al, 1981). Less directly, early serious bacterial infection, infective gastroenteritis, and early gut colonisation with gram negative bacteria are all associated with increased incidence and severity of graft versus host disease (Kay et al, 1983; Yolken et al, 1982; Schmeiser et al, 1984). Finally, at least two studies in which intravenous immunoglobulin has been given to BMT recipients for other purposes have found an unexpected reduction in the incidence of GvHD, perhaps due to antiendotoxin antibodies in the infused immunoglobulin (see below) (Winston et al, 1987; Occhini et al, 1986).

Recent devlopments in the understanding of the biochemistry of endotoxin and the availability of cell wall deletion mutants of gram negative bacteria offered us the opportunity of testing directly the hypothesis that endotoxin was the component of the gut flora responsible for these profound effects on GvHD. Extensive studies by Braude, Ziegler and others have provided persuasive evidence that the 'rough' Rc mutant of E.coli (J5 strain) could be used to induce an antibody which was widely protective against many of the immunological and toxic properties of endotoxin (reviewed in Proctor, 1986). We therefore undertook a series of experiments with E.coli J5 in models of allogeneic BMT in mice.

EXPERIMENTAL OBSERVATIONS

In initial experiments we used a conventional model of bone marrow transplantation across major histocompatability barriers (Lampert et al, 1987). Female C57BL/B6 ($H-2^b$) donors were transplanted in CBA/Ca ($H-2^k$) recipients, using standard techniques. Groups of mice received oral gut decontamination with neomycin, streptomycin and natamycin, or were controls. Typical features of graft versus host disease developed during the third week after transplantation , and we confirmed earlier observations that oral antibiotic decontamination appeared to protect the recipient mice from GvHD. As a quantitative measure of this protection, we developed a morphometric method of assessing tissue damage which involved detailed histological examination of the ear. We were able to obtain a ratio of abnormal/normal epidermidis and showed a clear cut and reproducible effect when comparing control to antibiotic decontaminated animals. These observations confirmed firstly, the role of the gut flora in influencing GvHD, and furthermore showed that a diffusable factor could induce a striking effect at a distant site, in this case the ear.

To investigate the potential role of endotoxin, we carried out further experiments using the same model in which recipient mice received E.coli J5 endotoxin, either as killed whole cells or purified lipopolysaccharide (Moore et al, 1987a). The animals were challenged intraperitoneally once a week for three weeks starting on day -28 (in the experiments with killed whole cells) or on day -42 in the case of the purified LPS. Appropriate controls were included. The administration of endotoxin exacerbated the effects of graft versus host disease as indicated by increased weight loss ($p=0.002$) and earlier death of the treated animals ($p=0.043$). In contrast, attempts to protect the recipient mice by passive administration of a rabbit anti E.coli J5 antiserum had no effect. We wondered if the immunological stimulus produced by fully allogeneic combinations may have been too great to allow the detection of any benefit conferred by the antiendotoxin antiserum. We therefore proceeded to repeat the experiments using a different model, this time a 'minor histocompatibility mismatch' system.

The model we chose has been described previously by Hamilton and others and uses LP/J ($H-2^b$) donors and C57BL/B6 ($H-2^b$) recipients (Moore et al, 1987b). Donor bone marrow is T cell depleted with anti-Thy-1,2 and complement, and injected intravenously with spleen cells to the irradiated mice. Animals were passively immunised with polyclonal rabbit anti E.coli J5 antiserum, receiving 1ml intraperitoneally on day 0, +7, +14 and +21. Control animals received normal rabbit serum in the same manner. The immune serum had a mean anti E.coli J5 titre of 1/2048, and the normal rabbit serum had a titre of 1/8. Graft versus host disease was significantly mitigated in the immunised group: weight loss was less severe ($p=0.01$) median survival time increased ($p=0.01$) overall survival was increased ($p<0.01$) and death was delayed ($p<0.01$). Thus for the first time, these results incriminated endotoxin directly in the pathogenesis of GvHD, and furthermore suggested that passive immunisation might be used to reduce the severity of GvHD in clinical practice.

To increase further the specificity of these observations we have now repeated the experiments using a monoclonal anti E.coli J5, designated WLP 15. This is an IgM class antibody which demonstrates broad reactivity in vitro against a range of gram negative bacteria (Fig 1), and isprotective in a mouse live infection model of gram negative sepsis. Full details of this work will be published elsewhere, but in brief, three groups of animals were studied: group 1 were transplanted but not otherwise manipulated, group 2 received four weekly intraperitoneal injections of 0.5ml of normal human serum on day 0, +7, +14 and +21, and group 3 received similarly WLP 15 ascites. The development of graft versus host disease was monitored by observation of the animals' weight, serologically by measuring the acute phase protein SAP (serum amyloid P) and by histology. Animals passively immunised with the monoclonal anti-endotoxin were protected substantially from graft versus host disease; in the NHS group weight loss on day 28 was -10.3% compared to weight gain of 11.8% in the test group. Similarly at day 40, weight loss in group 2 was -17.3% compared to a gain of 1.6%

in the antibody treated group (p=0.03 for both comparisons, one tailed t test) (Fig 2). Bi-weekly measurement of SAP levels showed an early peak in all groups related to the BMT procedure, and then a fall to the nadir between days 15-25. Thereafter both control groups demonstrated a brisk rise in SAP which proceded unchecked to the point at which the experiment was terminated on day +55 (Fig 3). In striking contrast, the WLP 15 treated animals demonstrated a delayed and considerably less intense rise in SAP. At day 35, the mean SAP level in group 1 was 110mg/ml and in group 2 115mg/ml, while in group it had risen to 55mg/ml from the nadir of 35mg/ml. At day 55 the control groups had SAP level of a mean 160mg/ml compared to antibody treated groups 70mg/ml (p<0.001 by analysis of variance). Experiments are in progress to characterise this result more closely, and further controls will be required to confirm these preliminary observation.

In parallel with these animal experiments, we carried out a clinical study on a cohort of patients undergoing allogeneic BMT for leukaemia (Cohen et al, 1987). The question we wished to address was the relationship, if any, between 'native' anti E.coli J5 antibodies in pre-transplant sera and the risk of subsequent GvHD.

Natural anti-endotoxin antibody is known to confer a degree of benefit in patients who develop septic shock. In one study of Pseudomonas septicaemia patients who had high antibody titres in serum samples taken before the episode of shock were less likely to die compared to those with low titres (Pollack et al, 1983).

We examined pre-transplant sera from 70 patients with leukaemia, and compared the titre of anti E.coli J5 antibody in these samples with that of normal controls, and a group of 'patient controls' (individuals undergoing surgery for relief of obstructive jaundice). Two findings emerged: firstly, the titres were reduced in pre-transplant sera, in respect of both IgG and IgM class antibody, compared to the control groups. Secondly, the IgM titre in the leukaemic group correlated inversely with the risk of GvHD; that is, patients with low anti-endotoxin titres in pre-transplant sera were significantly more likely to develop GvHD. It was particularly interesting that this protection from GvHD was associated with IgM antibody, since the same association has been made in relation to the benefit obtained from anti endotoxin antiserum in septic shock (McCutchan and Ziegler, 1983). Thus, this clinical study confirmed and extended the data gained from our animal experiments, implicating endotoxin in the expression of GvHD.

LPS			WHOLE BACTERIAL CELL					
LPS	E.coli J5 Lipid A	S.minn R595 Lipid A	E.coli	Klebsiella	Salmonella	Citrobacter	Serratia	Pseudomonas
Pos	Neg	Pos	Pos	Pos	Pos	Pos	Pos	Pos

Figure 1. Solid-phase binding characteristics of WLP 15, a monoclonal antibody against endotoxin. Radioimmunoassay, showing binding to both pure LPS and to whole cells.

Fig 3 →

Fig 2

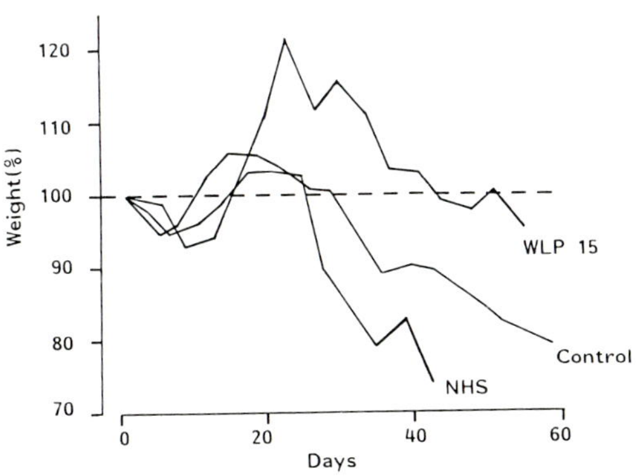

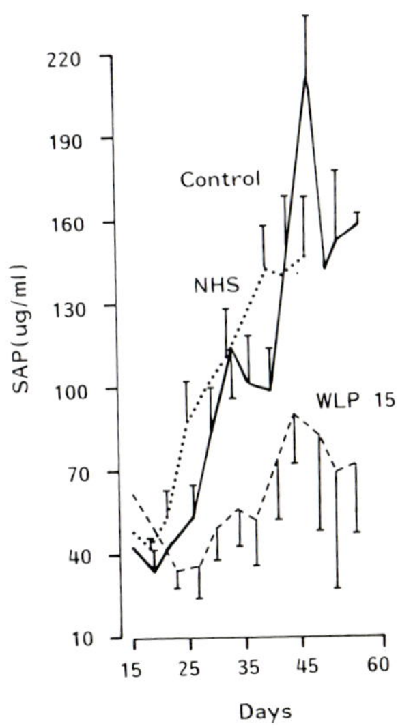

DISCUSSION

The clinical and experimental observations outlined above point very clearly towards bacterial endotoxin as having the ability to influence the course and severity of GvHD. What might be the mechanism of this effect? GvHD is, initially at least, an immunological phenomenon mediated by T lymphocytes in the donor marrow. It can be thought of in two phases: induction of the graft versus host response, dependent upon donor lymphocytes recognising alloantigens on host cells, and then an effector, or tissue-damage stage. Endotoxin might be involved in several aspects of these processes. It is known to be able to stimulate macrophages and T lymphocytes with the production and release of IL-1, IL-2, gamma interferon and other cytokines (Rosenstreich et al, 1977; Michalek et al, 1980; Friedman et al, 1986), and these powerful immuno-stimulatory effects could thereby enhance both induction and/or effector limbs of the graft versus host response. A particular cytokine which lately has received considerable attention is TNF/cachectin, a 17K protein produced by macrophages in response to exposure to endotoxin (Beutler and Cerami, 1987). TNF appears to have a multitude of biological effects which are currently being investigated actively, but one in particular is its ability to interact synergistically with gamma-interferon to enhance class II antigen expression on cell surfaces (Phillip and Epstein, 1986). With this in mind it is possible to assemble a plausible hypothesis to explain the role of endotoxin in GvHD.

The bowel microflora contains large number of coliform bacteria whose cell wall is composed of endotoxin. Under usual circumstances small amounts of endotoxin released from dying bacteria can enter the portal circulation; doubtless this accounts for the presence of measurable amounts of anti-endotoxin antibodies in normal sera. Various circumstances can cause a marked rise in the amount of gut-derived endotoxin entering the circulation, and one such is radiation,as occurs during conditioning for BMT. As endotoxin enters the circulation it stimulates macrophages to release a host of inflammatory mediators, notably gamma-interferon and TNF. These enhance class II antigen expression which serves to stimulate the induction phase of the graft versus host response. Cytotoxic mechanisms (the 'effector' response) then proceed, perhaps promoted by the other endotoxin induced mechanisms noted above. If this analysis is correct, passive immunisation of BMT recipients with either anti-endotoxins or anti-TNF antibodies might make a positive contribution to reducing the clinical consequences of GvHD. Such reagents are rapidly becoming available, and the results of clinical trials should be available in 2-3 years time.

Acknowledgements: Supported in part by the Wellcome Trust and a grant from the National Association for Colitis and Crohn's Disease. We thank Rabia Amin for typing the manuscript.

REFERENCES

Bekkum DW van, Roodenburg J, Heidt PJ, Waaij D van der (1974). Mitigation of secondary disease of allogeneic mouse radiation chimeras by modification of the intestinal microflora. J Natl Cancer Inst 52: 401-404.

Beutler B, Cerami A (1987). Cachectin: more than a tumor necrosis factor. N Engl J Med 316: 379-385.

Cohen J, Moore RH, Al Hashimi S, Jones L, Apperley JF, Aber VR (1987). Antibody titres to a rough mutant strain of Escherichia coli in patients undergoing allogeneic bone marrow transplantation. Evidence of a protective effect against graft-versus-host disease. Lancet 1: 8-11.

Deeg HJ, Storb R (1984). Graft-versus-host disease; pathophysiological and clinical aspects. Ann Rev Med 35: 11-24.

Forbes JT, Nakao Y, Smith RT (1975). T mitogens trigger LPS responsiveness in mouse thymus cells. J Immunol 114: 1004-1007.

Friedman H, Klein TW, Nowotny A, Blanchard K Stewart W, Szentivanyi A (1986). Lipopolysaccharide induction of gamma interferon production. J Imm and Immunopharm 6 (3) Suppl: 76-77.

Gans H, Matsumoto K (1974). Are enteric endotoxins able to escape from the intestine? Proc Soc Exp Biol Med 147: 736-739.

Gratama JW, Zwaan FE, Stijnen T, Weijers TF, Weiland HT, D'Amaro J, Hekker AC, The TH, DeGast GC, Vossen JM (1987). Herpes-virus immunity and acute graft-versus-host disease. Lancet 1: 471-474.

Heit H, Wilson R, Fliedner TM, Kohne E (1973). Mortality of secondary disease in antibiotic treated mouse radiation chimeras. In: Germfree Research, Biological Effect of Gnotobiotic Environments. Heneghan JB Ed. Academic Press, New York; 477-483.

Jones JM, Wilson R, Bealmear PM (1971). Mortality and gross pathology of secondary disease in germfree mouse radiation chimeras. Radiat Res 45: 577-588.

Kay HEM, Watson JG, Jameson B, Morgenstern GR, and Powles RL (1983). Infections after bone marrow transplantation using cyclosporine. Transplantation 36: 491-495.

Keast D, Walters MN-I (1968). The pathology of murine runting and its modification by neomycin sulphate gavages. Immunology 15: 247-262.

Lampert IA, Moore RH, Huby R, Cohen J (1987). Observations on the role of endotoxin in graft-versus-host disease. Prog Clin Biol Res; In press.

McCutchan JA, Ziegler EJ (1983). Treatment with antigram negative antibodies. Lancet 2: 802-803.

Mahmoud HK, Schaefer UW, Schuning F, Schmidt CG, Bamberg M, Haralambie E, Linzenmeier G, Hantschke D, Gross-Wilde H, Luboldt W, Richter HJ (1984). Laminar air flow versus barrier nursing in marrow transplant recipients. Blut 49: 375-381.

Maxwell A, Gaffin SL, Wells MT (1986). Radiotherapy, endotoxaemia and nausea. Lancet 1: 1148-1149.

Michalek SM, Moore RN, McGhee JR, Rosenstreich DL, Mergenhagen SE (1980). The primary role of lymphoreticular cells in the mediation of host responses to bacterial endotoxin. J Inf Dis 141: 55-63.

Moore RH, Lampert IA, Chia Y Aber VR, Cohen J (1987a). Influence of endotoxin on graft-versus-host disease after bone marrow transplantation across major histocompatability barriers in mice. Transplantation 43: 731-736.

Moore RH, Lampert IA, Chia Y, Aber VR, Cohen J (1987b). Effect of immunisation with Escherichia coli J5 on graft-versus-host disease induced by minor histocompatability antigen in mice. Transplantation 44: 249-253.

Narayanan PR, Kloehn DB, Sundharadas G (1978). Immune response to alloantigens in vitro, amplification of the development of cytotoxic T lymphocytes by lipopolysaccharide and polyadenylic: polyuridylic acid. J Immunol 121: 2502-2508.

Occhini D, Bacigalupo A, van Lint MT Corbetta G, Nardelli M, Frassoni F, Pittaluga PA, Marmont A (1986). Prophylactic high dose i.v. IgG in bone marrow transplantation: a randomized trial. Bone Marrow Transpl 1 (Suppl 1): 57.

Philip R, Epstein LB (1986). Tumor necrosis factor as immunomodulator and mediator of monocyte cytotoxicity induced by itself, gamma interferon and interleukin - 1 Nature 323: 86-89.

Pollack M, Huang AI, Prescott RK, Young LS, Hunter KW, Cruess DF, Tsai C-M (1983). Enhanced survival in Pseudomonas aeruginosa septicaemia associated with high levels of circulating antibody to Escherichia coli endotoxin core. J Clin Invest 72: 1874-1881.

Pollard M, Chang CF, Srivastava KK (1976). The role of microflora in development of graft-versus-host disease. Trans Proc 8: 533-536.

Proctor RA (1986). Role of antibody in the prevention and pathogenesis of endotoxin and gram-negative septic shock. In: Proctor RA Handbook of Endotoxin Vol 4: Clinical Aspects of Endotoxin Shock. Elsevier, Amsterdam; 161-184.

Rogers MJ, Moore RH, Cohen J (1985). The relationship between faecal endotoxin and faecal microflora of the C57BL mouse. J Hyg (Camb) 95: 397-402.

Rose WC, Rodey GE, Rimm AR, Truitt RL, Bortin M (1976). Mitigation of graft-versus-host disease in mice by treatment of donors with bacterial endotoxin. Exp Haematol 4: 90.

Rosenstreich DL, Glode LM, Mergenhagen SE (1977). Action of endotoxin on lymphoid cells. J Inf Dis 136 (Suppl): S239-S245.

Schmeiser Th, Kurrle E, Arnold R, Heit W, Krieger D, Kubanek B, Heimpel H (1984). Antimicrobial prophylaxis for prevention of infection and GvHD in allogeneic bone marrow transplantation. Proceedings of the Third International Symposium on Infection in the Immunocompromised Host. Toronto. Abstract 45.

Skopinska E (1972). Some effects of Escherichia coli endotoxin on the graft-versus-host reaction in mice. Transplantation 14: 432-437.

Smith WW, Alderman IM, Schneider C, Cornfield J (1963). Sensitivity of irradiated mice to bacterial endotoxin. Proc Soc Exp Biol Med 113: 778-781.

Storb R, Prentice RL, Buckner CD, Clift RA, Appellbaum F, Deeg J, Doney K, Hansen JA, Mason M, Sanders JE, Singer J, Sullivan KM, Witherspoon RP, Thomas ED (1983). Graft-versus-host disease and survival in patients with aplastic anaemia treated by marrow grafts from HLA-identical siblings. Beneficial effect of a protective environment. N Engl J Med 308: 302-307.

Sullivan KM, Parkman R (1983). The pathophysiology and treatment of graft-versus-host disease. Clin Haem 12: 775-789.

Vossen JM, Heidt PJ, Guiot HFL, Dooren LJ (1981). Prevention of acute graft-versus-host disease in clinical bone marrow transplantation: complete versus selective intestinal decontamination. Recent Advances in Germfree Research. Sasaki S et al. Eds Tokai Univ Press. 573.

Walker RI (1978). The contribution of intestinal endotoxin to mortality in hosts with compromised resistance: a review. Exp Hemat 6: 172-184.

Wick MR, Moore SB, Gastineau DA, Hoagland HC (1983). Immunologic, clinical, and pathologic aspects of human graft-versus-host disease. Mayo Clin Proc 58: 603-612.

Winston DJ, Ho WG, Lin C-H, Bartoni K Budinger MD, Gale RP, Champlin RE (1987). Intravenous immune globulin for prevention of cytomegalovirus infection and interstitial pneumonia after bone marrow transplantation. Ann Int Med 106: 12-18.

Yolken RH, Bishop CA, Townsend TR, Bolyard EA, Barlett J, Santos GW Saral R (1982). Infectious gastroenteritis in bone marrow transplant recipients. N Engl J Med 306: 1009-1012.

The Significance of Endotoxemia

Bacterial Endotoxins: Pathophysiological Effects, Clinical Significance, and Pharmacological Control, pages 209–212

THE COMPLEX CHALLENGE OF BACTEREMIA AND ENDOTOXEMIA

Lowell S. Young, M.D.

Director, Kuzell Institute for Arthritis & Infectious Diseases, Clinical Professor of Medicine, University of California, San Francisco

This review is an attempt to bridge and encompass a number of the remarks made during this conference. In his introduction, Dr. Jack Levin cited an example of a case that he felt represented endotoxemia without documented gram-negative bacillary bacteremia. Dr. J.P. Nolan reviewed the complex and controversial literature that implicates endotoxin or lipopolysaccharides (LPS) as a significant pathophysiologic factor in a variety of gastrointestinal and hepatic disorders. If I can paraphrase his comments, he indicated that there exists a wide variety of evidence suggesting but not proving that endotoxin is involved. The best available evidence, under the circumstances, suggests a role for this important pyrogenic moiety, but unequivocal proof remains lacking.

As we review the complex field of the pathophysiology of endotoxemia, it is clear that endotoxin is not an effector molecule but triggers a variety of important physiologic mediators (Young, 1984). This is perhaps the single most important concept relating to our understanding of the biological effects of bacterial endotoxins. In early studies carried out in my laboratory, we noted, as did many other earlier observers, that intravenous injection of gram-negative bacilli or LPS did not immediately precipitate hypotension, fever and death but some time elapsed before the evolution of these events (Young, 1975). It now seems clear that important mediators are generated during LPS administration and their effect is not coincident with the magnitude of bacteremia. The mediators that are shown to be liberated during gram-negative infections include gamma-interferon, tumor necrosis factor (TNF) (Beutler and Cerami, 1987), and granulocyte-monocyte colony stimulating factor (GM-CSF). It is likely that by the time that the pathophysiologic aspects of endotoxin are obvious clinically, it is the effects of mediators rather than the presence of LPS moiety that we are observing.

Much of this conference and those that have taken place previously have focused on the role of LPS assays and the difficulty in establishing a "gold standard" by which to validate many of the commonly used procedures, including limulus assay (Elin, 1985; Sturk and ten Cate, 1985; Thomas et al, 1984). Controversies regarding the standardization and interpretation of this assay are beyond the scope of this policy paper. Nevertheless, it seems clear that LPS is the only substance that consistently triggers the limulus gelatin assay. Furthermore, improvements in methodology have made the test more sensitive and reproducible.

What I would like to review, however, are important but often forgottenstudies reported by Postel and colleagues of experimental gram-negative bacteremia in dogs (Postel et al, 1975). The key question is whether organisms circulate in the blood as inert particles and can be quantitated in a reproducible manner. What these investigators found was that in order to maintain a fairly fixed number of organisms circulating per milliliter of blood (10^2/ml), a massive inoculum of 10^6 organisms per minute had to be administered during a "steady state" infusion into a peripheral limb vein. In other words, if one were to sample venous blood at a site distant from the one used to inject bacilli, in order to recover just a few hundred organisms, literally millions of <u>P. aeruginosa</u> would have to be injected over a sustained period of time. The implications are that effective clearance mechanisms must exist even in a "1 or 2 pass" system that will dramatically lower the number of circulating microorganisms. Furthermore, these organisms may shed their surface endotoxin at some point during the course of the infection and, if the organisms are serum sensitive, we might expect significant liberation of LPS while the number of organisms detected or quantitated by standard blood culture technique would be small or undetectable. Lethal bacteremia in a normal host, such as the canines studied by Postel and associates, resulted only after sustained delivery of 10^7 organisms per minute for 5 hours. Thus, while 10^3 or more bacilli may be necessary to trigger a positive limulus assay *in vitro*, a viable count of serum sensitive organisms does not always correlate with either the amount of LPS present or the number of organisms that may have entered the bloodstream.

What the clinician "sees" at the bedside are the end results of fever, hypotension and hemodynamic consequences of gram-negative bacteremia. We know that many bloodstream infections represent intermittent bacteremia. Following injection of organisms intravenously, the bacteremia has long cleared at the time when experimental animals are most febrile (Young, 1975). We are aware from the quantitative blood culture studies that even in clinical situations where there may be bacteremia not all blood cultures are positive. Thus, a single blood culture may be negative even while bacteremia is progressing.

The attention focussed on mediators has given us some interesting studies of individuals with meningococcemia in which very high levels of TNF have been measured. Our own studies indicate that TNF can be readily quantitated by *in vitro* systems following exposure of macrophages to different bacterial endotoxins. However, the measurement of free TNF in the plasma of patients with gram-negative bacteremia shows rates of positivity only of about 60%. These studies were carried out with stored plasma samples which were culture positive for gram-negative bacilli. Again, the same arguments surrounding the interpretation of the limulus assay would pertain to these data. An important issue is the collection of specimens in relationship to positive blood cultures. In our study, they were obtained simultaneously. On the other hand, TNF is a labile substance and scrupulous attempts at serum or plasma preservation may be necessary to document the presence of free TNF. Possibly a combination of measurements of assays with LAL test would be the most appropriate approach towards the solution to this problem.

Elsewhere during this symposium in depth attempts are made to refine the limulus assay. It is certainly improved in the chromogenic form, but this procedure does not resolve all of the issues. I suggest that in order to improve the detection of endotoxemia, concurrent assays for TNF and perhaps GM-CSF be carried out, as well as for limulus activity. Furthermore, one would hope that appropriate procedures would be undertaken in order to neutralize limulus positivity (e.g., with polymixin B) or TNF activity (with a specific antibody) so as to provide additional evidence for the presence of LPS.

We might predict that further inferential evidence will become available in the future about the role of endotoxemia in gram-negative infections. For instance, we all are aware that clinical trials are underway with monoclonals of the IgM isotype that stay intravascularly and bind lipid-A epitopes (Young, 1984). For non-bacteremic gram-negative infections, the efficacy and ability of such antibodies to neutralize the biological effects of endotoxin would be another indirect but important piece of evidence that would support a role for circulating endotoxemia in the absence of bacteremia.

REFERENCES

Young LS (1984). Gram-negative sepsis. In Mandell G, Douglas JG, Bennett JE (eds): "Principles and Practice of Infectious Diseases," New York: John Wiley and Sons, Second Edition, pp 452-475.

Young LS, Stevens P, Ingram J (1975). Functional role of antibody against "core" glycolipid of Enterobacteriaceae. J Clin Invest 56:850-861.

Beutler F, Cerami A (1987). Cachectin: More than a tumor necrosis factor. N Engl J Med 316:379-385.

Elin RJ, Hosseini J (1985). Clinical utility of the limulus amebocyte lysate test. In Liss A (eds): "Bacterial Endotoxins: Structure, Biomedical Significance, and Detection With the Limulus Amebocyte Lysate Test," New York: Alan R. Liss, Inc, pp 307-324.

Sturk A, ten Cate JW (1985). Endotoxin testing revisited. Eur J Clin Microbiol 4:382-385.

Thomas LL, Sturk A, Buller HR, ten Cate JW, Spijker RE, ten Cate H (1984). Comparative investigation of a quantitative chromogenic endotoxin assay and blood cultures. Am J Clin Path 82:203-206.

Postel J, Schloerb PR, Furtado D (1975). Pathophysiologic alterations during bacterial infusions for the study of bacteremic shock. Surg Gyn & Ob 5:683-692)

Young LS (1984). Immunoprophylaxis and serotherapy of bacterial infections. Am J Med 76:664-671.

Bacterial Endotoxins: Pathophysiological Effects, Clinical Significance, and Pharmacological Control, pages 213–224

ENDOTOXEMIA, BACTEREMIA AND UROSEPSIS

S.J.H. van Deventer, I. de Vries, L.W. Statius van Eps, W. Pauw, H.R. Büller, A. Sturk and J.W. ten Cate.

Department of Thrombosis and Hemostasis, Academic Medical Center, Amsterdam and Department of Bacteriology, Slotervaart Hospital, Amsterdam, The Netherlands.

Supported in part by a grant (no 28-704) from the Praeventiefonds, The Hague, The Netherlands.

INTRODUCTION

Endotoxins, lipopolysaccharides from the Gram-negative bacterial cell wall (Rietschel et al., 1985) have numerous biological effects (Morrison and Ryan 1987), including complement activation via both the classical and alternative pathway (Vukajlovitch et al., 1987), induction of platelet activation factor synthesis (Doebber et al., 1985), activation of the coagulation cascade (Maier and Hahnel, 1984; Shands, 1984) and induction of the synthesis and release of the macrophage mediators cachectin (Beutler and Cerami, 1987) and interleukin 1 (Dinarello, 1984). Therefore, endotoxins are important in the pathogenesis of Gram-negative septicemia (Ryan, 1986), and clinicians aimed to detect endotoxins in blood for many years to predict the clinical outcome of Gram-negative septicemia. This was however hampered by lack of sensitivity and specificity of the available assays, and high false positive rates. Recent modifications of the Limulus test for the detection of endotoxins in blood (Sturk et al., 1987) using chromogenic substrates, resulted in an improved sensitivity (detection limit 5 pg/ml), thereby enabeling clinical studies in patients with infections.

In a prospective study in consecutive febrile patients with a low prevalence of septicemia, this optimized Limulus assay was demonstrated to have clinically relevant positive and negative predictive values for the development of Gram-negative septicemia (Van Deventer et al., 1987). The sensitivity and specificity of this test were shown to be 79% and 96% respectively. In addition, it was shown that endotoxemia and Gram-negative bacteremia were separate clinical entities with different significance for the development of Gram-negative septicemia.

According to Bays theorem, the predictive values of a diagnostic test are dependent upon the pretest likelihood (prevalence) of disease in the population studied (Bays, 1763; Sackett, 1985), while sensitivity and specificity are virtually stable properties of the test. Consequently, we calculated that at a hypothetical prevalence of septicemia of 20%, the positive predictive value of the endotoxin assay would be 83%, whereas the negative predictive value would be as high as 95%.

Patients with Gram-negative infections of the urinary tract do have a significant risk to develop septicemia. The urinary tract is the most important source of Gram-negative bacteremia in hospitalized patients (Hamill and Maki, 1986). No exact data on the incidence of septicemia in these patients are available, but reported case-fatality rates ranged from 13.7% (Bryan and Reynolds, 1984a) to 30.8% (Bryan and Reynolds, 1984b). We therefore decided to investigate the predictive values of the chromogenic endotoxin test for septicemia in consecutive febrile patients with Gram-negative infections of the urinary tract.

PATIENTS AND METHODS

Study population

The study was performed in the Slotervaart Medical Center, a general teaching hospital in Amsterdam, The Netherlands. All consecutive febrile patients (body temperature >38°C) with a Gram-negative infection of the urinary tract (either pyelonephritis or an invasive bladder infection) were eligible for the study. This study was in part embodied in a larger trial investigating the clinical usefulness of the endotoxin test. A Gram-negative infection of the urinary tract was considered to be present if more than 15 white blood cells were seen on low magnification of a spun urine specimen, and more than 10^5 colonies of Gram-negative bacteria per ml of urine were cultured.

Study design

Upon request of the attending physician, blood for bacteriological culture and the endotoxin test was obtained thrice within one hour from eligible patients. In the subsequent 72 hours the occurrence of septicemia was prospectively determined by a single investigator who was unaware of the endotoxin test result. The diagnosis of septicemia was based upon the occurrence of either shock (systolic blood pressure <90 mm Hg), oliguria (urine production <500 ml/24 hours, in the absence of obstruction or hypovolemia), thrombocytopenia (blood platelets $<100 \times 10^9$/L), or metabolic acidosis (plasma bicarbonate <15mEq/L with high anion gap).

Patients were excluded from analysis if no samples for either the blood culture or the endotoxin assay had been obtained (incomplete sampling).

Bacteriological cultures and endotoxin assay

Quantitative cultures of urine were performed using a dip-slide method. Blood cultures were processed using routine bacterial techniques for culture of aerobic and anaerobic bacteria. Endotoxins were assayed with a Limulus test (detection limit: 5 ng/L) for determination of endotoxins in blood, which has been described in detail (Sturk et al., 1987). All assays were performed in duplicate. An endotoxin test result above 5 ng/L was considered to indicate endotoxemia (Van Deventer et al., 1987).

Calculation of accuracy indices

Sensitivity was defined as the number of patients with septicemia and a positive endotoxin test result (true positives), divided by all patients with septicemia. Specificity was defined as the number of patients without septicemia and a negative test result (true negatives) divided by all patients without septicemia. The number of patients with a positive endotoxin test and septicemia (true positives) divided by the total number of patients with a positive test result was designated the positive predictive value, whereas the negative predictive value was calculated by dividing the number of patients with a negative test result without septicemia (true negatives) by the total number of patients with a negative test result. The likelihood ratio of a positive test result was calculated using the formula:

$$\text{likelihood ratio pos. test} = \frac{\text{sensitivity}}{1-\text{specificity}}$$

For reasons of comparison with the endotoxin test, we also calculated the same accuracy indices of the blood cultures.

Adequate antibiotic therapy

Based upon antibiotic sensitivity patterns of bacterial isolates from blood in the Slotervaart Hospital, the administration of an animoglycoside or co-trimoxazol was considered adequate therapy for presumed Gram-negative bacteremia originating from the urinary tract.

TABLE I

Study population

Eligible patients	80
excluded (incomplete sampling)	4
Patients enrolled	76
Patients with septicemia	14 (18%)

RESULTS

Patients

Eighty consecutive febrile patients with Gram-negative infections of the urinary tract were enrolled in the study. Of these, 4 patients were excluded because of incomplete sampling. Of the 76 patients who were analyzed, 14 (18%) developed septicemia during the follow up period (table I). Three of the 14 patients with septicemia (21%) died within the follow-up period, whereas one of the 62 patients without septicemia died (2%). The mean age of the studied patients was 70 years (range 41-92) and the male:female ratio 1:1.3. One or more of the following risk factors for urinary tract infections were present in the studied patients: an indwelling urinary catheter (37 patients), major abdominal surgery (17 patients), hemiplegia (13 patients), and malignancy (5 patients). Of the 76 studied patients, 64 (including all septic patients), received adequate "blind" intravenous antibiotic treatment directly after sampling (co-trimoxazol, gentamicin, kanamycin or amikacin), and 12 were not, or inappropriately treated.

TABLE II

Bacteremic and non-bacteremic episodes

blood culture results	patients (n=76)
Gram-negative bacteria	38
Gram-positive contaminants	2
No bacteria isolated	36

Bacteremia and endotoxemia

In 40 of the 76 study patients (febrile: 29; septic:11), microorganisms were isolated from the blood sample (table II). In two patients isolates were considered contaminants. The spectrum of bacteremia is shown in table III.

TABLE III

Spectrum of bacteria isolated (from blood)

Microorganism	febrile patients	septic patients
Escherichia coli	17	7
Proteus mirabilis	4	3
Klebsiella pneumoniae	4	0
Klebsiella oxytoca	1	0
Serratia marcescens	1	1

Gram-negative bacteremia without septicemia was present in 27 febrile patients (table IV), 4 of whom had concomitant endotoxemia (table V). In 11 of the 14 septic patients Gram-negative bacteria were isolated, the remaining 3 patients received antibiotic treatment at the time of blood sampling. Sensitivity and specificity of blood cultures for septicemia were 79% and 56% respectively, whereas positive and negative predictive values were 29% and 92% (table IV).

TABLE IV

Sensitivity, specificity and predictive values of bacteremia for urosepsis (n=76)

	septicemia	febrile
Gram-negative isolate	11	27
Blood culture sterile	3	35

sensitivity = 79%
specificity = 56%
pos. predictive value = 29%
neg. predictive value = 92%

TABLE V

Sensitivity, specificity, and predictive values of endotoxemia (5 ng/L) for urosepsis (n=76)

	septicemia	febrile
endotoxin >5 ng/L	11	4
endotoxin <5 ng/L	3	58

sensitivity = 79%
specificity = 94%
pos. predictive value = 73%
neg. predictive value = 95%

Endotoxemia was observed in 15 patients (febrile: 4; septic: 11: table V). Eleven of the 14 patients with septicemia were endotoxemic. Endotoxin results in the septic and febrile patients are summarized in table VI and VII. In 8 of the 11 septic patients with endotoxemia, multiple samples for the endotoxintest had been obtained. Of these, three had single positive tests, whereas in five persistant endotoxemia was observed. Sensitivity and specificity of the chromogenic Limulus test for septicemia were 79% and 95%, the positive predictive value was 73% and the negative predictive value 95% (table V).

TABLE VI

Endotoxin results from septic patients (n=14)

number of assays performed	41
positive assays (> 5 pg/mL)	17 (41%)
mean positive test result	26 pg/ mL
range	6-100 pg/mL

TABLE VII

Endotoxin results from febrile patients (n=62)

number of assays performed	172
number of positive tests	6 (3%)
mean positive test result	7 pg/mL
range	5-11 pg/mL

DISCUSSION

The aim of this study was to determine the predictive values of endotoxemia and bacteremia in patients at high risk for septicemia. Of 76 consecutive febrile patients with deep-seated urinary tract infections, 38 (50%) had Gram-negative bacteremia, and 14 (18%) developed septicemia. Our results indicate that the chromogenic endotoxin test has high predictive values for the development of septicemia (positive predictive value 73%; negative predictive value 95%) in this population. Futhermore, the results also indicate that the presence of endotoxins in the bloodstream is a more important prognostic factor in the pathogenesis of septicemia, than is Gram-negative bacteremia. Indeed, Gram-negative bacteremia without endotoxemia was frequently observed, but in these patients septicemia was uncommon. On the other hand, an endotoxemic patient appears to be at great risk for the development of septicemia.

By definition, sensitivity and specificity are stable properties of a diagnostic test. Remarkably, in this study in high risk patients, sensitivity and specificity of the endotoxin test (79% and 95% respectively) were almost equivalent to the test characteristics we reported in a low risk population (sensitivity 79%, specificity 96%) (Van Deventer et al., 1987).

It should be noted that, despite appropriate antibiotic treatment some febrile patients developed septicemia (14 patients). Antibiotics do not have anti-endotoxin activities, with the notable exception of polymyxin, which is too toxic for clinical use. It has been speculated that antibiotics might liberate endotoxins, by killing bacteria (McCabe, 1975; Anonymous, 1985), and thereby inducing symptoms of septicemia.

In experimental septicemia in rats, gentamicin induced a rise in the serum endotoxin concentration (Shenep et al., 1985). To our knowledge however, there are no date that indicate that antibiotics induce endotoxin liberation in humans.

The major aim of the present investigation was to predict which patients, regardless of antibiotic treatment or the presence of bacteremia, went on to develop septicemia. The results indicate that the endotoxin test is indeed able to predict this outcome accurately.

Several investigators have successfully used anti-endotoxin-core (anti-J5) preparations, in order to lower the mortality of patients with septicemia (Ziegler et al., 1982), or to prevent septicemia in febrile surgical patients (Baumgartner et al., 1985). However, the anti-J5 preparations employed have been shown to be heterogenic (Appelmelk et al., 1986). Recently it has become clear that most pathogenic effects of endotoxins are mediated by endogenous molecules, such as cachectin (Beutler and Cerami 1987) and platelet activating factor (Doebber et al., 1985). Interestingly, anti-TNF antibodies and PAF-antagonists prevented some of the deletorious effects of experimental septicemia (Beutler et al., 198; Issekutz et al., 1986).

We conclude that the chromogenic Limulus assay has high predictive values for septicemia in febrile patients with Gram-negative infections of the urinary tract. Therefore, this assay could be employed in a clinical setting to select patients at high risk for septicemia, who might benefit from intervention treatment directed to either endotoxins, or it's endogenous mediators.

Acknowledgements

We thank dr. A. van de Ende for kind support of this study, and are indebted to Marian Janssen and the laboratory personnel of the Slotervaart Hospital for excellent technical assistance.

REFERENCES

Anonymous (1985). A nasty shock from antibiotics? Lancet ii: 594.

Appelmelk BJ, Rapson NT, Verwey-van Vught ANJJ, et al. (1986). Heterogeneity of Eschericia coli J5 vaccins. Lancet ii: 1273-4.

Baumgartner JD, Glauser MP, McCutchan JA, et al. (1985). Prevention of Gram-negative shock and death in surgical patients by antibody to endotoxin core glycolipid. Lancet ii: 59-63.

Bayes T (1763). An assay toward solving a problem in the doctrine of chance. Philosph Transact Royal Soc 53: 370-418.

Beutler B, Milsark IW, Cerami A (1985). Passive immunization against cachectin/tumor necrosis factor protects mice from lethal effect of endotoxin. Science 229: 869-71.

Beutler B, Cerami A (1987). Cachectin: More than a tumor necrosis factor. N Engl J Med 316: 379-85.

Bryan SC, Reynolds KL (1984a). Community-acquired bacteremic urinary tract infections: Epidemiology and outcome. J Urol 132: 490-3.

Bryan SC, Reynolds KL (1984b). Hospital-acquired bacteremic urinary tract infections: Epidemiology and outcome. J Urol 132: 494-8.

Calandra T, Schellekens J, Verhoef J (1986). A comparative study of standard (IGIV) versus E. coli J5-hyperimmune intravenous gammaglobulins (J5-IGIV) in the treatment of Gram-negative septic shock. Abstracts of the 4th International Symposium on infections in the Immuno compromized Host. Ronneby Brunn, Sweden.

Dinarello C (1984). Interleukin-1. Rev Infect Dis 6: 51-95.

Doebber TW, Wu MS, Robbins JC, Choy BM, Chang MN, Shen TY (1985). Platelet activating factor (PAF) involvement in endotoxin-induced hypotension in rats. Studies with PAF-receptor antagonist kadsurenone. Biochem Biophys Res Commun 127: 799-808.

Hamill RJ, Maki DG (1986). Endotoxin shock in man caused by Gram-negative bacilli. In: Proctor RA (ed). Handbook of Endotoxin vol 4, p. 55-126 Elsevier Amsterdam.

Issekutz AC, Szpejda M (1986). Evidence that platelet activating factor may mediate some acute inflammatory responses. Studies with the platelet activator antagonist CV3988. Lab Invest 54: 275-81.

Maier RV, Hahnel GB (1984). Microthrombosis during endotoxemia; potential role of hepatic versus alveolar macrophages. J Surg Res 36: 362-70.

McCabe WR (1975). Antibiotics and endotoxin shock. Bull NY Acad Med 51: 1084-95.

Morrison DC, Ryan JC (1987). Endotoxin and disease mechanisms. Ann Rev Med 38: 417-32.

Rietschel ET, Brade H, Brade L, et al. Newer aspects of the chemical structure and biological activity of bacterial endotoxins. In: Ten Cate JW, Büller HR, Sturk A, Levin J (eds.). Bacterial Endotoxins: Structure, Biomedical significance and Detection With The Limulus Amebocyte Lysate Test. pp. 31-50 Alan R. Liss, Inc. New York.

Ryan JL (1985). Microbial factors in pathogenesis: Lipopolysaccharides. In: Root RK, Sande MA, eds. Septic Shock, p. 13-25, Churchill Livingstone, New York.

Sackett DL, Haynes RB, Tugwell P (1985). Clinical Epidemiology. A basic science for clinical medicine. Little Brown, Boston.

Shands JW Jr. (1984). Macrophage procoagulants. Hemostasis 14: 373-77.

Shenep JL, Barton RP, Mogan KA (1985). Role of antibiotic class in the rate of liberation of endotoxin during therapy for experimental Gram-negative bacterial sepsis. J Infect Dis 151: 1012-8.

Sturk A, Janssen ME, Muylaert FR, et al. (1987). Endotoxin testing in blood. In: Watson SW, Levin J, Novitsky TJ (eds.) Detection of bacterial endotoxins with the Limulus amebocyte lysate test, pp. 371-85. Alan R. Liss, Inc. New York.

Van Deventer SJH, Pauw W, ten Cate JW, et al. (1987a). Clinical evaluation in febrile patients of an optimized endotoxin assay in blood. In: Watson SW, Levin J, Novitsky TJ. Detection of bacterial endotoxins with the Limulus lysate test, pp. 489-99. Alan R. Liss, Inc. New York.

Van Deventer SJH, Büller HR, ten Cate JW, et al. (1987b). Significance of endotoxemia and bacteremia for the development of septicemia in febrile patients. A prospective study. Submitted.

Vukajlovitch SW, Hoffman J, Morrison DC (1987). Activation of human serum complement by bacterial lipopolysaccharides: Structural requirements for antibody indendent activation of the classical and alternative pathways. Mol Immunol 24: 319-32.

Ziegler EJ, McCutchan JA, Fierer J, et al. (1982). Treatment of Gram-negative bacteremia and shock with human antiserum to a mutant Escherichia coli. N Engl J Med 307: 1225-30.

Bacterial Endotoxins: Pathophysiological Effects, Clinical Significance, and Pharmacological Control, pages 225–236

MEASUREMENT OF ENDOTOXIN IN THE ACUTE PHASE OF SEPTIC SHOCK

A. Christine McCartney, Boguslawa I. Piotrowicz
S. Edlin, IMcA. Ledingham.
University of Glasgow, Department of Bacteriology, Royal Infirmary, Glasgow G4 0SF and Intensive Therapy Unit, Western Infirmary, Glasgow G11 6NT, Scotland.

In the last three decades, the incidence of septicaemia caused by Gram-negative bacteria has substantially increased. This increase in Gram-negative sepsis appears to coincide with advances in surgery, the use of broad spectrum antibiotics, and the increased numbers of immunosuppressed patients.

All shock, including septic shock, is a state of systemic inbalance between supply and demand for oxygenated blood. In patients with septic shock, Gram-negative bacteria are responsible for the majority of cases (Ledingham et al, 1983).

The mortality due to septic shock varies with the underlying disease and may exceed 80% in some groups of patients (Ledingham et al 1980). The precise aetiology of shock in association with Gram-negative sepsis is uncertain but among other factors, the endotoxin of the cell envelope of Gram-negative bacteria may play a significant role.

The Intensive Therapy Unit (ITU) of Glasgow Western Infirmary is a referral centre for septic shock patients in the west of Scotland. The criteria required for the diagnosis of septic shock at this centre are circulatory collapse associated with mental confusion, oliguria and a positive bacteriological culture either from blood or a major primary septic focus. Treatment of these patients has particular emphasis on both prompt resuscitation of shock and subsequent aggressive surgical care (Ledingham and McArdle, 1978.)

In our initial study on endotoxaemia in septic shock,a semi-quantative gel-clot Limulus amoebocyte lysate (LAL) assay was used (McCartney et al, 1983). Thirty one patients, admitted to ITU with a diagnosis of septic shock, were studied.

In 21 of these the primary source of infection was gastrointestinal. There were 21 deaths in this group of 31 patients. Blood samples for all LAL assays were taken from all patients on admission to the ITU and thereafter as frequently as possible. Blood cultures were performed on all blood samples taken for LAL assays. Endotoxin was detected in the bloodstream of all patients during the 24 hours following referral. Three distinct patterns of serial LAL assays emerged in this group of patients. There was a good correlation between these patterns and the outcome in each patient.

Eighteen patients (58%) had a persistently positive LAL assay and all died. Nine patients (29%) had initially positive LAL assays which gradually became negative. All these patients survived. Four patients (13%) had intermittently positive and negative assays; three of them died, with positive LAL assays at death, and one survived, with a LAL assay eventually becoming persistently negative.

There was not a close relationship in any of these patients between positive LAL assays and positive blood cultures as approximately half of the patients had negative blood cultures with positive LAL assays. It was concluded from this study that endotoxin in the bloodstream was a feature of septic shock and that the continuous presence of endotoxin in the peripheral blood was associated with a poor prognosis.

With the development in our laboratory of a quantitative LAL assay which employed a chromogenic substrate (Piotrowicz et al, 1985 a; Piotrowicz et al, 1985 b), a further study was established in order to determine if there was any connection between the quantitative levels of plasma endotoxin and the severity of sepsis. For the assay, plasma extracts were obtained by 1:4 dilution with extraction buffer (Thomas et al, 1981) and subsequent heating at 100°C for 1 minute. The LAL assay method was a simple one-stage procedure (Table 1) with a threshold of sensitivity of 1 picogram of endotoxin per ml of plasma (0.012 EU/ml). QCL-1000 kits (Quantitative Chromogenic LAL, MA Bioproducts, Walkersville, MD 21793, USA), which were found to have a good performance (Piotrowicz and McCartney, 1987), were used.

TABLE 1 ONE-STAGE LAL ASSAY WITH A CHROMOGENIC SUBSTRATE

100 μl plasma extract
\+
100 μl mixture of 30μl LAL+70μl S-2423*
↓
Incubate at 37°C for 45 min
↓
Add 100 μl 50% (v/v) acetic acid

Measure absorbance at 410 nm in MICROELISA MiniReader MR 590

* Reagents from QCL 1000 kit, M.A. Bioproducts, Walkersville, MD 21793, USA.

Preliminary results with this quantitative assay in a group of critically ill but not necessarily septic patients demonstrated some association between endotoxin levels, blood cultures and the clinical course of several patients. These results also suggested that frequent monitoring of patients in septic shock was essential.

A further group of 8 septic shock patients (6 males and 2 females: age range 22-77 year, mean 57 years) have now been studied. These patients were monitored on admission to ITU, then at four-hourly intervals for 48 hours and subsequently daily for one week. In addition to the normal clinical and laboratory investigations (including blood cultures), plasma endotoxin, cardiovascular and respiratory parameters and plasma fibronectin were measured. Pulmonary and systemic vascular resistance, pulmonary shunt (Qs/Qt) and alveolar-arterial oxygen difference $(A\text{-}a)DO_2$ were also calculated. Formulae and calculations for these measurements have been detailed previously (McCartney et al, 1987). Plasma fibronectin was measured in a turbidimetric immunoassay (Boehringer Mannheim, West Germany).

TABLE 2 CLINICAL INFORMATION ON 8 SEPTIC SHOCK PATIENTS

PATIENT	AGE	PRIMARY PATHOLOGY	SEPTIC STATUS	OUTCOME
BL	49	Ileal conduit Surgery	+++	SURVIVED
AC	60	Peritonitis	++++	DIED
RMcD	47	Multiple trauma	++++	DIED
EB	67	Infected varicose leg ulcers	+++	DIED
AMcD	72	Transrectal prostate biopsy	+++	DIED
PH	22	Drug addict with *Strep. viridans* septicaemia	++	SURVIVED
JW	77	Perforated gastric ulcer	++	SURVIVED
JM	65	Thoracic spinal abscess	++	SURVIVED

Details of the eight patients are given in Table 2. Septic status is a subjective clinical score (Elebute and Stoner, 1983). The time course of the LAL assays of all eight patients is shown in Figure 1. This figure shows the marked variation in endotoxin levels observed in individual patients during the first 48 hours of acute septic shock. Of the four patients (BL, AC, RMcD, EB) with the highest endotoxin levels in the first 48 hours, three died (AC, R McD, EB) within that period. In all four patients, there was a correlation between LAL assays, disturbances in pulmonary gas exchange (Qs/Qt and $(A-a)DO_2$) and plasma fibronectin levels. The other four patients (AMcD, PH, JW, JM) all presented clinically with acute septic shock but none had markedly elevated endotoxin levels. Their pulmonary shunt and alveolar-arterial oxygen difference were only moderately elevated with the exception of AMcD who had grossly elevated cardio-respiratory parameters. The three survivors in this

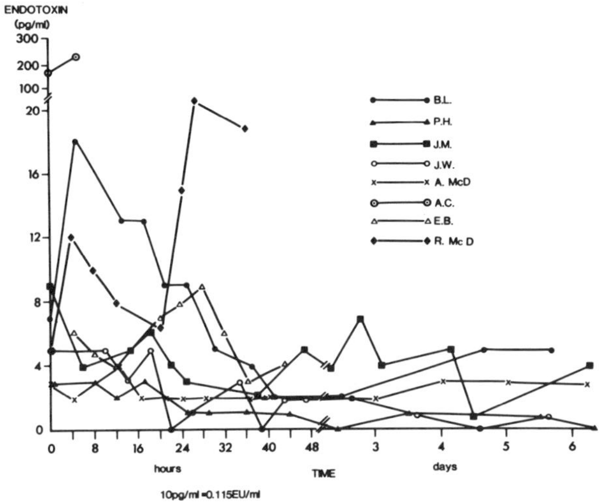

Figure 1 The time course of the LAL assays in eight septic shock patients.

group (PH, JW, JM) had the least septic score of all patients studied. None of the three had bacteriological evidence of a Gram-negative source of sepsis, and in fact one (PH) had Streptococcus viridans isolated from several blood cultures. The fourth patient (AMcD), had a positive blood culture with Escherichia coli prior to admission to ITU and died later (10 days after admission).

The case histories of three of these eight patients have been selected for detailed presentation, to illustrate some points discussed above.

Patient BL. This patient was a 49 year old woman who developed septic shock 24 hours after ileal conduit surgery. She had had no antibiotic prophylaxis for surgery. On admission to ITU she had a temperature of 39.2°C, a systolic blood pressure of 60 mm Hg and a pulse of 153 beats per minute. Resuscitation was started with plasma, mechanical ventilation and an inotropic infusion of dopamine. She was commenced on ampicillin, gentamicin and metronidazole. Figure

2 shows her endotoxin levels, alveolar-arterial oxygen difference and pulmonary shunt. There was a close temporal relationship between endotoxin and pulmonary shunt ($r=0.9211$) and endotoxin and alveolar-arterial oxygen difference ($r=0.8852$). There was an inverse relationship between changes in plasma fibronectin and endotoxin levels, with $r=-0.6899$ ($p \leq 0.05$). There was also good correlation between endotoxin and septic score, and fibronectin and septic score (Figure 3). Plasma fibronectin returned to within the normal range (275-300 µg/ml) as the endotoxaemia resolved. In this patient, Staphylococcus aureus was isolated from six of 10 blood cultures taken within the first 48 hours, which was the period corresponding with endotoxaemia. Similarly, Harris and colleagues, in two of their patients with Gram-positive bacteraemia, also found systemic endotoxaemia (Harris et al 1984). We have confirmed this finding on several occasions in a study on endotoxaemia in immunosuppressed patients (McCartney et al, in press). This may be explained in two ways. Firstly, this patient's Gram-positive septicaemia could have caused abnormal permeability of her gut. Endotoxin detected in the LAL assay would in this case be of intestinal origin. Secondly, this patient could have had a cryptic septic focus outside the bloodstream. Endotoxin released by bacteria from this source could have been a reason for the endotoxaemia in this patient (Levin 1979; Levin 1982). It is interesting to speculate that had the LAL assays in this patient been monitored only on a daily basis, her septic episode would probably have been missed!

Patient AC. This patient was a 60 year old chronic alcoholic admitted to hospital with an 11 hour history of vomiting, rigors and persistent lower abdominal pain. He was pyrexial with a temperature of 39°C, a pulse of 120 beats per minute and a systolic blood pressure of 70 mm Hg. Investigative laparotomy did not reveal perforation of any viscus but it was observed that he had an enlarged cirrhotic liver and a collection of peritoneal fluid, from which Escherichia coli was subsequently cultured. The patient died nine hours after admission. He had the highest endotoxin levels recorded in our septic shock patients (180 pg/ml on admission to ITU and 220 pg/ml 4 hours later), despite only moderately elevated pulmonary shunt and alveolar-arterial oxygen difference, and negative blood cultures. (Figure 4) The plasma fibronectin levels in this patient were almost undetectable at 40pg/ml.

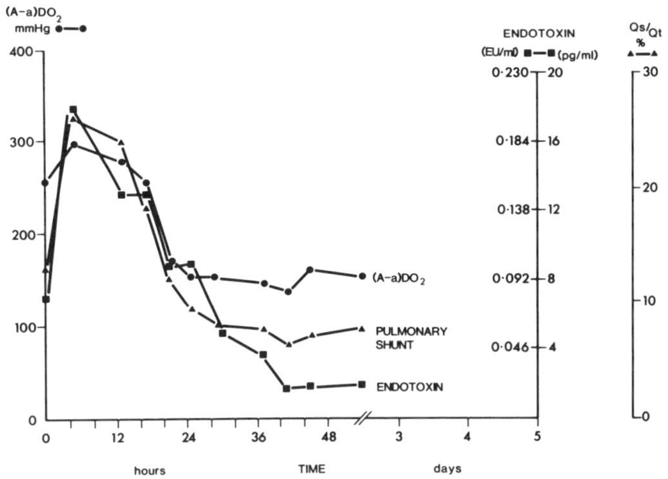

Figure 2 Endotoxin levels, alveolar-arterial oxygen difference and pulmonary shunt in BL, 49 year old female patient, admitted to ITU with septic shock 24 hours after ileal conduit surgery.

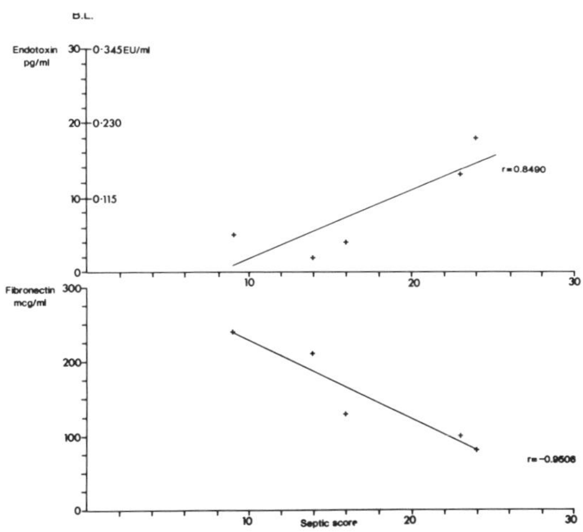

Figure 3 Relationship between endotoxin, fibronectin and septic score in patient BL.

It may be postulated that this patient might have died as a result of direct toxic action of endotoxin, while the condition of patient BL was caused by the indirect action of endotoxin via host mediators of shock ie. various enzymatic systems activated by endotoxin present in the circulation (Morrison and Ulevitch, 1978, Cumming et al, 1984).

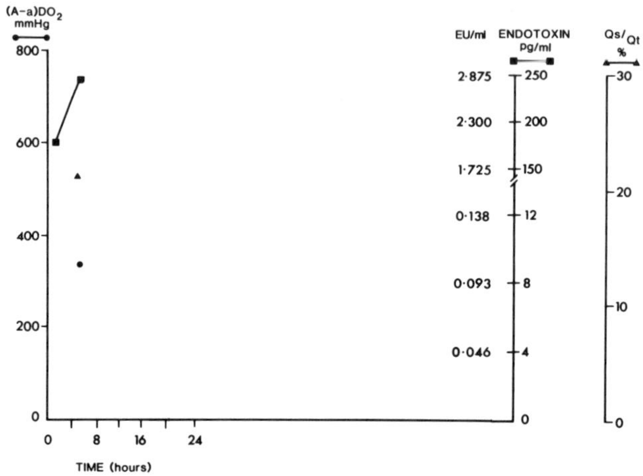

Figure 4 Endotoxin levels, alveolar-arterial oxygen difference and pulmonary shunt in AC, 60 year old male patient with peritonitis.

It is also possible that patients at the greatest risk of developing septic shock may be hypersensitive to endotoxin. In these patients even low levels of endotoxin could cause septic shock. This hypersensitivity may be genetic and/or result from their clinical condition and/or treatment. Impaired clearance of endotoxin and continuous flow of endotoxin from a septic source may further enhance this sensitivity.

Patient AMcD. This patient was a 72 year old man admitted to a peripheral hospital six hours after having a transrectal prostate biopsy performed as an outpatient. He was in septic shock, with a systolic blood pressure of 80 mm

Hg, a pulse of 120 beats per minute and a temperature of 40°C. Escherichia coli was grown from blood cultures taken on admission. His condition deteriorated and 48 hours later he was transferred to ITU at the Western Infirmary. He was by now in severe septic shock. Despite intensive therapy, he died 10 days later.

Patient AMcD illustrates the opposite situation to patient AC , as his endotoxin levels were negligible despite clinical evidence of overwhelming sepsis, grossly elevated pulmonary shunt and alveolar-arterial oxygen difference consistent with early Adult Respiratory Distress Syndrome (ARDS) (Figure 5). The initial level of plasma fibronectin in patient AMcD was 80 μg/ml and remained very low during this patient's course prior to his death in ITU.

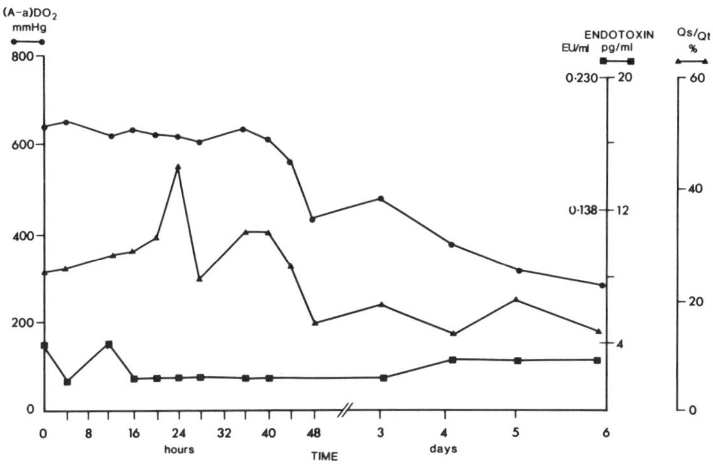

Figure 5 Endotoxin levels, alveolar-arterial oxygen difference and pulmonary shunt in AMcD, 72 year old male patient, admitted to ITU with severe septic shock after a transrectal prostate biopsy.

In this patient, it is likely that an endotoxin peak might have occurred prior to his admission to ITU and gone undetected because of the failure to sample blood for endotoxin during this time. This would explain the clinical situation at death, when, despite low endotoxin levels, cardio

-respiratory indices were very high, possibly as a result of the secondary effects of endotoxin.

Our studies demonstrate that there are marked differences in plasma endotoxin levels when patients are monitored frequently during the acute phase of septic shock. We have also observed this in patients with severe sepsis given Taurolin infusion (McCartney and Browne, this volume). In the group of patients with elevated endotoxin levels (BL, AC, RMcD, EB), the LAL assay appeared to relate appropriately to the changes in the pulmonary gas exchanges and plasma fibronectin levels during accute septic shock. However, in the group of patients (AMcD, PH, JW, JM), which was clinically quite similar, this relationship was not as apparent. The intermittent nature of endotoxaemia, variability in reticuloendothelial cell function, and differences in sensitivity to endotoxin in individuals may have contributed to the fact that markedly elevated endotoxin levels were not observed in this group.

Conclusions, at present, suggest that there are probably no "clinically significant" levels of endotoxin in critically ill patients with septic shock. However, our studies on this intensively monitored group of patients do not preclude a central role for endotoxin in septic shock.

ACKNOWLEDGEMENT

The authors are grateful to Dr. Paul Newman for the fibronectin assays, and to Miss M. McChristie for secretarial help.

REFERENCES

Cumming AD, Robertson CE, Jeffrey, SS, Robson JS, Ledingham IMcA (1984). The plasma kallikrein-kinin system in severely ill and traumatised patients. Arch Emerg Med 3: 135-142.

Elebute EA, Stoner HB (1983). The grading of sepsis. Brit J Surg 70: 29-31.

Harris RI, Stone PCW, Evans GR, Stuart J (1984). Endotoxaemia as a cause of fever in immunosuppressed patients. J Clin Pathol 37: 467-470.

Ledingham IMcA, McArdle CS (1978). Prospective study of the treatment of septic shock. Lancet 1:1194-1197.

Ledingham IMca, McArdle CS, MacDonald RC (1980). Septic shock. In Taylor S (ed): "Recent advances in surgery", No 10, Edinburgh: Churchill Livingstone, pp 161-199.

Ledingham IMcA, Bradley JA, McCartney AC, Redding PJ (1983). Infection and septic shock. In Bain WH, Taylor KM (eds): Handbook of Intensive Care, London: John Wright, pp 196-237.

Levin J (1979). The Limulus test: a status report. In Cohen E (ed): "Biomedical Applications of the Horseshoe Crab (Limulidae)", New York: Alan R Liss, pp 235-244.

Levin J (1982). The Limulus test and bacterial endotoxins: some perspectives. In Watson SW, Levin J, Novitsky TJ (eds): "Endotoxins and Their Detection With the Limulus Amebocyte Lysate Test", New York: Alan R Liss, pp 7-24.

McCartney A Christine, Banks JG, Clements GB, Sleigh JD, Tehrani M, Ledingham I McA (1983). Endotoxaemia in septic shock: clinical and post mortem correlations. Int Care Med 9: 117-122.

McCartney, AC, Robertson MRI, Piotrowicz BI, Lucie NP. Endotoxaemia, fever and clinical status in immunosuppressed patients - a preliminary study. J Infect, in press.

McCartney A Christine, Browne M K. Clinical studies on administration of Taurolin in severe sepsis: a preliminary study. In Levin J (ed): Bacterial Endotoxins: Pathophysiological Effects, Clinical Significance and Pharmacological Control, New York: Alan R Liss, in press.

Morrison DC, Ulevitch RJ (1978). The effects of bacterial endotoxins on host mediation systems. Am J Pathol 93: 527-618.

Piotrowicz BI, Watt I, Edlin S, McCartney AC (1985 a). A micromethod for endotoxin assay in human plasma using Limulus amoebocyte lysate and a chromogenic substrate. Eur J Clin Microbiol 4: 52-54.

Piotrowicz Boguslawa I, Edlin SE, McCartney A Christine (1985 b). A sensitive chromogenic Limulus amoebocyte lysate micro-assay for detection of endotoxin in human plasma and water. Zbl Bakt Hyg A 260: 108–112.

Piotrowicz Boguslawa I, McCartney A Christine (1987). Chromogenic Limulus amoebocyte lysate assay for endotoxin: comparison of three commercial products. Med Lab Sci 44: 89–91.

Bacterial Endotoxins: Pathophysiological Effects, Clinical Significance, and Pharmacological Control, pages 237–246

ENDOTOXAEMIA IN MULTIPLE ORGAN FAILURE DUE TO SEPSIS

G. Ramsay, P.M. Newman, A.C. McCartney and I.McA. Ledingham.

University Department of Surgery, Western Infirmary Glasgow G11 6NT, SCOTLAND. (G.R., P.M.N., I.McA.L.)
Department of Bacteriology, Royal Infirmary, Castle Street, Glasgow, SCOTLAND (A.C.McC)

SUMMARY

Patients with multiple organ failure secondary to intra-abdominal sepsis are often blood culture negative despite exhibiting the features of septic shock. This study examined the possible central role of endotoxin in such patients.

In 15 consecutive intensive care patients with the above clinical picture endotoxin was measured by a chromogenic limulus (LAL) assay; on admission and thereafter 4 hourly. Regular blood cultures and cultures of any primary septic focus were also performed and liver function was assessed by measurement of indocyanine-green clearance from plasma (ICGC).

All 15 patients had significant endotoxaemia at least intermittently. No significant difference was observed between survivors (n=5) and non-survivors (n=10) in either initial or peak endotoxin levels, although the pattern of endotoxaemia differed with non-survivors exhibiting consistently high or steadily increasing levels. Of 5 patients with an intra-abdominal (I/A) septic focus only one had a positive blood culture while 5 of 10 patients with extra-abdominal (E/A) infection had positive cultures. Despite this the I/A group had higher initial and peak endotoxin levels. 3 patients with Gram-positive septicaemia had significant endotoxaemia in the absence of any gram-negative infection. Changes in ICGC appeared to be of useful prognostic significance. ICGC was significantly lower in the I/A group and in both groups there was a significant negative correlation between ICGC and the level of endotoxaemia.

These results suggest that endotoxin may play a central role in the syndrome of multiple organ failure and further suggest that the endotoxin is endogenous (gut-derived) secondary to failure of hepatic filtration.

INTRODUCTION

It has been reported that patients with high post-operative fevers, who subsequently prove to have intra-abdominal infections, tend to be blood culture negative (Le Gall et al, 1982). Patients with multiple organ failure (MOF)secondary to sepsis often exhibit the clinical features of septic shock, although they too are often blood culture negative. The term "abacterial sepsis" has been coined to describe this apparent syndrome. As long ago as 1954 (Fine, 1954) it was postulated that failure of the liver reticulo-endothelial system and detoxication capacity may allow passage of gut-derived toxins into the systemic circulation during shock. More recently it has been confirmed that impaired liver function, in particular obstructive jaundice, predisposes to the development of systemic endotoxaemia (Pain et al, 1985; Cahill, 1983). One possible theory to explain the clinical features in patients with "abacterial sepsis" is that endotoxin released into the splanchnic circulation may pass into the systemic circulation, as a result of liver impairment. The presence of such endogenous endotoxins in patients with multiple organ failure or sepsis would also explain the finding that patients diagnosed clinically as septic shock are blood culture negative in at least 50% of cases (Ledingham et al, 1980).

The aim of this study was to investigate the frequency with which endotoxaemia occurred in septic patients with multiple organ failure, in an attempt to look for evidence supporting the hypothesis that endogenous endotoxin plays a role in producing the clinical features in such patients. In addition hepatic clearance of injected indocyanine green was assessed as a test of liver function. It has previously been stated that clearance of dyes such as indocyanine green (ICG) is the most sensitive, practical and reliable index of liver injury (Waife and Platcow, 1974). Bromsulfalein has previously been used in this way but the properties of ICG make it a superior drug for use as a test of liver function. In comparison to bromsulfalein ICG produces fewer side effects, is cleared from plasma by the liver with no extra hepatic clearance, follows a monoexponential plasma clearance and in health has a very high first pass extraction ratio making it a useful estimate of liver blood flow; though this does not hold true in disease.

PATIENTS AND METHODS

A total of 15 patients were studied, these being consecutive admissions to a single intensive care unit, with a diagnosis of septic shock together with multiple organ failure (MOF). Our criteria for diagnosing septic shock were as follows; septicaemia or a bacteriologically confirmed major source of sepsis together with systolic blood pressure <90 mmHg, and oliguria (<30 mls/hr), and core peripheral temperature gradient >4°C. In addition the patients were required to have evidence of failure in at least one other major organ system; the definitions of organ failure being those used by Knaus (Knaus et al, 1985a). In order to allow quantification of the degree of sepsis and severity of illness two scoring systems were applied to each of the patients - the APACHE II (Knaus et al, 1985b) and the sepsis score (Elebute and Stoner, 1983).

In addition to routine intensive care monitoring, plasma endotoxin levels were measured on admission and thereafter four hourly. Regular blood cultures and cultures of any primary septic focus were also performed (blood cultures were performed in triplicate at the time of admission and at least twice daily thereafter) and liver function was assessed by measurement of indocyanine green clearance from plasma, at the time of admission and twice daily thereafter.

Endotoxin levels were measured by a modified chromogenic limulus (LAL) assay, the use of which has been previously described in work from our department (Piotrowicz et al, 1985). The kit used for our endotoxin assays was the QCL 100 (MA Bio Products) coupled with chromogenic substrate S 2423 (Kabi diagnostica). ICG clearance was measured by the method of Caesar et al (1961). ICG was administered in a dose of 0.5 mg/kg body weight as an intravenous bolus, usually through a central venous catheter, 1.5 ml blood samples were withdrawn before the injection of ICG and at 3 minute intervals after the injection, for 21 minutes. These samples were placed in

Eppendorf tubes primed with EDTA. Following centrifugation the ICG concentrations were estimated spectrophotometrically at 805 nm. Plasma from the base line blood sample formed a plasma blank, which was used to zero the spectrophotometer. The absorbance of all samples was read against the plasma blank, as was the absorbance of a standard solution prepared by adding 1 μl of dye (5 mg/ml) to 1 ml of plasma blank. The calculated plasma ICG concentrations can then be plotted on semilogarithmic paper; the resulting plot being used to calculate the initial concentration and the half life of the dye. The initial concentration of ICG (C) and the injected dose (D) can be used to calculate the volume of distribution (V), $V=D/C$. The observed half life ($T\frac{1}{2}$) is related to the volume of plasma cleared per minute (CL), $T\frac{1}{2}=0.693 \times V/CL$. The volume of plasma cleared per minute can be converted to the corresponding blood clearance by using the hematocrit value. All statistical analyses were by Student's t-test, paired where appropriate.

RESULTS

The patients had a mean age of 53 years (range 26-74). The mean APACHE II score at the time of admission was 20.3 (range 12-31) while the mean sepsis score was 21.4 (range 9-31). In each patient the primary focus of infection was identified - in five patients this primary septic focus was intraabdominal while in 10 patients the primary septic focus was extraabdominal. Bacteriological culture of these primary septic foci revealed Gram-negative organisms only in 8 patients, Gram-positive organisms only in 3 patients, mixed Gram-negative and Gram-positive infection in 2 patients and no positive bacteriology in the remaining 2 patients. The results of regular blood culturing revealed Gram-negative bacteraemia in 3 patients and Gram-positive bacteraemia in 3 patients, while in the remaining 9 patients no bacteraemia was detected at any stage during their stay on intensive care, despite regular sampling. Of the 5 patients with an intra-abdominal septic focus only one (20%) had a positive blood culture while in the 10 patients with extraabdominal infection 5 had positive blood cultures.

All 15 patients had significant levels of endotoxin present in plasma, at least intermittently. This includes the 3 patients with Gram-positive bacteraemia, who each had significant endotoxaemia at the time of the positive blood culture, in the absence of any focus of Gram-negative infection.

Table 1 shows the mean initial and peak levels of endotoxin observed with the values seen in survivors compared to non-survivors and those in patients with intraabdominal infection compared with levels seen in patients with extraabdominal infection.

TABLE 1. Plasma Endotoxin Levels.

	Plasma Endotoxin Levels (pg/ml)	
	Initial*	Peak*
Survivors (n=5)	33 (0-131)	81 (15-200)
Non-survivors (n=10)	38 (0-139)	82 (8-200)
Intra-abdominal focus (n=5)	48 (0-139)	93 (15-165)
Extra-abdominal focus (n=10)	31 (0-131)	86 (8-200)

* Values are mean (range)

Table 2 shows the values of indocyanine green clearance (ICGC) observed at the time of admission. Initial ICGC values were not obtained in 2 patients but the table shows the mean values in 5 survivors and 8 non-survivors; no significant difference being observed between these groups. In contrast the initial ICGC values were significantly lower in patients with an intraabdominal septic focus than in patients with an extra-abdominal septic focus (p=0.02). Although the initial values for ICGC were not significantly different in survivors and non-survivors, the trend in ICGC values from 0 - 24 hours showed a striking difference in these two groups. In 5 survivors ICGC was higher by 24 hours in each case while, in 8 non-survivors, ICGC was lower by 24 hours in all but one case.

Table 2. Indocyanine Green Clearance.

	Initial values (ml/min)*	
Survivors (n=5)	547 (125-1100)	
Non-survivors (n=8)	708 (290-1800)	
Intra-abdominal focus (n=5)	496 (290- 950)	p = 0.02
Extra-abdominal focus (n=8)	841 (125-1800)	

*Values are mean (range)

An apparent link between ICGC values and plasma endotoxin levels was observed. Figure 1 shows a correlation plot of endotoxin against ICGC, each point representing the values measured concurrently (within 30 minutes of each other). There appears to be a significant negative correlation between ICGC values and endotoxin values ($r = 0.502$).

DISCUSSION

The group of patients investigated in this study formed an extremely sick group, suffering from septic shock together with multiple organ failure. Evidence that the protocol had indeed selected an extremely sick group of patients is provided by the mean scores for the APACHE II and Sepsis Scoring Systems; both of which would be expected to be associated with a high mortality. This is an important point since the patients studied constituted a more severely ill group than those described in earlier published work from our department, in which the incidence of endotoxaemia in the acute phase of septic shock was studied (McCartney et al, 1987).

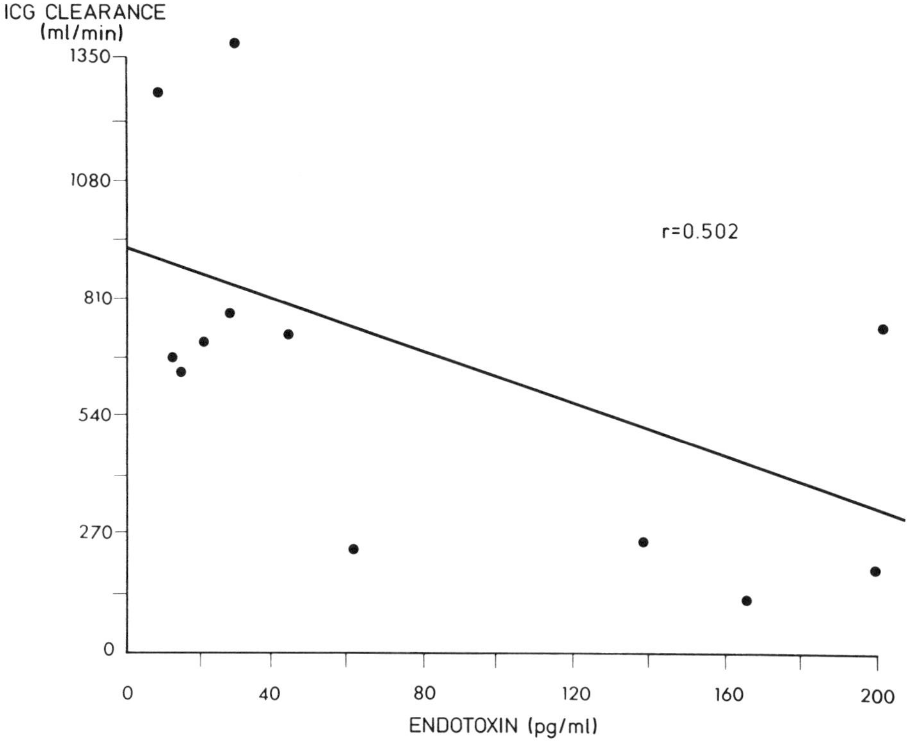

Figure 1. Correlation between simultaneously measured indocyanine green clearance and plasma endotoxin.

Jacob et al (1977) found that endotoxin was present in the portal venous blood of 97% of 34 patients in whom samples were taken during elective abdominal surgery. Although subsequent authors have not found endotoxin in such a high proportion of samples, the balance of evidence in the literature would support the assertion that portal endotoxaemia occurs, at least intermittently, even in individuals without gastrointestinal or liver disease. Furthermore, surgical patients with obstructive jaundice have been shown to have high levels of portal endotoxin and, following surgery, often develop systemic endotoxaemia which has been implicated in producing the renal failure that has long been recognised as a complication of surgery in jaundiced patients (Pain et al, 1985). Certain bile salts are known to bind and inactivate endotoxin and their absence from the gut, in jaundiced patients, might reasonably be expected to lead to higher levels of endotoxin in the portal blood; the evidence for this has been reviewed by Pain (1985). In addition the presence of cholestatic jaundice is known to lead to impaired Kupffer cell function (Katz et al, 1984: Tanaka et al, 1985) and this would explain the relatively common occurrence of systemic endotoxaemia in jaundiced patients and patients with liver failure.

An analogy can be drawn between the patients with multiple organ failure described in this study and surgical patients with obstructive jaundice, since marked cholestatic jaundice is an almost invariable accompaniment of multiple organ failure (Banks et al, 1982). However, in addition to the reduced binding of endotoxin by bile salts within the gut and the reduced Kupffer cell function that are seen in jaundiced patients, other additional mechanisms are likely to be operating in the multiple organ failure group, increasing the risk of portal and systemic endotoxaemia. Patients with shock of any aetiology will have a degree of splanchnic vasoconstriction as a compensatory mechanism, leading to both intestinal ischaemia (and therefore increased mucosal permeability) and also reduced hepatic blood flow (and therefore impaired Kupffer cell function). The reduction in splanchnic blood flow in even very minor degrees of shock can be very marked as shown by earlier work from our department (Gilmour et al, 1980); which showed that a 10% reduction in plasma volume, while producing no change in pulse or blood pressure, resulted in a 40% decrease in colonic blood flow. Other authors have shown that endotoxin shock results in a 60% decrease in gut mucosal capillary blood flow (Rayner et al, 1960).

When all of the above predisposing factors are considered it is not surprising that endotoxaemia was observed in all of the 15 patients investigated in this study. The level of endotoxaemia did not appear to relate directly to outcome, since both initial and peak levels were similar in survivors and non-survivors. However, persistent endotoxaemia was associated with an unfavourable outcome:- 7 patients had steadily increasing or persistent high levels of endotoxin and all 7 died; 4 patients had steadily decreasing levels and all 4 survived; 4 patients had intermittent peaks of endotoxaemia and of these 3 died and 1 survived.

Although ICG clearance is not a measure of Kupffer cell function it was interesting to note that there was an apparent negative correlation between ICG clearance and systemic endotoxin levels. This may be because ICG clearance is largely dependent on hepatic blood flow. Our department has previously published work showing that standard liver function tests appeared to be insufficiently sensitive to accurately evaluate the degree of hepatic impairment in septic shock (Banks et al, 1982) and in addition none of the standared liver function tests appeared to be of any prognostic value in such patients. In this study although ICG clearance was not significantly different in survivors and non-survivors, the trend in ICG clearance from the time of admission to 24 hours post-admission appeared to be extremely useful as a prognostic indicator. Such a trend would have correctly predicted outcome in all but one of the 15 patients, with an increasing ICG clearance being associated with survival while a decreasing ICG clearance was associated with non-survival.

Taken together the results presented in this study provide circumstantial evidence to suggest that endotoxin may play a central role in the syndrome of multiple organ failure acting either directly or, more likely, indirectly via secondary mediators. The results further suggest that much of the endotoxin is likely to be endogenous (gut derived) spilling over into the systemic circulation as a result of failure of hepatic filtration. This concept of endogenous endotoxaemia in such patients would of course explain the finding that many patients with septic shock have negative blood cultures and in addition provides an explanation for the finding of endotoxaemia during Gram-positive bacteraemia.

REFERENCES

Banks JG, Foulis AK, Ledingham IMcA, Macsween RNM (1982). Liver function in septic shock. J Clin Pathol 35:1249-1252.

Caesar J, Shaldon S, Chiandussi L, Guevara L, Sherlock S (1961). The use of indocyanine green in the measurement of hepatic blood flow and as a test of hepatic function. Clin Sci 21:43-57.

Cahill CJ (1983). Prevention of postoperative renal failure in patients with obstructive jaundice - the role of bile salts. Br J Surg 70:590-595.

Elebute EA, Stoner HB (1983). The grading of sepsis. Br J Surg 70:29-31.
Fine J (1954). "The bacterial factor in traumatic shock". In Page IM, Corcoran AG (eds): American Lecture Series, Illinois: Thomas.
Gilmour DG, Aitkenhead AR, Hothersall AP, Ledingham IMcA (1980). The effect of hypovolaemia on colon blood flow in the dog. Br J Surg 67:82-84.
Jacob AI, Goldberg PK, Bloom N, Degenshein A, Konzinn PJ (1977). Endotoxin and bacteria in portal blood. Gastroenterol 72:1268-1270.
Katz S, Grosfeld JL, Gross K, Plager DA, Ross D, Rosenthal RS, Hull M, Weber TR (1984). Imparied bacterial clearance and trapping in obstructive jaundice. Ann Surg 199:14-20.
Knaus WA, Draper EA, Wagner DP, Zimmerman JE (1985a). Prognosis in acute organ system failure. Ann Surg 202:685-693.
Knaus WA, Draper EA, Wagner DP, Zimmerman JE (1985b). APACHE II: A severity of disease classification system. Crit Care Med 13:818-828.
Ledingham IMcA, McCardle CS, Macdonald RC (1980). Septic Shock. In Taylor S (ed): "Recent Advances in Surgery - 10", Edinburgh: Churchill Livingstone, pp 161-199.
LeGall JR, Fagniez PL, Meakins J, Brun Buisson C, Trunet P, Carlet J (1982). Diagnostic features of early high post-laparotomy fever: a prospective study of 100 patients. Br J Surg 69:452-455.
McCartney AC, Piotrowicz BI, Edlin SE, Ledingham IMcA (1987). Evaluation of the chromogenic Limulus Lysate Assay in septic shock. In Watson SW, Levin J. Novitsky TJ (eds): "Detection of Bacterial Endotoxins with Limulus Amebocyte Lysate Test", New York: Alan R. Liss, pp459-474.
Pain JA, Cahill CJ, Bailey ME (1985). Perioperative complications in obstructive jaundice: therapeutic considerations. Br J Surg 72:942-945.
Piotrowicz BI, Edlin SE, McCartney AC (1985). A sensitive chromogenic Limulus Amoebocyte Lysate micro-assay for detection of endotoxin in human plasma and in water. Zbl Bkt Hyg A 260:108-112.
Rayner R, McLean LD, Grim E. (1960). Intestinal tissue capillary flow in shock due to endotoxin. Circ Res 8:1212-1217.
Tanaka N, Bergqvist L, Christensen P, Bengmark S (1985). Reticulo-endothelial function in rats with obstructive jaundice. Br J Surg 72:946-949.
Waife SO, Platcow EL (1974). "Evaluation of liver function". Indianapolis: Lilly Research, pp 23-31.

Bacterial Endotoxins: Pathophysiological Effects, Clinical Significance, and Pharmacological Control, pages 247–262

INCREASED PLASMA LEVELS OF ENDOTOXIN AND CORRESPONDING CHANGES IN CIRCULATORY PERFORMANCE IN A PORCINE SEPSIS MODEL:THE EFFECT OF ANTIBIOTIC ADMINISTRATION.

Røkke.O,Revhaug.A, Østerud.B *,Giercksky.K.E.
Departments of Surgery and Biochemistry *,
N-9012 University Hospital of Tromsø, Norway.

ABSTRACT.

Changes in endotoxin levels and cardiovascular performance during antibiotic therapy in septicemia were investigated in a porcine model.One group of animals (n=9) received gentamicin 2 mg/kg intravenously infusion two hours after induction of sepsis with live E.coli bacteria. Another group (n=7) served as non-treated septic controls.
Plasma-levels of endotoxin increased significantly after antibiotic administration from 0.26 ±0.02 ng/ml before treatment (0 hrs), to 1.1 ±0.3 ng/ml after two hours ($p<0.01$) and 2.1 ±0.98 ng/ml four hours after treatment ($p<0.01$). In the control group no significant increase occurred in the observation period.No difference could be demonstrated between the groups with regard to the number of live bacteria in blood, either before or after treatment. When the data from all the animals were taken together for the first two hours following antibiotic administration a significant negative correlation ($p<0.05$) was demonstrated between changes in endotoxin levels and cardiac output.This correlation was significant for animals in which the levels of endotoxin increased above 0.5 ng/ml ($p<0.05$).

The present study indicates that endotoxin is liberated after antibiotic administration during bacteremia, and that this increase correlates with cardiac performance.

INTRODUCTION.

Modern appropriate use of antibiotics in gram negative septicemia reduces death rates with 50-60% (2,14). Still this maladie remains a major cause of death in surgical departments with an average mortality rate of 30% in most hospital materials (5).

The outcome of septicemia is believed to be partly dependent on the responses of the cardiovascular system. In a review of 612 patients with gram negative bacteraemia, Kreager et al. demonstrated a mortality rate of 47% in patients having gram negative sepsis complicated with shock; compared to 7% in septic patients without shock (5). In another clinical study, Nisihijima et al. reported a correlation between cardiac output and survival during septicemia (9). These studies indicate a prognostic value of changes in circulatory parameters for the outcome in septicemia.

A huge body of evidence indicates that endotoxin directly or indirectly damages the organism,and that pathophysiological changes are partly dose dependent.In this context, a sudden massive killing of bacteria could increase the level of endotoxin to levels of clinical importance.Already in 1935, Reilly et al. warned about this possible danger in antibiotic therapy (9).With the introduction of the Limulus Amebocyte Lysate test for detection of endotoxin, some in vitro studies have confirmed this hypothesis (1).In addition, casuistic reports have been published demonstrating rapid unexpected deterioration in septic patients after antibiotic therapy (3).

Yet,others have been unable to reproduce these results in clinical studies, finding no correlation between plasma endotoxin levels, other parameters of clinical importance, and the outcome of gram-negative septicemia (5,12). These authors report low plasma levels of endotoxin, in spite of antibiotic therapy and deteriorating clinical state, thus suggesting that the biologic effect of endotoxin is independent of the plasma levels.

In the present study we have investigated the possible changes in plasma endotoxin levels following antibiotic therapy and evaluated the corresponding circulatory changes in a large animal model.

MATERIALS AND METHODS.

Experimental procedure.
16 Norwegian landrace piglets, weight 20-30 kg were used in this study.

Treatment group. (n=9). Septicemia was induced by i.v. E.coli infusion. After two hours, gentamicin(2mg/kg) dissolved in 100 ml saline was administered (time=0 hours) intravenously over a 20 min period, and the animals observed for 6 hours thereafter.
Blood samples were drawn before induction of septicemia-(time= -2hours),before antibiotic infusion(time=0 hours), and 2 hours(2 h),4 hours(4 h) and 6 hours(6 h) after the antibiotic infusion.

Control group. (n=7).Septicemia was induced as in the treatment group. At T=2 hours, placebo(100 ml saline) was infused. The observation period and blood-sampling procedures were performed as in the treatment group.

Animals and preparation.
All animals were fasted 20 hours prior to the experiment with free access to water. Anesthesia was induced by halothane (Fluothane, ICI, Great Britain) 5% / O2 on mask,and maintained with 0.7-0.9% halothane vaporized in O2/N2O (50%/50%) controlled by a servo ventilator (Selema Schønander, Stockholm, Sweden). Liquid (0.9% NaCl / 50% glucose) was continuously administered through an i.v ear vein cannula; 20 ml/kg/h the first hour, and 10 ml/kg/h thereafter. A Swan-Ganz thermodilution catheter (5 F, Edwards lab., Santa Ana, Cal., USA) was inserted into the pulmonary artery through the right external jugular vein. Another Swan-Ganz catheter (7 F) was introduced into the femoral artery and advanced into the decending aorta. A Foley-catheter was placed in the bladder through a cystotomy for urine collection. Another catheter (Connecta 3 way stopcock, Viggo prod., Helsingborg, Sweden) was placed into the superior caval vein through the left jugular vein for subsequent blood sampling. All operative procedures were performed under sterile conditions.

E.coli preparation and induction of septicemia. An E.coli isolate from a severe septic patient was used in this study. The same batch, stored on Brain-Heart medium (Oxoid CM 225,Basing Stoke,England)at -70° C, was used in all experiments. Bacteria from this batch were cultured for 24 hours at 37° C on lactose. Two hours before infusion, the microbes were harvested and dissolved in saline. Bacterial counts were determined by photometry (Spectonic 20, Bauch & Lamb) at a wavelength of 475 nm. A bacterial suspension with optic density (OD) = 0.25 corresponding to $4*10^9$ bact/ml was prepared.A control of the procedure was performed by a 24 hour culture on lactose of a sample from the suspension.

After a one-hour stabilization period, the animals received a potentially lethal infusion of live E.coli-bacteria through the ear vein cannula over a 20 min period with a dose of $4*10^9$ bacteria/kg, ($1*10^{11}$ /25 kg).

Antibiotic therapy. Gentamicin (Gensomycine,Roussel,Roussel lab., London, England) was selected as antibioticum because of its bactericide action. The selected E.coli strain was highly sensitive to gentamicin as demonstrated by antibiogram.

Circulatory parameters were monitored continuously. Temperature (TEMP), mean pulmonary artery pressure (PAP) and cardiac output (CO) were determined every half hour. Registrations were done using a multi-channel recorder (Recorder 2800-S,Gould,Cleveland,Oh,USA). The Swan-Ganz catheter placed in the pulmonary artery, was connected with a cardiac output computer (9520 A, Edwards Laboratories, Santa Ana, Cal., USA), and core temperature was also recorded with this catheter. Cardiac output was measured with the thermo-dilution technique. Systolic and diastolic arterial blood pressures were monitored by the 7 F Swan-Ganz catheter placed in the aorta.Aorta mean pressure was calculated as diastolic pressure +(syst-diast)/3.Mean pulmonary artery pressure was calculated similary,using the registrations from the 5 F Swan-Ganz catheter placed in the pulmonary artery.Urinary output was collected every half hour.

Blood bacterial counts were estimated in 1.6 ml heparinized samples and put into plastic tubes containing 0.2 ml glycerol. Bacterial samples from blood were cultured 24 hrs. at 37° C on blood culture plates after appropriate dilution. The number of cultures on two plates were counted and the mean taken as the number of colony forming units (CFU).

Plasma endotoxin concentration. 5 ml blood samples were collected in pyrogen free heparinized tubes. Samples were centrifuged immediately at 3500 g for 15 min at 4 C. Plasma was divided into two pyrogen-free plastic tubes at each time point, and immediately frozen at -70 C. Analysis was performed using a Limulus Amoebe Lysate chromogenic substrate method (Coatest,S-2423, KabiVitrum, Stokholm,Sweden) according to the description of the manufacturer.The mean of two analyses was taken as endotoxin concentration.Results are given as ng/ml.

Gentamicin level in serum was determined in 3 ml serum aliquots prepared from central venous blood samples, aspirated 5 min after the termination of infusion (peak concentration), and regularly for six hours after the infusion,using spectrophotometer(Stasar III,Syva)according to current laboratory spesifications(Tromsø,Norway).

Blood gas analysis.Heparinized arterial blood samples of 0.5 ml were drawn each hour, and Ph, pCO2, pO2, HCO3, and base exess were determined using an auto-analyzer (ABL3, Radiometer, Copenhagen, Denmark).

Statistics:Students unpaired two-tailed T-test with the Bonferroni correction for repeated measurements was used to test differences within groups.The time point before administration of antibiotics (t=0 hours) was compared to 3 time-points,namely 2 hours (2h), 4 hours (4h) and 6 hours (6h) after antibiotic administration. Mann Whitney U Test was applied to test differences between groups. The Fisher exact test was used to estimate significance of survival.Correlation between changes in important parameters were evaluated. Values above the critical two-tailed f-value corresponding to $p>0.05$ were taken as significant. All values are expressed as mean$\pm$ SEM.
The data analyses were performed using a standard statistical package (Microstat 4.1, Ecosoft,Inc.,Ind. USA) on an Olivetti M 24 computer(Olivetti, Ivrea, Italy).

RESULTS

Survival. (Table 1) The administration of gentamycine did not improve survival in the 6-hour observation period after antibiotic treatment. (Fisher exact test=0.34). Of the 9 animals treated with gentamicin, 3 died after 4, 4.5, and 5 hours, and one animal in this group died shortly thereafter(6.5 hours). In the control-group, 4 died at 3, 5, 5.5 and 6 hours respectively.

Circulatory parameters.The changes in aortic mean pressure, urinary output,and temperature did not differ significantly between the antibiotic treated group and the control group.

Heart rate.(Fig.1) Administration of the antibiotic did not influence the heart rate. In both groups mean values peaked at 2 hours, from basal level in the treatment group of 117 $\pm$ 4 beats pr min.(-2 hrs), to 205 $\pm$ 10 beats pr. min. at two hours. The values in the control group were 127 $\pm$ 8 beats pr.min.(2 hrs.) and 240 $\pm$ 5 beats pr. min. (2 hrs.).The response of the heart rate exhibited a parabola for all animals, but the heart rate of the animals which died, increased more (252 $\pm$4 beats pr.min) than that of the animals which survived (216 $\pm$ 8 beats pr. min), $p<0.01$.The increase in heart rate in the surviving animals was also lower in the antibiotic treated group(max.value 202 $\pm$ 5 beats pr. min) compared to surviving animals in the control group(max.value 245 $\pm$ 3 beats pr. min),$p<0.001$.

Mean Pulmonary Artery Pressure(PAP). The initial PAP value was 13 $\pm$ 1 mmHg in the treated group and 12 $\pm$ 1 mmHg in the control group.E.coli infusion induced a rapid increase in both groups, stabilizing after one hour at a level of 23.6 $\pm$ 5 mmHg in the treatment group and at 18 $\pm$ 4.3 mmHg in the control group.The difference is significant($p<0.05$).

Aorta mean pressure.(AM)(Fig 1).The initial value was 82 $\pm$ 2 mmHg in the antibiotic group and 86 $\pm$ 4 mmHg in the non-treated group. Thereafter, a gradual decrease occurred in both groups following E.coli infusion. After antibiotic administration, the AM stabilized with no further significant decrease. Surviving animals actually experienced an increase in AM. In the control group, AM continued to deteriorate, and 6 hours after antibiotic administration, the difference between the groups was significant (78 $\pm$ 4 mmHg versus 42 ± 7 mmHg, $p<0.05$).

Cardiac output (CO).(Fig 1). Initially, the values were 3.6 ± 0.1 l/min and 4.0 ± 0.4 l/min in the antibiotic group and the non-treated group respectively. Following E. coli

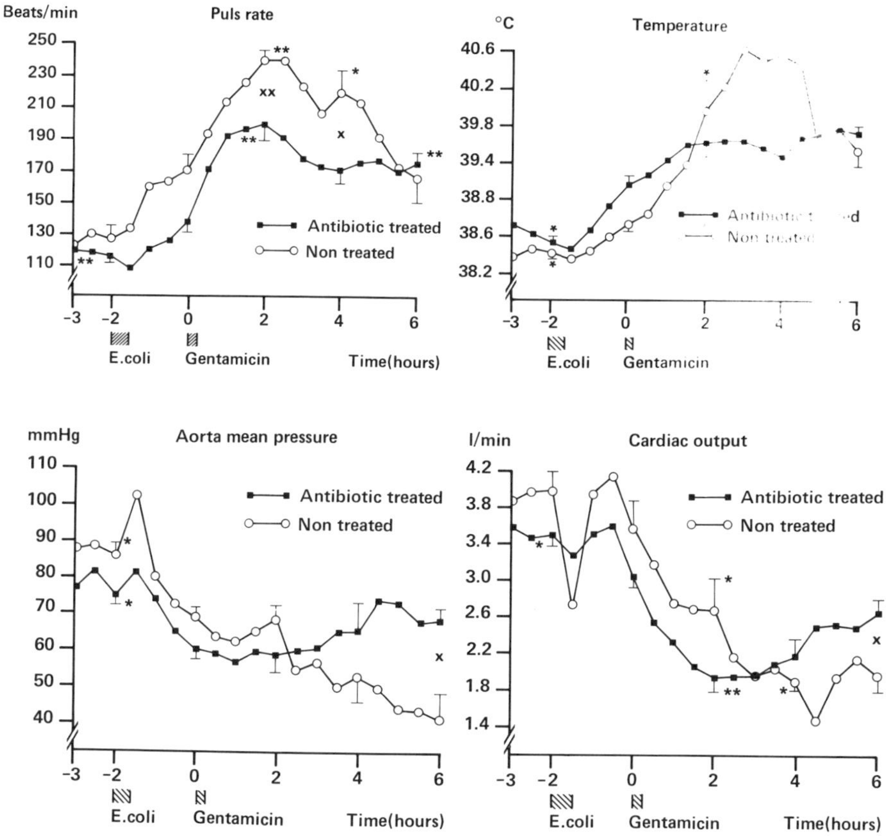

Fig.1. Heart rate, temperature, mean aorta pressure and cardiac output mean values in antibiotic treated(■—■) and non treated (○ ○) animals. Mean values ± SEM defines the timepoints for statistical analysis. Asterix(*) denotes significant changes within groups, * = $p<0.05$, ** = $p<0.01$. Differences between groups are denoted (**X**), **X** = $p<0.05$, **XX** = $p<0.01$.

infusion, CO decreased in both groups. After gentamicin infusion in treatment group, mean CO values stabilized,then started to rise, and 6 hours after gentamycine administration, the difference between the groups was significant.- (2.8 ± 0.2 l/min in the treatment group versus 1.7 ± 0.3 l/min in the control group, $p<0.05$).

Urinary output.The urine production decreased in both groups,from 33 ± 6 ml/h before bacterial infusion in the treatment group,and from 95 ± 11 ml/h in the control group to values below 20 ml/h one hour after antibiotic administration.Dying animals exibited a more rapid decrease in urine production (anuria at range 1/2-2 h) than in surviving animals(range 1/2-4h).

Temperature.(Fig 1)The body core temperature started to rise one hour after E. coli infusion. From basal levels of 38.5 ± 0.1 C in the treated group and 38.4 ± 0.1 C in the control group,the temperature reached maximal values at 3 hrs. In both groups this rise was significant.Even if no statistical difference could be demonstrated at the selected time points, the temperature in antibiotic treated animals did not reach a high values as the animals in the control group(fig.1).In the treatment group no difference was observed between animals which died and surviving animals. The animals in the control group which died,always reached higher temperatures than animals which survived.

Blood gas analysis. In both the treatment group and the control group,animals developed a metabolic acidosis, with decreased values of pH,base exess and HCO3. No significant change occurred in O2 or CO2 levels. The changes were similar in both groups(data not shown).

Concentration of gentamicin in serum reached recommended peak values (5-10 mg/l); with peak concentrations of 7.0 ± 0.55 mg/l 15 min after infusion,decreasing exponentially to 2.4 ± 0.4 mg/kg 6 hours after infusion (Fig 3).

Bacterial counts in blood.(Fig 3). The bacteria were rapidly cleared from the circulation,with minimal counts at the time of antibiotic administration (0 hours),78 ± 19 CFU/ml in treatment group, 57 ± 16 CFU/ml in the control group.Thereafter the number of CFU increased in both groups. The increase in the treatment group was similar to the increase in the control group, with no significant differences between the groups in the observation period.

Plasma endotoxin levels. (Fig.2) The endotoxin was rapidly cleared from the circulation; lowest levels were demonstrated two hours after E.coli infusion. Thereafter, plasma endotoxin levels increased in both groups. After antibiotic administration, levels of endotoxin increased significantly at both two hrs (1.1 ± 0.3 ng/ml) and four hrs (2.1 ± 0.3 ng/ml) compared to levels before treatment (0.3 ± 0.1 ng/ml). Both at two and four hours after antibiotic administration, the levels of endotoxin in the treated group were significantly higher than in the non-treated group (2h:$p<0.02$, 4h:$p<0.02$).In non-treated animals, endotoxin values increased at 2hrs (0.4 ± 0.05 ng/ml),4hrs (1.0 ±0.2 ng/ml) and 6hrs (1.3 ± 0.4 ng/ml),but without reaching significance with the Bonferroni correction.

To further investigate the relationship between plasma endotoxin levels and circulatory parameters, the first two hours following antibiotic administration were more closely examined.The relationship was examined using statistical correlation analyses. This time period was chosen because animals began to die after this time point. In vitro studies have shown that the study of this time period after administration of antibiotics is appropriate, as liberation of endotoxin, if any, is detected within two hours (7). In this period, the levels of gentamicin were within therapeutic levels in all animals.

The individual values of central parameters before antibiotic administration(0h) and two hours after antibiotic administration(2h) are shown in table 1.The changes in endotoxin levels were correlated to corresponding changes in circulatory parameters.The assosiation between both the relative changes(the ratio:value at 2h/value at 0h) and the absolute changes(the difference:value at 2h-value at 0h) were calculated.Even though mean endotoxin levels increased twice as high in the treated group as in the non- treated animals, no significant correlation could be demonstrated between the relative changes in plasma levels of endotoxin and circulatory parameters.
When the changes were expressed in terms of differences, a significant correlation between changes in endotoxin levels and changes in CO was demonstrated when all animals in

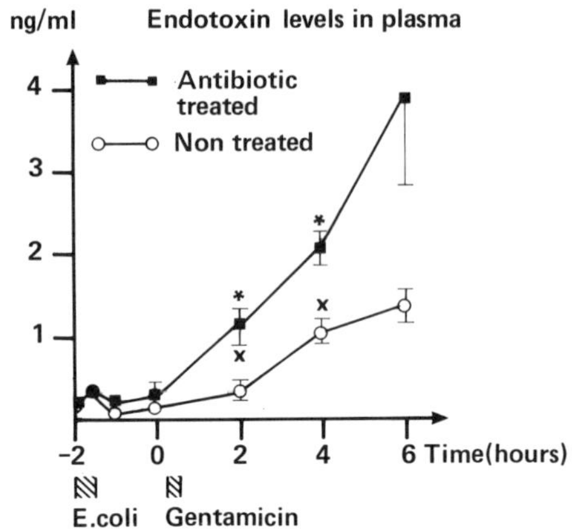

Fig.2. Plasma endotoxin levels before E.Coli infusion (-2hrs), before antibiotic administration (0hrs), and evry two hours thereafter in antibiotic treated animals(■—■) and non treated animals(○—○),mean values ± SEM. Asterix(*) denotes significant differences within group; * = $p<0.05$. Differences between groups are denoted (**X**), **X** = $p<0.05$.

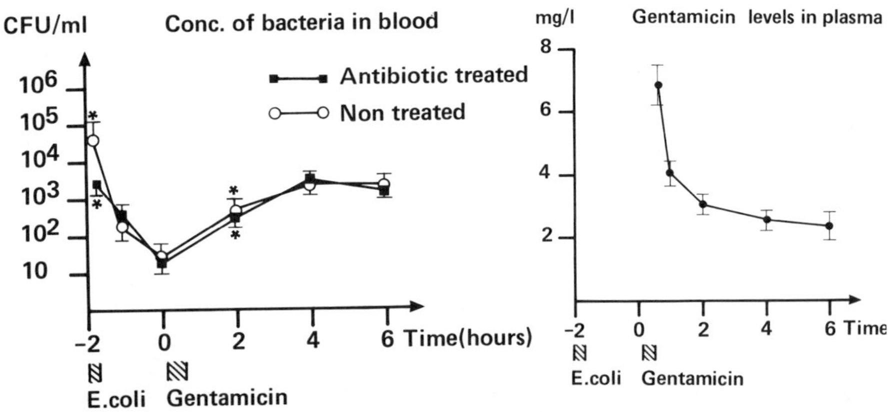

Fig.3. Number of colony forming units of E.Coli bacteria (CFU) determined in whole blood from animals treated with antibiotic treated (■—■), and non treated(○—○) animals. Plasma levels of gentamicin after treatment is also shown,mean values ± SEM. Asterixes(*) denote significant differences from 0 hrs * = $p<0.05$.

INDIVIDUAL VALUES

Animal nr.and group.		AM 0	AM 2	CO 0	CO 2	END 0	END 2	
	1	56	38	2.71	1.18	0.80	2.74	*
	2	70	54	3.52	1.64	0.70	2.10	*
	3	67	38	3.21	1.20	0.09	1.46	*
	4	69	88	2.86	2.34	0.09	0.61	
A	5	81	44	2.55	1.87	0.09	0.76	
	6	60	84	3.03	2.54	0.12	1.05	+
	7	62	70	3.36	2.49	0.12	0.65	
	8	56	60	2.96	2.10	0.25	0.28	
	9	54	82	2.77	2.10	0.11	0.37	
	10	67	66	3.30	1.85	0.34	0.54	
	11	68	63	3.05	2.41	0.23	0.42	
	12	60	45	2.45	1.17	0.27	1.02	*
B	13	75	89	3.17	2.33	0.12	0.28	
	14	80	74	3.65	3.45	0.08	0.16	*
	15	76	71	4.89	4.52	0.06	0.13	*
	16	67	72	4.62	3.06	0.02	0.15	*

Table 1. Individual values of aorta mean pressure(AM), cardiac output(CO) and plasma levels of endotoxin(END) before antibiotic administration (0 hrs),and two hours thereafter(2 hrs) are shown for animals treated with gentamycin,(1-9);group A, and non-treated animals,(10-16); group B. Asterix(*) denotes animals dying within observationperiod. One animal dying 15 min after observation period is denoted(+).

treatment group and control group were considered (fig 4)(corr quot. -0.59;critical val.$\pm$ 0.50 , $p< 0.05$). Also, when only the animals treated with antibiotics were considered, a significant correlation was found (corr.quot. = -0.73,crit.val.$\pm$0.66 $p<0.05$.). In the non-treatment group, no such significant correlation was demonstrated. Then, all the 16 animals in treatment group and control group were stratified according to increase in endotoxin levels, and the 8 animals with highest increase were compared to the 8 animals with lowest increase. Seven out of the 8 animals with the highest increases belonged to the treatment group, and endotoxin values at 2hrs were consistently above the 0.5 ng/ml level. A significant negative correlation was found between endotoxin concentration and cardiac output in these animals (corr.quot.= -.073, crit.value $\pm$.07,p<0.05). In the animals with the lowest increases, no correlation was demonstrated between endotoxin concentration and cardiac output.

The mean CO values at 2hrs in the animals with the highest endotoxin levels (1.8 $\pm$ 0.2 l/min) were significantly lower than CO values at 2hrs in the animals with lower endotoxin levels (2.7 $\pm$ 0.3 l/min), p<0.05. High endotoxin levels at two hours were also associated with mortality. Of 5 animals reaching endotoxin concentration of 1 ng/ml at two hours, 4 died in the obs.period; the remaining animal died 15 min. later.

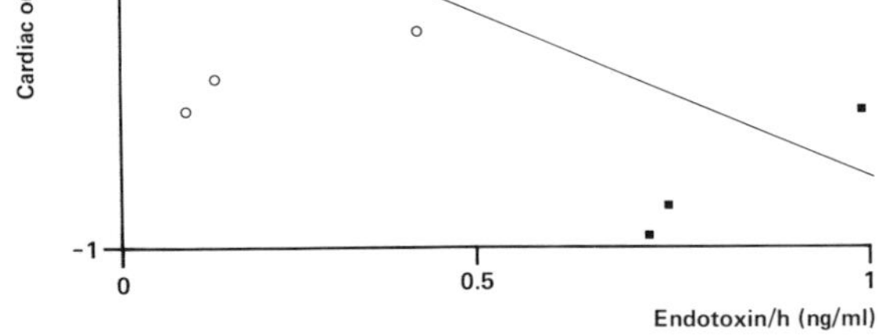

Fig.4. Relation between changes in plasma endotoxine level and cardiac output the first two hours after gentamicin infusion for treated animals(■)and non treated animals(O) Corr.quot. -.59,critical two-tailed value $\pm$.50, p<0.05.

DISCUSSION

The results of this study support the hypothesis that administration of a bactericide antibioticum increases levels of plasma endotoxin.Previously, this effect has been documented in in vitro (1), and Shenep et al. demonstrated an increase in free endotoxin after antibiotic treatment in rabbits with E.coli peritonitis(10). In the present study, total levels of endotoxin in centrifuged plasma was considered,and an increase was demonstrated two and four hours after gentamicin administration.

The source of this plasma endotoxin in Gram negative septicemia is usually thought to be the cell walls of intruding bacteria. In our study, however, we were unable to demonstrate any decrease in the concentration of viable bacteria in blood after antibiotic treatment, in spite of documented therapeutic gentamicin levels. Several mechanisms might account for this apparently contradictory increase.

a) Gentamicin kills bacteria in the circulation, with a corresponding increase in plasma levels of endotoxin, but the bacterial killing is concealed by the substantial-number of bacteria and the exponential growth rate of bacteria during this time period.

b) Gentamicin kills bacteria in tissues other than-blood, thereby increasing the circulating levels of endotoxin.

c) Gentamycin induces changes in the immunesystem (7, 11), resulting in altered endotoxin processing (12).

Endotoxin is generally thought to give rise to serious clinical symptoms in concentrations of 0.5 -5 ng/ml. The present study also indicates an association between endotoxin concentration and clinical state. Animals with levels of endotoxin above 1.0 ng/ml at an early stage in the experiment (2 hrs) all died in the observation period or shortly after (Tab 1). The finding of a significant correlation between changes in endotoxin level and cardiac output in animals reaching endotoxin levels of 0.5 ng/ml and not in animals with lower endotoxin levels, also indicate the existence of a certain level, above which endotoxin becomes a dominant factor. Finally, the demonstration of a significant correlation between endotoxin levels and cardiac output when actual levels were compared and not when relative values (ratios) were compared, also indicates importance of absolute endotoxin levels. This finding may support the case reports of rapid deterioration of the state of patients after administration of antibiotics. This could be even more important in the treatment of more fragile bacteria,i.e. N.Meningitidis.

Conflicting results from previous studies with regard to the clinical significance of actual endotoxin levels could be related to a difference in the timing of the determinations. The number of bacteria in circulation at the time of treatment might also be of importance. Some in vitro studies also demonstrate differences between bacteria strains with regard to the liberation of endotoxin upon antibiotic challenge(1).

Even if antibiotic therapy alone did not improve survival, surviving animals managed better after antibiotics than without, supporting the clinical studies. Antibiotic-therapy is important in the treatment of septicemia, but measures for elimination of endotoxin, such as administration of endotoxin core antibody (14), or inhibition of the endotoxin activity should be further evaluated. We believe that this study indicates that such interventions might be of clinical relevance.

<u>In conclusion:</u> The results from this study indicate that antibiotic therapy in bacteremia increases plasma levels of endotoxin to levels of clinical importance. The study supports the theory that high endotoxin levels correlate with impaired cardiovascular performance.

ACKNOWLEDGEMENTS.

The authors want to thank Helen Sletten, Knut Hansen, Eli Berg, Karin Myhr, Jorunn Støvne Pettersen and Jan Ole Olsen for technical assistance. This work was supported by grants from the Norwegian Council for Science and Humanities.

REFERENCES.

1.Andersen.B.M,Solberg.O.(1980) The Endotoxin-liberating effect of Antibiotics of Meningococci in vitro. Acta Path.Microbiol.Scand.88;231-236.

2.Bryan C.S,Reynolds K.L,Brenner E.R.(1983).Analysis of 1186 Episodes of Gram-negative Bacteremia in Non-University Hospitals:The effects of antimicrobial therapy.Rev. Infect. Dis.5,4;629-638.

3.Hopkin.D.(1977)Too rapid destruction of gram-negative organisms.Lancet,2;603-604.

4.Kreger B.E,Craven D.E,McCabe W.R.(1980).Gram-Negative Bacteremia.IV.Re-evaluation of Clinical Features and Treatment in 612 Patients.Am.J.Med,68,344-355.

5.Lewis.L.S.(1979) Prognostic factors in acute meningococcemia.Arch.Dis. Child.54;44-48.

6.Morrison D.C,Duncan R,Gouthwan S.A.(1985).In Vivo biological activity of endotoxin. In ten Cate J.W, Büller H.R,Sturk A,Levin J.(eds):"Bacterial endotoxins. Structure,biomedical significance,and detection with the Limulus Amebocyte Lysate test." Progress in clinical and biological research,Vol 189.Alan R. Liss,INC,New York, USA,pp 81-98.

7.Munster A.M,Loadholdt C.B,Leary C.L.T,Barnes M A.(1977) The effect of antibiotics on cell-mediated immunity. Surgery. 81,6;692-695.

8.Nishijima H,Weil M.N,Shobin H,Cavaniles J.(1973)Hemodynamic and metabolic studies on shock associated with gram-negative bacteriemia. Medicine.52;287-294.

9.Reilly J,Rivalier E,Campagnon A,Laplane R,Du Buit H. (1935).Sur la pathogenic de la dothienenteric.La fievre typhoide experimentale.Ann de Med,1935,37,182&241.

10.Shenep.J.L and Mogan K.A.(1984). Kinetics of Endotoxin Release During Antibiotic Therapy For Experimental Gram-Negative Bacterial Sepsis.J.Infect.Dis.150,3;380-388.

11.Simone C,Pugnaloni L,Cilli A,Forastieri E.M.A.(1984). Pharmacokinetic assessment of immunosuppressive activity of antibiotics in human plasma by a modification of Mixed lymfocyte reaction.Crit. Care. Med.12;483-485.

12.Tubbs.H.R.(1980)Endotoxin in meningococcal infections. Arch.Dis.Child.55;808-810.

13.Weinstein M.P,Murphy J.R,Reller L.B,Lichtenstein K.A. (1983).The clinical significance of positive blood cultures:a comprehensive analyses of 500 episodes of bacteremia and fungemia in adults.II.Clinical observations,with special reference to factors influencing prognos is.Rev.Infect.Dis. 5,1;54-70.

14.Ziegler.E.J,McCutchan.J.A,Fierer.J,Glauser.M,Sadoff.J C, Douglas.H,Braude.A.I.(1982).Treatment of gram-negative bacteremia and shock with human antiserum to a mutant Escherichia Coli.N.Engl.J.Med.307,1225-1230.

Bacterial Endotoxins: Pathophysiological Effects, Clinical Significance, and Pharmacological Control, pages 263–271

SYSTEMIC MENINGOCOCCAL DISEASE: A MODEL INFECTION TO STUDY ACUTE ENDOTOXINEMIA IN MAN.

Petter Brandtzaeg, Peter Kierulf, Peter Gaustad, Jan Dobloug, Tom Eirik Mollnes and Kari Sirnes
Department of Infectious Diseases (P.B.,J.D.),
Department of Clinical Chemistry (P.K.),
Department of Microbiology (P.G.),
Bloodbank and Department of Immunology (T.E.M., K.S.),
Ullevål University Hospital, Oslo, Norway.

ABSTRACT

Plasma LPS were quantitated in 40 patients with bacteriological verified systemic meningococcal infection (SMD). Twenty-two patients (55%) demonstrated LPS > 25 pg/ml. Fourteen patients with initial plasma LPS levels > 700 pg/ml developed a severe septic shock, impaired renal function and extensive coagulopathy compared to 2 out of 26 patients with LPS levels below 700 pg/ml. Seven patients died due to the circulatory collapse and multiple organ failure all with inital LPS levels >1,000 pg/ml. Initiation of antibiotic therapy did not result in further increase of plasma LPS concentration. LPS were cleared from the circulation with a half-life of 1.5-3 hours after initiation of antibiotic treatment with penicillin and chloramphenicol. The results suggest that plasma LPS are quantitatively related to the development of septic shock, multiple organ failure and death from these complications in SMD, making LPS measurement a direct prognostic marker.

INTRODUCTION

Fulminant meningococcal septicemia is characterized by circulatory collapse, multiple organ failure, extensive consumption coagulopathy and a high mortality rate. Many of the features observed in these patients are identical to those seen in animals during LPS infusion. Since LPS are the only known toxic components produced by the Neisseria meningitidis, meningococcal LPS may initiate different cascade reactions and play an important role in the pathogenesis of fulminant meningococcal septicemia. Meningococci may cause other and less dramatic clinical syndroms (meningitis and meningococcemia) without multiple organ failure. By using the latter two syndroms as reference group, SMD may serve as a human model to study a rapidly evolving gram-negative infection in young, previously healthy persons and to test the hypothesis that LPS are involved in the development of multiple organ dysfunction.

MATERIAL AND METHODS

Patients: 40 patients consecutively admitted to Ullevål University Hospital were studied prospectively. All patients had Neisseria meningitidis cultivated from blood and/or cerebrospinal fluid (CSF) and were initially treated with benzylpenicillin (3oo mg/kg/24 hours, maximum dosis 18g/24 hours) and chloramphenicol succinate (100 mg/kg/24 hours, maximum dosis 3g/24 hours) devided in 4 doses. When the etiological agens was bacteriologically confirmed, penicillin treatment alone was given for at least 5 days.

Clinical definitions:

Fulminant septicemia: A patient suffering from a profound septic shock i.e. systolic blood pressure < 70 mm Hg in children (< 12 years) or < 85 mm Hg in adults needing plasma, crystalloids and vasoactive drugs (dopamine or dopamine and epinephrine) for at least 24 hours. In addition all patients showed signs of impaired renal function (creatinine clearance < 80 ml/min./1,73m^2 or pathologically elevated serum creatinine without signs of concomitant dehydration) and dissiminated intravascular coaculation (DIC).

Meningitis: A patient with clinical signs of meningism, > 100x10^6 leucocytes per l CSF, normal blood pressure after initial rehydration and normal renal function.

Meningococcemia: A patient with hemorrhagic skin rash, normal blood pressure after initial rehydration, < 100x10^6 leucocytes per l CSF, normal renal function and Neisseria meningitidis cultivated from blood culture(s).

Lipopolysaccharide assay in plasma: Blood was collected sequentially in heparinized (15 U/ml) vacuum tubes tested to be LPS-free, centrifuged at 1400 g for 10 minutes, platelet poor plasma pipetted off and stored at -70° C. Pooled plasma from healthy blood donors was diluted with H_2O 1:10 and used to prepare standards with LPS (E. coli B5505 Mallinckrodt) in concentrations varying from 2.5 to 100 pg/ml. All patient plasmas were initially diluted 1:10 with H_2O. If further dilutions were needed, pooled plasma from healthy donors diluted 1:10 with H_2O was used. Samples to be assayed were heat-inactivated for 12 minutes at 75° C, cooled to room temperature and 10 ul transferred to a 96 well microtiter plate. Fifty ul diluted LAL (Mallinckrodt, 0.063 EU/ml) was added, and the microtiter plate incubated at 37° C for 30 minutes. Subsequently, 50 ul prewarmed (37° C) chromogenic peptide substrate S-2423 (Kabi Vitrum, S-2423 5.3 mmol/l and TRIS-HCl 0.05 mol/l mixed 1:1) and the reaction terminated 180 seconds later with acetic acid (50%). The reaction was read automatically at 405 nm. The detecton limit was 25 pg/ml. The day to day variation coefficient was 16% at 250 pg/ml. All samples were assayed in duplo. Samples were always retested in separate runs and the mean value used. Forty healthy blood donors and 30 consecutively patients admitted to the Medical department all showed LPS values < 25 pg/ml.

RESULTS

Sixteen patients fulfilled the criteria of fulminant septicemia. The plasma LPS levels in the first available sample are seen in Fig.1.

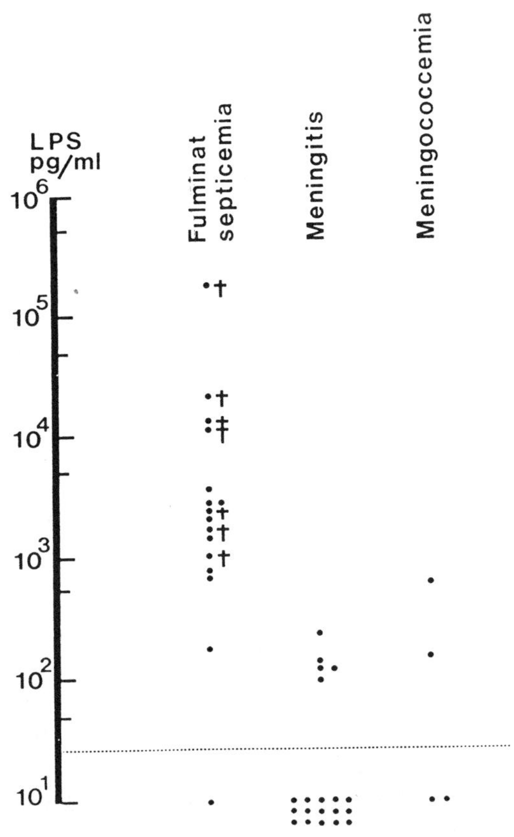

Figure 1. LPS plasma levels in the first available plasma sample in 40 patients with systemic meningococcal disease. The dotted line indicates lower detection limit (25 pg/ml). Non survivers are marked with a cross.

Fifteen patients had LPS levels ranging from 210 to 170,000 pg/ml all with positive blood culture(s). One patient had negative blood cultures on admission and negative plasma LPS samples collected 7 hours after initiation of antibiotic therapy. Neisseria meningitidis was, however, cultivated from CSF. The clearance of LPS was calculated to be 1,5 - 3 hours during the first hours of therapy. Initiation of antibiotica therapy never gave rise to plasma LPS. Eight patients with fulminant septicemia were treated with plasmapheresis (n=6) or blood exchange (children with body weight < 30 kg, n=2). The selection-criteria were the same as used by Bjorvatn et al.(3 positive clinical criteria out of 6). After Jan 1. 1986, plasmaphersis or blood exchange treatment was stopped in children (<13 years).

Seven patients died within 39 hours of hospitalization due to progressive circulatory collapse often combined with cardiac arrhythmias and in one case with cerebral herniation. Three patients treated with plasmapheresis or blood exchange died versus 4 not given this treatment. Two patients suffered severe sequelae: a 15 years old girl treated with plasmaphersis developed amaurosis due to intraocular hemorrhage and extensive skin necrosis and a 15 years old boy given conventional treatment, needed amputation of all 10 fingers due to gangrene. He was discharged with persistent reduction of renal and pulmonary function after 3 months of hospitalization.

Twenty patients fulfilled the criteria of meningitis of whom only five had detectable LPS (>25 pg/ml) The LPS plasma levels ranged from 85 to 260 pg/ml(Fig. 1), all 5 having positive blood culture(s). Five additional patients had positive blood cultures but LPS negative plasma samples collected immediately before initiation of antibiotic therapy. Two out of 4 patients with meningococcemia had detectable LPS ranging from 150 to 600 pg/ml (Fig.1). In both patients the symptoms emerged very rapidly, and they might have developed a fulminant septicemia if left untreated.

All together 227 plasma samples were collected serially. In the septicemia-group 90 samples out of 118 were LPS positive (76%) versus 14 out of 84 (17%) in the meningitis-group and 7 out of 23 (30%) in the meningococcemia-group.

DISCUSSION

The role of LPS in the symptomatology of in SMD is still controversal. In 2 studies from Africa on fulminant meningococcal septicemia, Lewis (1979) and Tubbs (1980) using a semiquantitative method, did not find LPS quantitation useful to predict the severity or fatal outcome in Nigerian children. In 2 small Norwegian studies on SMD using the more senitive LAL and chromogenic substrate method (Harthug et al 1983, Bjorvatn et al 1984), 7 patients had LPS plasma levels > 3,000 pg/ml, all developing multiple organ failure of whom 4 died. However no control group with less severe cases were included in their studies.

This study comprises 40 patients with bacteriologically verified SMD studied prospectively. Using the LAL and chromogenic substrate method 22 patients (55%) had detectable LPS in plasma sample while 29 (73%) had positive blood culture(s). A plasma LPS level > 700 pg/ml was found in 14 out of 16 patients defined as fulminant septicemia and in none of 24 with meningitis or meningococcemia ($p=0.0006$ Fisher's exact test). Seven patients died due to septic shock in the fulminant septicemia group all with LPS plasma levels >1,000 pg/ml. None in the other two clinical syndroms died ($p=0.0002$ Fisher's exact test).

"Naturally" occuring LPS is eliminated rapidly from the circulation with a half-life varying from 1.5 to 3 hours after initiation of antibiotic therapy in SMD. Increasing LPS values were never observed. This findings contrast data from previous experimental E. coli infections in rabbits suggesting that betalactam antibiotics may rise the the free plasma LPS level during the first hours of treatment (Shenep et al 1984).

Fulminant meningococcal septicemia evolves rapidly. Within 24 hours most patients develop circulatory collapse and multiple organ failure. The very high LPS levels observed in these patients may indicate a rapid bacterial proliferation. In addition meningococci are known to shed LPS containing outer membrane blebs during growth (DeVoe et al 1973) and very easily undergo lysis. Endotoxin spill over from a diseased bowel is a less likely explanation due to the following observations: 1) Patients with detectable LPS showed decreasing values in spite of a deteriorating general condition. 2) Four patients of whom 3 died, with fulminant gram-positive septicemia (Streptococcus pneumoniae, n=3 and Streptococcus pyogenes, n=1) with septic shock, renal and pulmonary failure, skin hemorrhages and DIC, all had LPS negative sequentilly collected plasma samples.

The LPS clearance seems to be fairly constant in different patients and independent of the initial LPS level, a phenomenonobserved in animals under experimental conditions (Freudenberg et al 1985). Very high LPS levels generated in short time, may indicate the lacking immune competence of the host towards the insulting strain of Neisseria meningitidis.

Plasmapheresis or blood exchange have been proposed as therapeutic procedures that might reduce the high mortality rate in fulminant meningococcal septicemias (Bjorvatn et al 1984, Brandtzaeg et al 1985). Our recent findings are not in accordance with the first optimistic results. Three out of 8 patients in the plasmapheresis/blood exchange group died versus 4 out of 8 given conventional intensive care treatment. One patient in each group suffered severe sequelae. So far, no prospective, randomized clinically controlled trial has been conducted and the role of plasma exchange in the treatment of fulminant meningococcal septicemia remains unsettled.

Our results suggest that LPS, being the only known toxic component of the meningococci, are quantitatively related to the development of septic shock, multiple organ failure and death from these complications in SMD. In spite of a possible difference in individual responsiveness to LPS (Caroff et al,1986), a plasma level of 700 pg/ml seemed to be critical in this study. All patients passing this level developed a severe septic shock and renal dysfunction. These observations suggest that LPS quantitation is of prognostic value and may give us insight in the dynamics of the host bacteria interaction in SMD.

REFERENCES

Bjorvatn B, Bjertnaes L, Fadnes HO, Flaegstad T, Gutteberg TJ, Kristiansen BE, Pape J, Rekvik OP, Østerud B, Aanderud L (1984). Meningococcal septicaemia treated with combined plasmaphersis and leucapheresis or with blood exchange. Br Med J 288:439-441.

Brandtzaeg P, Sirnes K, Folsland B, Godal HC, Kierulf P, Bruun JN, Dobloug J (1985). Plasmapheresis in the treatment of severe meningococcal or pneumococcal septicemia with DIC and fibrinolysis. Scand J Clin Lab Invest Vol 45, Suppl 178:53-55.

Caroff M, Cavaillon JM, Fitting C, Haeffner-Cavaillon N (1986). Inability of pyrogenic, purified Bordetella pertussis lipid A to induce interleukin-1 release by human monocytes. Infect Immun 54:465-471.

DeVoe I.W, Gilchrist J.E (1973). Release of endotoxin in the form of cell wall blebs during in vitro growth of Neisseria meningitisis. J Exp Med 138:1156-67

Freudenberg MA, Kleine B, Freudenberg N, Galanos C (1984). Distribution, degradation and elimination of intravenously applied lipopolysaccharide in rats. In: Homma JY, Kanegasaki S, Luderitz O, Shiba T, Westphal O, eds. Bacterial endotoxin. Chemical, biological and clinical aspects.

Harthug S, Bjorvatn B, Østerud B (1983). Quantitation of endotoxinaemia in meningococcal disease using a Limulus lysate test in combination with chromogenic substrate. Infection 11:192-195.
Lewis LS (1979). Prognostic factors in acute meningococcaemia. Arch Dis Child 54: 44-48.
Shenep JL, Mogan KA (1984). Kinetics of endotoxin release during antibiotic therapy for experimental gram-negative bacterial sepsis. J Infect Dis 150:380-388.
Tubbs HR (1980). Endotoxin in meningococcal infections. Arch Dis Child 55:808-10.

Bacterial Endotoxins: Pathophysiological Effects, Clinical Significance, and Pharmacological Control, pages 273–282

SPONTANEOUS ENDOTOXINAEMIA AND ONSET OF ORAL FEEDING IN 80 PRETERM NEONATES

Alistair P J Thomson, Colin M M Stern, Chris West (1), Jon W Scopes. Dept of Paediatrics, St Thomas' Hospital, London SE1, and (1) Dept of Community Health, University of Liverpool, ENGLAND

INTRODUCTION

The origin and significance of endotoxinaemia in Very-Low-Birth-Weight (VLBW) neonates remains unclear. A firm relationship between endotoxinaemia and Gram-negative infection has not been demonstrated (Scheifele et al, 1981b); even those babies with culture proven Gram-negative septicaemia are not invariably endotoxinaemic (Adhikari et al, 1985). Recent work has suggested that neonates may become spontaneously endotoxinaemic whilst well, and that endotoxinaemia is more pronounced after oral feeding (Scheifele et al, 1985).

Moreover, the Limulus test is difficult to perform in plasma compared to other body fluids, (Kelsey et al, 1982), and there is no generally accepted method for detection of endotoxin in neonatal plasma, with its special problems of small sample size, and apparent false positives in jaundiced babies (Goldberg et al, 1982).

In this prospective study with serial sampling of Very-Low-Birth-Weight (VLBW) babies, the relationship of endotoxinaemia to infection and oral feeding has been examined using a chromogenic Limulus amebocyte lysate assay.

AIMS

Our aims were to:

1. develop a rapid, reliable, quantitative chromogenic Limulus amebocyte lysate assay for endotoxin in neonates,
2. document endotoxin levels in healthy VLBW babies,
3. document endotoxin levels at the time of infection screens and relate these to the presence or absence of infection, and
4. relate endotoxinaemia to the onset of oral feeding.

METHODS

Population and Protocol

Eighty babies of birthweight less than 1500 gm admitted to the St Thomas' Hospital Special Care Baby Unit between 1/10/83 and 16/2/85 were studied. These babies were consecutive admissions; 50 were inborn, and 30 transferred from local maternity units which had no neonatal intensive care facilities. Six inborn babies, who showed signs of life at birth, but were not offerred active resuscitation because of extreme prematurity, were excluded. Three other live births, who all died before two hours of age despite active measures, were not sampled by the time of death.

Birthweight ranged from 620-1500 gm (median 1140 gm). Gestation ranged from 24-36 weeks (median 29.0 weeks). Stay in the neonatal intensive care unit ranged from 1-165 days (median 51.5 days).

Blood samples for endotoxin were obtained daily, when sampling necessary to the baby's routine management was taking place, and at the time of 165 of a total of 188 infection screens. Sampling continued until the baby was discharged from the intensive care unit. Sterile technique was used throughout. The amount of blood taken, approximately 150 µl, was a negligible proportion of the baby's total blood volume of 85 ml/kg. Permission for the study was obtained from the hospital Ethical Committee.

Blood was collected from umbilical arterial catheters, indwelling arterial cannulae, percutaneous radial artery punctures, venous punctures, or heel-stick capillary sampling, after skin cleansing with alcohol (Mediswab, or surgical spirit). Each sample was taken into an endotoxin-free 1ml disposable syringe (Gillette Surgical), via a needle (Gillette Surgical), or a butterfly (Abbott Laboratories Ltd). The syringe was washed out with sterile heparin obtained from multidose vials (1000 IU per ml) by a no-touch sterile technique before sampling.

If several samples were being collected from different babies they were stored at 4°C to minimise endotoxin degradation or production prior to separation of the plasma. Separated plasma was stored at -20°C until assay.

Assay Method

The assay method was based on the Whittaker MA Bioproducts chromogenic method for detection of endotoxin, using kit no 50-640U and chromogenic substrate S2423, though this needed modification to enable accurate endotoxin determination in very small initial quantities (0.025 ml) of neonatal plasma.

The assay was performed in a 96-well microtitre plate, 16 samples at a time, each sample being measured in duplicate to minimise errors. The samples were diluted 1:10 in sterile water and heated at 75°C for ten minutes to remove activators and inhibitors of the assay.

Each microtitre plate contained material for a standard curve (run in duplicate) consisting of doubling dilutions of standard endotoxin (E coli 0111:B4) in freshly drawn adult plasma, diluted and heated in the same way as the samples.

Fifty µl of the diluted, heated sample was added to each of 4 microtitre plate wells. Fifty µl of chromogenic Limulus lysate was added to all wells at time = 0, and the plate preincubated at 37°C for 45 minutes. Fifty µl chromogenic substrate S2423 was then added to half of the sample wells, and 50 µl of pH 9.0 buffer without chromogenic substrate was added to the other 2 wells. Glacial acetic acid was added to terminate the reaction after a further 3 minutes. The plate was read 55 minutes after the start of the reaction, at 405 nm in a microtitre plate reader (Titertek Multiskan MC).

The standard samples were used to derive a straight

line graph of optical density versus endotoxin concentration, usually between 0 and 2.88 Endotoxin Units per millilitre of plasma (EU/ml).

Readings from the 2 sample wells without chromogenic substrate were subtracted from those of the 2 wells containing S2423, to eliminate inaccuracy due to jaundiced plasma. The mean of the differences was compared with that plate's standard curve to give an endotoxin concentration. Sensitivity of the assay was between 0.01-0.05 EU/ml plasma.

Data Analysis

An episode of endotoxinaemia was defined as any sequence of one or more positive values for endotoxin in plasma, occurring between two negative values. One or more high levels from birth followed by a zero value were also counted as an episode, as were one or more high levels before discharge preceded by a zero value.

Data were analysed on the Liverpool University IBM 3083 mainframe computer using the Statistical Package for the Social Sciences (SPSS-X).

RESULTS

The 80 neonates had a total of 322 episodes of endotoxinaemia. Mean endotoxin concentration during an episode was 0.799 EU/ml. Median endotoxin during an episode was 0.340 EU/ml. Range was 0-7.2 EU/ml.

Of the 322 episodes of endotoxinaemia, 57 coincided with infection screens. A further 129 occurred in 'ill' babies, a group defined as: those with symptoms suggesting necrotising enterocolitis; on mechanical ventilation; with a pneumothorax; or an intraventricular haemorrhage on ultrasound scan during the previous 48 hours; with shock; or who were otherwise unwell. 136 episodes occurred in the absence of any concomitant illness. These episodes therefore appear to be 'spontaneous' endotoxinaemia.

Onset of oral feeding was defined as the first day of nasogastric milk feeds. Of the 80 study babies, 70 were orally fed, (the other 10 died before oral feeding could be commenced). Babies were fed first from 0 - 39 days of age, with a median of 4.5 days, and a mode of 0 days.

Forty-three of the 322 episodes of endotoxinaemia were before introduction of oral feeds and 279 afterwards. Median endotoxin levels during an episode before feeding were 0.270 EU/ml: median levels during episodes after feeding were 0.350 EU/ml. There is no significant difference between these using the Mann-Whitney U-test.

However, this analysis ignores zero endotoxin values between episodes, so comparison should be made between all endotoxin levels before and after feeds: before feeds the levels were 0.213 +/- 0.042 EU/ml (mean +/- standard error of mean on 531 samples), and 0.322 +/- 0.055 EU/ml (1242 samples) afterwards. This is highly significant ($p < 0.005$ by 2-tailed t test).

Endotoxinaemia occurred before oral feeding in 47 of 63 (75%) babies sampled, and after feeding in 67 of 69 (97%) babies sampled ($p < 0.002$).

		AFTER FEEDING		
		LPS NEG	LPS POS	TOTAL
BEFORE	LPS NEG	0	13	13
FEEDING	LPS POS	1	38	39
	TOTAL	1	51	52

p = 0.0018 (sign test on paired data)

Table 1: Number of babies with endotoxinaemia: 52 babies sampled both before and after feeding (LPS NEG = Number of babies endotoxin-free, LPS POS = Number of babies endotoxinaemic).

Fifty-two babies were sampled both before and after feeding (see table 1). Thirteen babies who were endotoxin-free before feeds became endotoxinaemic after feeds; 38 neonates were endotoxinaemic both before and after feeds. Only one neonate endotoxinaemic before feeds became endotoxin-free after feeds. More babies were therefore endotoxinaemic after feeds than before ($p = 0.0018$ by sign test on paired data).

To correct for the effects of illness on endotoxinaemia, values for endotoxin obtained at the time of an infection screen, or when the neonate was 'ill' (vide supra for definition) have been excluded from analysis. This leaves 136 episodes of endotoxinaemia, when these babies were well.

In these well babies before feeds the levels were 0.187 +/- 0.066 EU/ml (mean +/- standard error of mean in 73 samples), and 0.321 +/- 0.024 EU/ml (969 samples) afterwards. This is highly significant ($p < 0.005$ by 2-tailed t test). Non-parametric statistical analysis, to allow for the skewed distribution, confirms this, ($p = 0.006$ by Mann-Whitney U test).

	Before feeds	After feeds
Endotoxin-free babies	16	3
Endotoxinaemic babies	17	66
TOTAL	33	69

$p = <0.001$ (Chi-squared test)

Table 2: Number of babies with endotoxinaemia whilst well, before and after feeding.

Endotoxinaemia occurred, (see table 2), before oral feeding in 17 of 33 babies sampled (52%), and after feeding in 66 of 69 babies sampled (96%) ($p = <0.001$ by Chi-squared test).

The paired data are shown in table 3.

		AFTER FEEDING		
		LPS NEG	LPS POS	TOTAL
BEFORE FEEDING	LPS NEG	1	14	15
	LPS POS	0	17	17
	TOTAL	1	31	32

p = 0.00012 (sign test on paired data)

Table 3: Number of babies with endotoxinaemia: 32 babies sampled whilst well, both before and after feeding (abbreviations as table 1).

Thirty-two babies were sampled both before and after feeding. Of the 15 babies who were endotoxin-free before feeds 14 became endotoxinaemic after feeds; there were 17 babies who were endotoxinaemic both before and after feeds, and no baby who was endotoxinaemic before feeds became endotoxin-free afterwards. Thus, endotoxinaemia was more frequent in babies after feeding, even when all samples taken while the babies were ill were excluded, (p = 0.00012 by sign test on paired data).

DISCUSSION

Studies of endotoxinaemia in neonates have mostly been directed towards the diagnosis of Gram-negative infection. However, these have produced conflicting information and conclusions.

Scheifele et al (1981b), using a sensitive version of the Limulus amebocyte lysate gelation test, found a large proportion of babies had positive Limulus tests in the first week, without evidence of infection. Endotoxinaemia was detected in 7 out of 8 neonates who had Gram-negative septicaemias and 8 out of 11 infants with necrotising enterocolitis. It was thought that the positive tests in the first week reflected true endotoxinaemia, and concluded that the test would be of limited use for the diagnosis of infection in babies less than 7 days of age. Other reports also reflect the lack of a firm association between Gram-negative infection and endotoxinaemia, and demonstrate apparent false positive and false negative results (Goldberg et al, 1982, Goldstein et al, 1976, Kelsey et al, 1982).

Is spontaneous intestinal endotoxinaemia a possible explanation for these false positive and negative results? That intestinal endotoxinaemia occurs in ill babies is shown in a report of an infant with a sepsis-like syndrome, (Scheifele et al, 1981a). Endotoxin levels were high, but declined promptly after polymyxin B treatment was started. A group of similar patients and findings were subsequently described (Cooperstock et al, 1982). These reports suggested that some neonates might have lower levels of endotoxinaemia while well.

In a large study using serial sampling in 47 VLBW babies, (Scheifele et al, 1985), 28 of the 47 (60%) were endotoxinaemic at some stage; only 2 of 58 episodes of infection were accompanied by endotoxinaemia. This provides good evidence of spontaneous endotoxinaemia. Before feeding 6 out of 45 (13%) babies were endotoxinaemic, as compared with 56% after feeding (a difference significant at the $p<0.001$ level). Onset of oral feeding was followed by a rise in stool endotoxin-like activity in most babies, suggesting that the endotoxin was gut-derived. Intestinal wall permeability changes in the first days of life of VLBW neonates, (Beach et al, 1982), might facilitate endotoxin entry. One possible function of this endotoxinaemia is that it may be involved in early acquisition of immunity (Nolan, 1985).

Although our results appear to strongly suggest that higher levels of endotoxin and more episodes of endotoxinaemia tend to occur after onset of oral feeding, we feel that they should be treated with caution. Firstly, 85% of babies were orally fed by 15 days of age, thus leaving little time for endotoxin sampling before feeding. Secondly, with advancing age in each baby more frequent samples were obtained from heel-prick capillary sites rather than the more reliably sterile indwelling cannulae. Thirdly, the rise in endotoxin levels appeared to occur several days after feeding; indeed there was no significant difference between median endotoxin levels in the episode immediately before feeding and the episode immediately afterwards. Fourthly, we did not perform stool endotoxin studies.

Nevertheless, this present study lends support to the conclusions drawn by Scheifele and his colleagues. Our results show that neonates may become spontaneously endotoxinaemic without concomitant illness, that VLBW neonates can become endotoxinaemic after feeding, and that higher endotoxin levels may follow oral feeding. This endotoxinaemia is likely to be intestinal in origin.

REFERENCES

Adhikari M, Coovadia H M, Gaffin S L, Brock-Utne J G, Marivate M, Pudifin D J (1985). Septicaemic low birthweight neonates treated with human antibodies to endotoxin. Arch Dis Child 60:382-384.

Beach R C, Menzies I S, Clayden G S, Scopes J (1982). Gastrointestinal permeability changes in the preterm neonate. Arch Dis Child 57:141-145.

Cooperstock M S (1981). Neonatal intestinal endotoxinaemia (letter). Lancet i:721.

Goldberg P K, Kozinn P J, Kodsi B, Lewitis R, Ackerman B, Aschenbrand L, Feldman F (1982). Endotoxemia and hyperbilirubinemia in the neonate. Am J Dis Child 136:845-848.

Goldstein J A, Reller L B, Wang W-L L (1976). Limulus amebocyte lysate test in neonates. Am J Clin Path 66:1012-1015.

Kelsey M C, Lipscomb A P, Mowles J M (1982). Limulus amoebocyte lysate endotoxin test: an aid to diagnosis in the septic neonate? J Infection 4:69-72.

Nolan J P (1985). Spontaneous endotoxinemia (editorial). J Pediatr Gastroenterol Nutr 4:7-8.

Scheifele D W, Melton P W, Ebelt V (1981a). Endotoxinaemia per se as cause of neonatal morbidity (letter). Lancet i:337.

Scheifele D W, Melton P, Whitchelo V (1981b). Evaluation of the Limulus test for endotoxemia in neonates with suspected sepsis. J Pediatr 98:899-903.

Scheifele D W, Olsen E, Fussell S, Pendray M (1985). Spontaneous endotoxinemia in premature infants: correlations with oral feeding and bowel dysfunction. J Pediatr Gastroenterol Nutr 4:67-74.

Bacterial Endotoxins: Pathophysiological Effects, Clinical Significance, and Pharmacological Control, pages 283–292

ENDOTOXEMIA IN RATS - INFLUENCE OF LIPOXYGENASE BLOCKER VERSUS LEUKOTRIENE RECEPTOR ANTAGONIST (BW755C VS LY171883)

Soheyl Bahrami, Eva Paul, Heinz Redl, Günther Schlag, Fred Mihm*

Ludwig Boltzmann Institute for Experimental Traumatology, Vienna, Austria

* Stanford University Hospital, California, U.S.A.

INTRODUCTION

The cardiopulmonary sequelae of endotoxin occur concomitantly with an increase in arachidonic acid metabolites, such as thromboxane A_2(TXA_2) (Demling et al. 1983)and leukotrienes(Brigham et al. 1979). The TXA_2 is produced in the cyclooxygenase pathway of arachidonic acid metabolism (Ogletree 1983), whereas the leukotrienes (LT) are produced in the 5-lipoxygenase pathway (Borgeat et al. 1979). The thromboxane A_2 as a vasoconstrictor substance is supposed to be responsible for pulmonary hypertension (Demling et al. 1983). The leukotrienes have several properties. The increased-lung vascular permeability has been attributed to the LTD_4 (Brigham et al. 1979). The formation of thromboxane was also found to be stimulated by leukotrienes (Puustinen et al. 1984, Piper et al. 1981).

Inhibition of prostaglandin formation by different inhibitors was found to be beneficial in a number of endotoxin studies in sheep (Begley et al. 1984, Ogletree 1983, Ogletree and Brigham 1982, Schuette et al. 1984). Therefore BW755C, an inhibitor of both the cyclooxygenase and lipoxygenase pathways of arachidonic acid metabolism (Morgan 1971), could be expected to protect against endotoxin. The results however, are controversial (McKechnie et al. 1985).

The compound LY171883 was found to antagonize an LTD_4-induced increase in total pulmonary resistance and to reduce the vascular leakage caused by this LT (Fleisch et al. 1985). A protective effect of LY171883 in traumatic shock has been reported (Hock et al 1985). To our knowledge, the effect of BW755C and LY171883 in endotoxin shock has not yet been extensively studied.

Therefore it seemed reasonable to employ the dual lipo-cyclooxygenase inhibitor BW755C and the LTD_4 antagonist LY171883 in our standardized rat model (LD_{75}), which was successfully used e.g. to evaluate the effect of corticosteroids in the treatment of endotoxin shock (Bahrami et al. 1986). In the model of conscious, unrestrained rats anesthesia- and stress-induced complications are minimized. We attempted to evaluate the inhibitory effect of BW755C or LY171883 on endotoxin-induced changes in arachidonic acid metabolism, blood cells and mortality.

MATERIAL AND METHODS

Male Sprague Dawley rats weighing 230 - 270 g were used (eight in each group) after being kept in animal units on a standard commercial rat chow and water ad libidum for a 5-day adaptation period.

Twenty-four hours before the experiment the jugular vein was catheterized as previously described for introduction of endotoxin under halothane anesthesia (Bahrami 1984). The catheter was attached to a swivel arrangement and allowed unrestricted movements during the experiment. Endotoxin (DIFCO E. coli 026:B6) was dissolved in 0.9 % W/V sodium chloride solution and administered as a 5 mg/kg dose over a 30-minute period. Without treatment this resulted in an overall mortality of about 75 % within 48 hours (LD_{75}). Blood samples (a total of 2 ml, which was replaced with the same amount of Ringer's solution) for chemical analysis were withdrawn from the jugular vein catheter before starting the infusion and then at 30 minutes, four hours, 24 hours and 48 hours post-endotoxin from surviving animals (Fig.1).

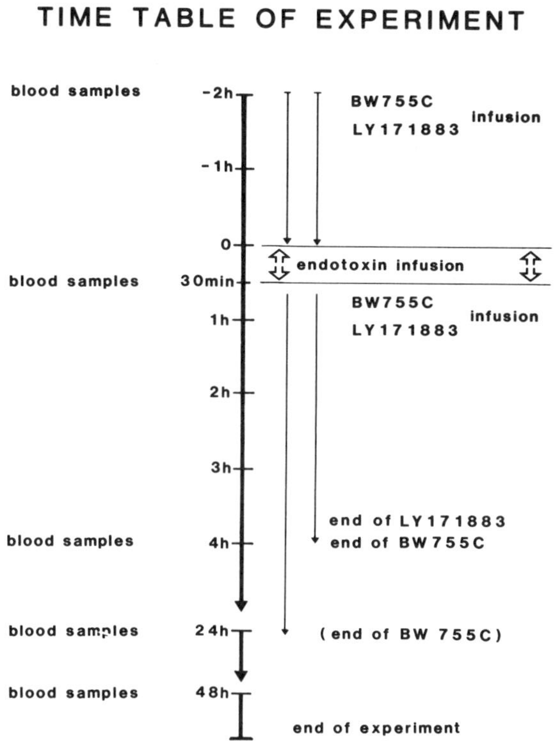

Fig. 1: Experimental procedure.

BW755C was dissolved (10 mg/ml) in 0.9 % W/V sodium chloride solution and administered i.v. as a 25 mg/kg dose over two hours pre-endotoxin, followed by 2.5 mg/kg/hr either up to four hours (BW group) or up to 24 hours (BT group) post-endotoxin.

LY171883 was initially dissolved in 0.5 M sodium bicarbonate to a concentration of 22 mg/ml. LY171883 was administerd i.v. (2 ml/hr, diluted with 0.9 % W/V sodium chloride solution) as a 4 mg/kg dose over two hours pre-endotoxin followed by 4 mg/kg up to four hours post-endotoxin (LY Group).

Two drug control groups of four animals each received equal amounts of BW755C (BO group) or LY171883 (LO group) drug without endotoxin infusion.

Two vehicle control groups of four animals each were treated with an equal volume of BW755C (BV group) or LY171883 (LV group) vehicle and endotoxin.

Seven animals received endotoxin alone without any teatment (CO group).

Statistical analysis was done with chi-square, analysis of variance and unpaired t-tests (with logarithmical transformation if data were heterogenous). Significance level = 95 %.

RESULTS

The time course of mortality was influenced by BW755C administered according to our protocol. There was a marked but not significant reduction of mortality during the first 8 hours either to one out of eight, when BW755C was infused for 4 hours, or to Ø if BW755C was infused for 24 hours post-endotoxin, as compared to three out of seven in the control group. This reduction of mortality was not seen after 24- or 48-hours post-endotoxin, even if treatment was continued for 24 hours (Table 1). LY171883 treatment resulted in 100% mortality within four hours post-endotoxin (Table 1).

TABLE 1: Time course of mortality (death/total) (ET = endotoxin)

	4^h	8^h	24^h	48^h
Control ET alone (CO)	1/7	3/7	3/7	5/7
4 hours BW755C + ET (BW)	0/8	1/8	6/8	7/8
24 hours BW755C + ET (BT)	0/8	0/8	4/8	6/8
LY171883 + ET (LY)	8/8	8/8	8/8	8/8
BW755C vehicle + ET (BV)	0/4	1/4	2/4	2/4
LY171883 vehicle + ET (LV)	1/4	2/4	4/4	4/4
BW755C alone (BO)	0/4	0/4	0/4	0/4
LY171883 alone (LO)	0/4	0/4	0/4	0/4

Neither BW755C nor LY171883 alone induced mortality during 48 hours of observation period, whereas their vehicle combined with endotoxin (ET) show a similar course of mortality as seen in animals that received endotoxin and saline (Table 1).

The general fall in blood platelets following endotoxin infusion was not prevented by BW755C or LY171883 treatment. At 30 minutes post endotoxin there were even more circulating platelets in the control group ($p < 0.05$) (Fig.2).

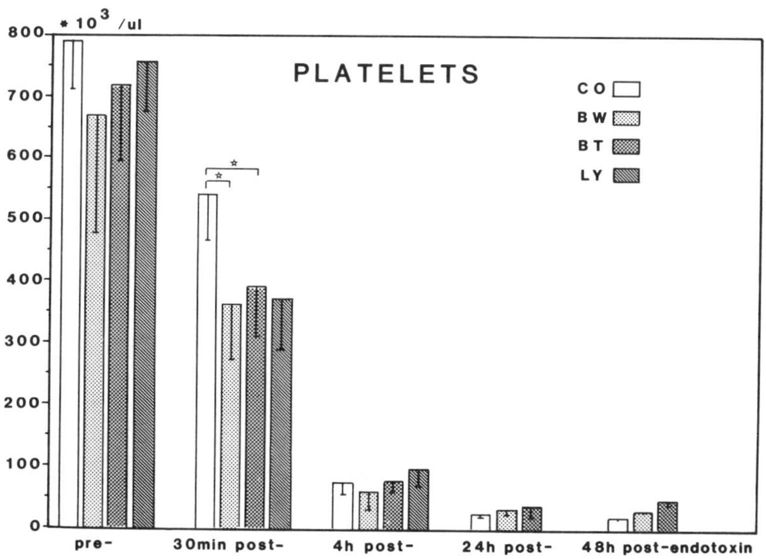

Fig. 2: Platelets in rat blood specimens (counts x 10^3/ul). Co - control (endotoxin without treatment). BW - BW755C administration up to four hours post endotoxin. BT-BW755C administration up to 24 hours post-endotoxin. LY-LY171883 administration up to four hours post-endotoxin.
* - significant ($p < 0.05$) difference.

Also, there was no influence on the drop of leukocyte counts due to our treatment protocol, but reoccurrence was faster in surviving BW755C-treated animals. This was not statistically significant because of the small number of animals caused by a high mortality (Fig. 3). When supplied according to our protocol, the drug itself had no effect except for a slight fall in platelets and leukocyte counts during the first hours, which completely recovered within 24 hours (data not shown). Each of the vehicles combined with endotoxin (BV, LV groups) revealed similar effects as endotoxin (CO group) alone (data not shown). Immediately after endotoxin infusion we found a significant reduction of initially increased thromboxane levels in the plasma of rats that received BW755C (Fig. 4). In contrast, at 4 hours post-endotoxin there was no difference between the groups that received BW755C and those who did not. The plasma levels of TXB_2 showed a delayed development similar to that in animals without any treatment (Fig.4). The plasma TXB_2 levels in rats treated with LY171883 was found to be increased tenfold compared to the pre-endotoxin levels within four hours post-endotoxin, during which all animals died (Fig.3).

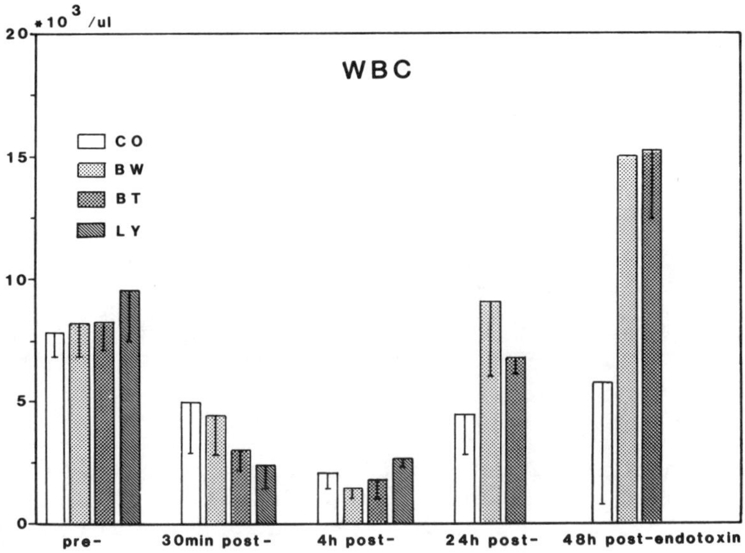

Fig. 3: Leukocyte counts (WBC) (counts x 10^3/ul) Groups as in Fig .2.

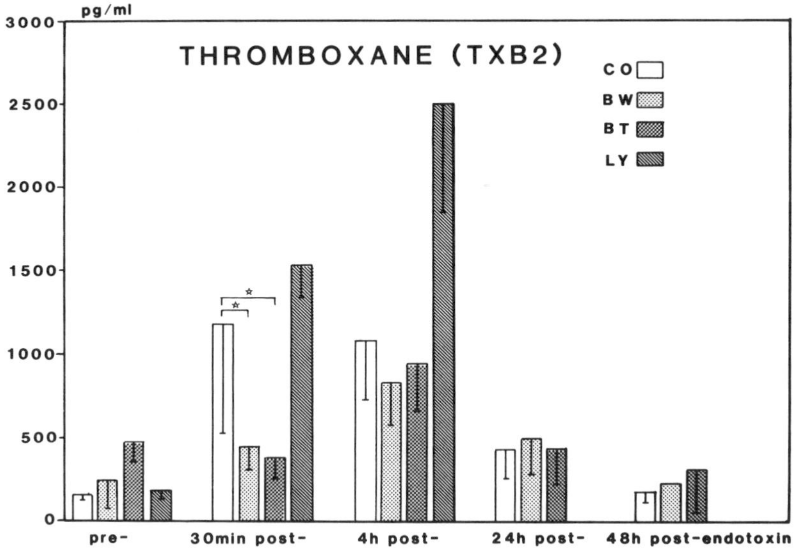

Fig. 4: Plasma thromboxane B_2 levels (TXB_2) (pg/ml). Groups as in Fig. 2. * - significant ($p < 0.05$) difference.

DISCUSSION

Among the primary events following endotoxin administration is the release of arachidonic acid derived prostaglandins and leukotrienes, also a possibly decisive event in endotoxin lethality. Accordingly, it has been shown that either a) inhibition of arachidonic acid release (phospholipase A_2 inhibition), b) of leukotriene formation (inhibition on lipoxygenase) and c) blocking the action of peptidoleukotrienes (LT antagonists) were curative.

ad a) Similar to others, we have recently demonstrated that application of corticosteroids prevents lethality in an endotoxin model with LD_{90} (Bahrami et al. 1986). Whether this is the effect of induction of phospholipase A_2 inhibitors,the influence on leukocytes (Redl et al. 1986) or rather a prevention of cachectin formation (Beutler et al. 1986) remains unclear.

McKechnie has demonstrated that TXB_2 formation was not prevented by corticosteroids in a similar model of endotoxin shock, but was effective in the reduction of mortality.

ad b) In contrast to this, McKechnie was able to block or probably only delay TXB_2 formation by BW755C, but without an effect on survival. We only found a significant delay in the rise of plasma thromboxane and improved survival for the first eight hours. However, the overall mortality (48 hours) was not reduced with BW755C in our rat model . This demonstrates that it is not sufficient to limit the observation period in drug studies to few hours only, as previously performed (Hock and Lefer 1985).
However, Hagmann who used an observation period of five days found 100 % protection by BW755C in mice with a 100 % lethal endotoxin model. Besides the different species, BW755C was administerd orally. In contrast in our study, the intravenous rout was chosen. Furthermore, other lipoxygenase inhibitors have been shown to be protective, e.g. diethylcarbamazine (Hagmann et al. 1986).

ad c) The formation and biliary elimination of peptidoleukotrienes has been proven by direct measurements in rats (Hagmann et al. 1986). Again, this group has shown that application of antagonists, e.g. FPL 55712, was fully protective in mice, while in our rat studies LY171883 even aggravated endotoxin action, both with regard to mortality and thromboxane formation. Possibly due to a feedback mechanism, as was suggested by Piper and Samhoun (1981), higher levels of circulating LT occurred because of the antagonist blocked receptor sites, which might have increased plasma thromboxane B_2.

The application of BW755C in a lethal model of endotoxin shock only moderately blocked the detrimental effects of endotoxin infusion, while LY171883 aggravated the situation. However, this should not lead to the conclusion that cyclooxygenase and lipoxygenase products of arachidonic acid are not related to the development of endotoxic shock.

ACKNOWLEDGEMENT

This study was in part supported by "Medizinisch-Wissenschaftlicher Fonds des Bürgermeisters der Bundeshauptstadt Wien".

REFERENCES

Bahrami S (1984) Untersuchung der Beeinflußbarkeit des Surfactantsystems der Lunge durch parenterale Ernährung unter Anwendung spezieller Phospholipidbestimmungsmethoden. Dissertation, Technische Universität Wien

Bahrami S, Schiesser A, Redl H, Schlag G (1986) Influence of different corticosteroids on cellular and humoral parameters in rat endotoxemia. In: Schlag G, Redl H (eds) "Progress in clinical and biological research", New York: A.R.Liss Inc., in press

Begley CJ, Ogletree ML, Meyrick BO, Brigham KL (1984) Modification of pulmonary response to endotoxemia in awake sheep by steroidal and nonsteroidal anti inflammatory agents. Amer. Rev. Resp. Dis. 130: 1140 - 1146

Beutler B, Krochin N, Milsark IW, Leudke C, Cerami A (1986) Control of cachectin (tumor necrosis factor) synthesis: mechanism of endotoxin resistance. Science, 232: 977-979

Borgeat P, Samuelsson B (1975) Arachidonic acid metabolism in polymorphnuclear leukocytes: Unstable intermediate in formation of dihydroxy acids. Proc Natl Acad Sci USA 76: 2994-2998

Brigham KL, Bowers R, Haynes J (1979) Increased sheep lung vascular permeability caused by Escherichia coli endotoxin. Circ. Res. 45: 292 - 297

Demling RH, Smith M, Gunther R, Flynn JT, Gee MH (1983) Pulmonary injury and prostaglandin production during endotoxemia in conscious sheep. Amer. J. Physiol. 240: H348 - H353

Fleisch JH, Rinkema LE, Haisch KD, Swanson-Bean D, Goodson T, Ho PK, Marschal WS. (1985) LY171883, 1-<2-Hydroxy-3-Propyl-4-(1H-Tetrazol-5-yl) Butoxy>Phenyl>Ethanone,an orally active Leukotriene D_4 Antagonist. J. Pharmacl. Exp. Ther. 233: 148-157

Hagmann W, Denzlinger C, Keppler D (1984) Role of peptide leukotrienes and their hepatobiliary elimination in endotoxin action. Circ. Shock 14: 223 - 235

Heflin ACjr, Brigham KL (1981) Prevention by granulocyte depletion of increased vascular permeability of sheep lung following endotoxemia. J. Clin. Invest. 68: 1253 - 1260

Hock CE, Lefer AM (1985) Protective effects of a new LTD4 antagonist (LY-171883) in traumatic shock. Circ. Shock 17: 263 - 272

McKechnie K., Furman L., Parratt J.R. (1985) Metabolic and cardiovascular effects of endotoxin infusion in conscious unrestrained rats: Effects of methylprednisolone and BW755C. Circ. Shock 15: 205 - 215

Morgan TE (1971) Biosynthesis of pulmonary surface active lipid. Arch. Intern Med. 127: 401 - 407

Ogletree ML (1983) Roles of arachidonic acid metabolites in endotoxin induced pulmonary edema. Semin. Resp. Med. 4: 303 - 307

Ogletree ML, Brigham KL (1982) Effects of cylooxygenase inhibitors on pulmonary vascular response to endotoxin in unanesthetized sheep. Prostagl. Leuk. Med. 8: 489 - 502

Piper PJ and Samhoun MN (1981) The mechanism of action of leukotienes C_4 and D_4 in guinea pig isolated perfused lung and parenchymal strips of guinea pig, rabbit and rat. Prostaglandins 21: 793-803

Puustinen T, Uotila P (1984) Thromboxane formation in human polymorphonuclear leukocytes is inhibited by prednisolone and stimulated by leukotrienes B_4, C_4, D_4,and histamine. Prostaglandins Leukotrienes Med 14: 161-167

Schuette AH, Hüttemeier PC, Hill RD, Watkins WD, Wonders TR, Kong D, Zapol WM (1984) Regional blood flow and pulmonary thromboxane release after sublethal endotoxin infusion in sheep. Surgery 95: 444 - 453

Anti-Endotoxin Mechanisms and Treatment

Bacterial Endotoxins: Pathophysiological Effects, Clinical Significance, and Pharmacological Control, pages 295–308

HYPERSENSITIVITY TO ENDOTOXIN AND MECHANISMS OF HOST-RESPONSE.

C. Galanos, M.A. Freudenberg, M. Matsuura, A. Coumbos.

Max-Planck-Institut für Immunbiologie, Stübeweg 51, 7800 Freiburg, F.R.G.

INTRODUCTION

Endotoxins are factors of pathogenicity present in gram-negative bacteria. Administration of endotoxin to experimental animals leads to the development of pathophysiological activities (fever, leucopenia and leucocytosis hemodynamic changes and lethal shock) which are also known to develop during infections with gram-negative bacteria. Endotoxins also exhibit a number of beneficial effects like adjuvant activity, stimulation of antibody producing cells, induction of protection to infection and to their own toxic effects as well as anti-tumour activity.

Endotoxins are chemically lipopolysaccharides (LPS) consisting of a sugar part, the polysaccharide, linked covalently to a lipid portion, the lipid A.

The toxic properties of LPS are expressed by the lipid A which is structurally similar and serologically cross-reacting among many gram-negative bacteria. (For reviews see Galanos et al., 1977 and Rietschel et al., 1982).

The lethal activity of LPS may be enhanced in a variety of experimental systems. It will be shown here that sensitization to endotoxin proceeds also during gram-negative infection. We believe that this is a general phenomenon and that the manifestation of endotoxin lethality resulting from gram-negative infections always takes place in a state of hypersensitivity which the host acquires during the infection. Experimental models of sensitization may help to understand the mechanisms leading to hypersensitivity. A number of sensitization models developed recently in our laboratory, which we believe to be relevant to natural sensitization will be presented here. Regarding the mechanism of host of responses to endotoxin, it will be shown that macrophages are the cells mediating endotoxin lethality and that the direct interaction of LPS with macrophages is the first step in the triggering of endotoxicity. Finally, evidence will be presented that tumour necrosis factor (TNF) a macrophage product inducible by LPS is a mediator of the lethal activity of endotoxin.

RESULTS AND DISCUSSION

Muramyl dipeptide (MDP) sensitization to the lethal activity of endotoxin

When mice are administered 100 µg of MDP intravenously they show an enhanced reactivity to a subsequent challenge with LPS. Hypersensitivity appears as two timely distinct toxic manifestations, an early and a late phase toxicity (Takada H and Galanos C, 1987).

The late phase toxicity is seen with small amounts (1 to 5 µg) of endotoxically active LPS. S- and R-form LPS, as well as free lipid A, are all highly potent in inducing late phase toxicity in MDP-treated mice. Lethality resulting from late phase toxicity commences usually 15h after LPS injection and is completed about 72h later, the time-course of lethality being similar to that induced in normal mice. Late phase lethality proceeds in endotoxin sensitive mice and is absent in endotoxin resistant C3H/HeJ mice. Tab. 1.

Tab. 1 Induction of early- and late-phase toxicity by S-and R-form LPS in MDP-treated C57Bl/6 and C3H/HeJ mice.

		C57Bl/6 deaths/total in phase:		C57Bl/6 deaths/total in phase:	
MDP (µg)	LPS (µg)	early	late	early	late
100	100	6/6	*	6/6	*
100	10	0/6	6/6	1/6	0/6
100	100	0/6	6/6	0/6	0/6
100	10	0/6	1/6	0/6	0/6

S-form LPS = S.abortus equi; R-form LPS = S.minnesota R595.
The LD_{50} of the LPS of S.abortus equi in normal C57Bl/6 mice is approx. 150 µg that of the R595 approx. 600 µg.
*No survivals from early-phase toxicity.

The early phase toxicity is obtained in MDP treated mice following administration of S-form LPS in amounts of 50-100 µg. The reaction may develop within minutes of LPS injection and lethality is usually completed within 1h. It is characterized by severe convulsions, unconsciousness and lethality. The induction, of this anaphylactic-like reaction is a property not shared by all S-form LPS. Only a few of a large number of LPS preparations tested so far were found to exhibit this activity. Among these, interestingly enough, was the LPS of Bacteroides gingivalis, a preparation of very low toxicity compared to LPS of Salmonella or E.coli.

The most important features of the early phase toxicity are that it is not inducible by R-form LPS or free lipid A, even when administered in mg amounts. Further, early phase toxicity is inducible also in MDP-treated endotoxin resistant, C3H/HeJ mice.

Analysis of the above findings leads to the conclusion that the early- and late-phase toxicities seen in MDP treated mice after endotoxin challenge are timely distinct toxic manifestations based on different mechanisms. Late-phase toxicity represents an enhanced reactivity of the host to the toxic activity of LPS, i.e. a sensitization to the lethal effects of lipid A. The absence of late-phase toxicity in endotoxin resistant mice is in agreement with this hypothesis. In contrast, the early-phase toxicity does not seem to be primarily a lipid A activity, because it is not inducible by R-form LPS or lipid A. This is supported further by the finding that early phase toxicity is inducible also by non-toxic LPS preparations and that it proceeds in endotoxin resistant mice.

Therefore the development of this anaphylactic-like reaction in MDP treated mice is not dependent on the endotoxic properties of the LPS. Other factors may be involved, such as an immune-mediated sensitization to the polysaccharide or to very small amounts of protein that may be present in the LPS.

A role of the physicochemical properties of the LPS in MDP sensitization may also not be excluded as shown by (Takada H and Galanos C. 1987).

Protein-dependent sensitization

Mice administered bacterial protein and challenged 7 to 14 days later with the same protein, become sensitized to the lethal effects of LPS Tab. 2. When high amounts of LPS are used (e.g. 100 µg), an anaphylactic-like reaction occurs, the animals dying within 10 to 20 min. of injection. The sensitization is obtained with different proteins including non-bacterial proteins such as bovine serum albumin. Sensitization is not seen however, when, for the second protein treatment a different protein is used,

Tab. 2 Lethal toxicity of LPS/bacterial protein in protein-sensitized NMR/mice.

Sensitization	LPS + Protein (μg, i.v.)		Lethality %
100 μg Protein subcutaneously on day -7	25	100	100
	12.5	100	80
	none	100	0
	25	none	0
none	25	100	0

suggesting that the induction of sensitization is related to the antigenic specificity of the protein. This is supported by the finding that serum from protein-primed mice or specific anti-protein antiserum may substitute for the first protein injection. Thus mice receiving antiserum to a protein and challenged with the same protein up to several hours later become hypersensitive to endotoxin. This suggests that anti-protein antibodies may be mediating the above sensitization. It should be mentioned however that the sensitizing activity of a given antiserum is not related to its anti-protein antibody titre. This suggests that a given class of antibody may be involved or that the sensitizing effect of the antiserum is related to other, non-antibody serum factors which however must be specific for the protein employed.

The protein sensitizaton to endotoxin and generation of anaphylactic-like reaction also proceeds in endotoxin resistant C3H/HeJ mice.

D-galactosamine sensitization

Treatment of different experimental animals with D-galactosamine, a known hepatotoxic agent, increases their sensitivity to the lethal activity of endotoxin many-thousand-fold (Galanos C, Freudenberg MA and Reutter W, 1979) (Tab. 3). The duration of sensitization is short, being highest during the first hour after galactosamine administration. Thereafter it decreases and is no longer present 4h after D-galactosamine treatment.

Administration of D-galactosamine, leads to a depletion of UTP and consequent inhibition of RNA synthesis in hepatocytes. These early biochemical alterations in the liver are prerequisite for the development of sensitization because their reversion by uridine inhibits the development of sensitization to endotoxin.

Tab. 3 Sensitization to LPS by D-galactosamine in C57Bl/6 mice

D-galactosamine (mg)	LPS (µg)	Lethality %
18	0.01	100
18	0.001	80
18	-	0
-	50	0

D-galactosamine and LPS injected together i.p.

Although D-galactosamine is a hepatotoxic agent causing necrosis of hepatocytes, these effects are usually seen after repeated administration of the amino sugar. A single dose of D-galactosamine, as used here for sensitizing to endotoxin, leads to only minimal histological alterations and only in some of the animals treated. Moreover, such histological alterations when present are detectable at least 24h after administration of D-galactosamine, i.e. at a time when the state of hyperreactivity to endotoxin has long subsided. Therefore it becomes evident that the sensitization to endotoxin by D-galactosamine is not related to a histologically damaged liver but to the early biochemical effects of the hexosamine.

Sensitization by growing tumours

Sensitization to endotoxin was found to develop in mice bearing certain types of tumour (Bartoleyns J, Freudenberg MA and Galanos C, 1987). Two such tumours, a metastatic Lewis lung carcinoma (LLC) growing intramuscularly in C57Bl/6 mice and a EMT6 sarcoma growing subcutaneously in BALB/C mice were found to cause a steady increase in the susceptibility of the animals to the lethal activity of LPS. The time-course of sensiti zation is shown in Fig. 1. An increase of the susceptibility of the animals to endotoxin is detectable already on day 3 after tumour inoculation, increasing thereafter and reaching maximum by day 15. At this time a sensitization factor of 10,000 and 200 for the LLC and EMT6 bearing mice respectively is obtained.

The property to enhance the lethal activity of LPS is not shared by all tumours. Thus, a EL4 lymphosarcoma growing in C57Bl/6 mice and a methyl cholanthren induced fibrosarcoma in BALB/C mice failed to induce any significant sensitization to endotoxin.

Since the strain of mice (C57Bl/6 and BALB/C) in which these two tumours are growing are the same as those for the LLC carcinoma and EMT6 sarcoma, it becomes evident that the induction of sensitization to endotoxin by growing tumours is related primarily to properties of the tumour rather than of the host.

Sensitization to the lethal Effects of LPS by growing Tumours.

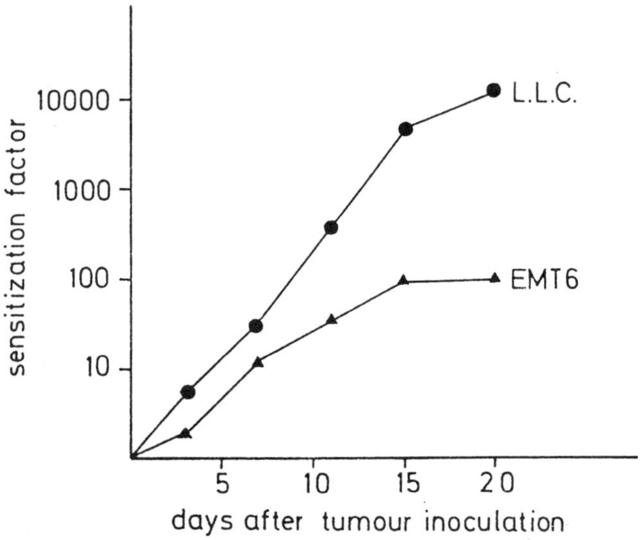

Induction of hyperreactivity to endotoxin by gram-negative infection

Hyperreactivity to endotoxin may be established during gram-negative infections in experimental animals. The property to induce sensitization has been demonstrated so far for representative pathogens of 3 different genera, C.burnetii, K.pneumoniae and S.typhimurium and may therefore be a property shared widely by at least pathogenic gram-negative bacteria in general.

Coxiella burnetii sensitization (Schramek, S et al., 1984): Mice administered live or killed C.burnetii bacteria show enhanced susceptibility to the lethal activity of endotoxin. Sensitization may be induced by both phase I or phase II cells, but it is more pronounced with the former. The sensitization lasts for a long time (several months), and is accompanied by a marked hepato-

Tab. 4 Induction of sensitization to endotoxin and hepato-splenomegaly in Dobra Voda mice infected with C.burnetii phase I.

Infection	LPS treatment on day	LPS LD_{50} (µg)	Average wt. spleen (mg)	Liver (g)
C.burnetii phase I 10^6 E LD_{50}	7	5	650	3.2
	14	3	1100	4.1
none		?500	340	1.9

LPS (S.abortus equi) was administered i.p.

splenomegaly Tab. 4. The property of C.burnetii phase II to induce sensitization to endotoxin also proceeds in endotoxin-resistant C3H/HeJ mice. Also here the sensitization is accompanied by a marked hepatosplenomegaly.

Salmonella typhimurium sensitization: Lethal toxicity tests in C57Bl/6 mice infected with S.typhimurium revealed that the sensitivity of mice to the lethal activity of endotoxin increases progressively up to the time when the animals succumb to the fatal effects of the infection (Krajewska D and Galanos C, unpublished). Already on day 2 after infection with 30 micro-organisms a detectable increase in sensitivity to endotoxin is present. It continues to increase on day 3, 4 and 5, by which time a precise measurement of the toxicity of LPS becomes difficult because lethality due to LPS overlaps with the lethality caused by infection, which just begins at this time.

Recently we adopted a different infection model using C3H/Tif mice. These are more resistant to S.typhimurium infection than the C57Bl/6 used above. In these animals lethality begins about 2 weeks after infection which allows sensitization to endotoxin to be better evaluated. With this model it could also be shown that hyperreactivity to endotoxin may also be established with sub-lethal infection. This is shown in Fig. 2. C3H/TifF mice were infected with $2x10^4$ viable S.typhimurium cells and at different times afterwards challenged with S.abortus equi LPS i.p. While no lethality occurred in the controls, in infected animals the toxicity of LPS increased. Maximum sensitization was on day 5-6 where the LD_{50} of the LPS was less than 1 µg (LD_{50} in normal mice = 150 µg). The toxicity of TNF also depicted in Fig. 2 is discussed below in a following chapter.

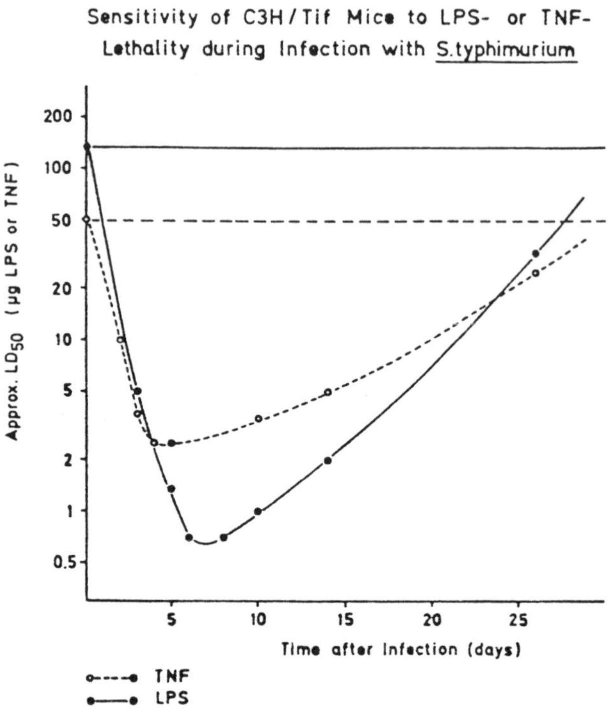

MECHANISMS OF HOST-RESPONSE TO ENDOTOXIN

Today it is generally accepted that the pathophysiological activities of LPS are not induced by LPS directly but through the action of endogenous mediators which are produced following interaction of LPS with cellular and humoral targets.

In the past it was shown that cells of the lymphoreticular system mediate the toxic activity of LPS (Rosenstreich DL et al., 1977). It has also been shown indirectly that tumour necrosis factor (TNF), a mediator produced by macrophages upon stimulation with LPS is responsible for the lethal toxicity of LPS (Beutler B, Milsark JW and Cerami AC, 1985).

Here, direct evidence will be presented that macrophages mediate the lethal activity of endotoxin. In direct toxicity tests it will be shown further that TNF is a mediator initiating the lethal activity of endotoxin.

Macrophages, the mediators of endotoxin lethality

In the previous chapters it was shown that when mice are treated with D-galactosamine their sensitivity to the lethal activity of endotoxin is drastically increased. In contrast, a similar treatment of endotoxin resistant C3H/HeJ mice did not enhance their susceptibility to endotoxin to any comparable extent. It could be shown further that, in C3H/HeJ mice, the early biochemical effects of D-galactosamine which are pre requisite for the development of sensitization (hepatocellular UTP depletion), were identical to those obtained in endotoxin sensitive mice. This suggested that C3H/HeJ mice are also sensitized by D-galactosamine, the sensitization however does not become evident because these mice lack the targets of endotoxin action.

The presence of sensitization to LPS by D-galactosamine became evident when endotoxin sensitive macrophages were administered.

Tab. 5 Lethal toxicity of C3H/HeN macrophages treated with LPS in vitro in D-galactosamine sensitized C3H/HeJ mice.

LPS/2 x 10^7 macrophages µg	Lethality %
0.5 (0.05)	100
0.1 (*)	80
0.02 (*)	60
0.004	0
control macrophages	0

Pure cultured macrophages derived from bone-marrow precursors of C3H/HeN mice were incubated with different amounts of S.abortus equi LPS at 37°C/90 min. The cells were washed and $2x10^7$ macrophages transferred into D-galactosamine-treated (20 mg/mouse) C3H/HeJ mice, i.v. Animals receiving $2x10^7$ macrophages treated under identical conditions but without LPS served as controls.

Values in parenthesis are µg of LPS actually associated with macrophages after incubation and washing. This was measured using ^{14}C LPS.

* = not detectable.

Transfer of $2x10^7$ cultured macrophages from endotoxin sensitive C3H/HeN mice, rendered C3H/HeJ animals highly susceptible to the lethal activity of sub-microgram amounts of LPS (Freudenberg MA, Keppler D and Galanos C, 1986).

LPS lethality was also achieved by macrophages that had been pre-activated with LPS in vitro. Thus C3H/HeN macrophages incubated in vitro with LPS, washed and transferred into D-galactosamine treated C3H/HeJ mice, were highly lethal without further LPS treatment of the animals (Freudenberg MA and Galanos C, in preparation) (Tab. 5).

The above findings allow the following conclusions to be made: Macrophages are the cells mediating the lethal activity of LPS. The direct interaction of LPS with macrophages is the first step in the initiation of toxicity. In the D-galactosamine model, sensitization to LPS and triggering of lethal toxicity are developed independently and are based on different mechanisms. Therefore, D-galactosamine, by as yet unknown mechanisms involving the liver sensitizes to endotoxin by rendering the host more susceptible to the lethal action of toxic macrophage products.

Tumour necrosis factor, a mediator of endotoxin lethality

Mice treated with Propionibacterium acnes and challenged 7 to 14 days later with LPS contain in their serum large amounts of TNF (Carswell E et al., 1975). Such TNF-rich sera were found to be highly lethal for D-galactosamine-treated C3H/HeJ mice (Coumbos, Freudenberg and Galanos, in preparation).

The lethal activity of TNF has been demonstrated directly using human recombinant TNF (rTNF) obtained from BASF/Knoll, Ludwigshafen, FRG (Lehmann V, Freudenberg MA and Galanos C, 1987). Tab. 6, compares the toxicity of LPS and rTNF in normal and D-galactosamine-treated endotoxin-sensitive and -resistant mice. The toxicity of rTNF which in normal mice is of the order of a few hundred micrograms is considerably enhanced by D-galactosamine, submicrogram amounts being sufficient to cause lethality. Unlike LPS the toxicity of rTNF is expressed equally well in endotoxin resistant mice.

The lethal toxicity of TNF was demonstrated also in a number of other sensitization models where it could be shown that enhancement of LPS toxicity was always paralleled by an enhancement of the toxicity of TNF. Sensitization to endotoxin by P.acnes pretreatment (Coumbos, Freudenberg and Galanos, in preparation) and by S.typhimurium infection (see Fig. 2) (Matsuura and Galanos, in preparation) were found to also sensitize mice to the lethal effects of TNF. Also in the hypersensitivity to endotoxin induced by growing tumours, TNF is a mediator of lethality.

Tab. 6 Lethal toxicity of rTNF or LPS in D-galactosamine-treated endotoxin-sensitive C57Bl/6 and -resistant C3H/HeJ mice

D-GalN (mg)	TNF or LPS (µg)		Lethality (dead/total) C57Bl/6	C3H/HeJ
18	TNF:	1.0	6/6	6/6
18		0.1	3/6	4/6
18		0.01	0/6	1/6
18		-	0/6	0/6
-		150	0/6	0/6
-		250	3/6	1/4
-		500	6/6	3/4
18	LPS:	0.01	6/6	0/6
18		500.0	ND	0/6
-		100.0	3/6	0/6

Groups of mice received D-galactosamine and TNF or LPS as mixture intraperitoneally in 0.5ml PBS. Controls received, D-galactosamine, TNF, or LPS alone.
ND = not determined

This is shown in Tab. 7 which compares the toxicity of TNF and LPS in LLC and EMT6 tumour-bearing mice on day 15 after tumour inoculation. Also the time-course of sensitization to endotoxin by the above tumours (Fig. 1) is identical to that obtained with TNF (not shown).

Induction of tolerance to endotoxin

As shown in the previous chapters, in the D-galactosamine sensitization model, mice administered LPS together with, or up to 3 h after D-galactosamine become hypersensitive to endotoxin. It was recently found that administration of LPS 1 to 24 h before D-galactosamine induces tolerance to a subsequent lethal dose of LPS and D-galactosamine. The tolerogenic activity of LPS, like its lethal activity, is mediated directly by macrophages (Freudenberg and Galanos, in preparation).

Tab. 8 shows that mice pretreated with LPS become tolerant also to TNF. Vice versa pretreatment with TNF induces tolerance to LPS and TNF lethality i.e. LPS and TNF are cross-reacting in their tolerance-inducing properties.

Tab. 7 Induction of hypersensitivity to endotoxin by growing tumors.

	LD_{50} (µg/mouse)	
Mouse	LPS	TNF
Normal C57Bl/6	400	300
C57Bl with carcinoma (15 day tumor)	0.1	0.01
Normal Balb/C	300	300
Balb/C with sarcoma (15 day tumor)	2	4

Female C57Black/6 and BALB/C mice were inoculated respectively with 3LL cells intramuscularly or EMT6 cells subcutaneously. Fifteen days later the animals were challenged intraperitoneally with increasing doses of recombinant TNF or of Salmonella a. equi LPS. Each group was formed of 6 animals receiving increasing doses of the test compound. The mortality was recorded during 48 h.

Tab. 8 Induction of tolerance by LPS or rTNF to the lethal toxicity of rTNF and LPS in D-galactosamine-treated C57Bl/6 mice.

Pretreatment with:		Lethality with:	
LPS	TNF	TNF (1 µg)	LPS(0,1µg)
0.10	-	0/6	0/6
0.01	-	0/6	n.d.
-	4.0	0/6	0/6
-	1.0	3/30	7/17
-	0.1	10/16	6/9
-	0.01	5/6	6/6
control (Pi/NaCl pretreatment)		10/10	10/10

Mice were pretreated intravenously with different amounts of LPS or TNF and 4 h later were challenged with a mixture of D-galactosamine (20 µg) and TNF (1 µg) or D-galactosamine (20 µg) and LPS (0.1 mg) intraperitoneally.

SUMMARY AND CONCLUSIONS

It was shown here that the lethal activity of endotoxin may be considerably enhanced in experimental animals treated with different agents. Some of these agents represent killed bacteria, bacterial products or hepatotoxic agents and their sensitizing effects may be relevant to the sensitization seen during gram-negative infections.

This study provided direct evidence that macrophages are cells mediating the lethal activity of endotoxin and that the direct interaction of LPS with macrophages is the first step in its initiation. Further, TNF is a macrophage mediator responsible for the induction of lethality.

In normal animals as well as in the various sensitization models investigated, TNF alone was sufficient to induce lethality. Therefore, the stimulation of macrophages by LPS to produce TNF is a mechanism common to diverse lethality models. It would seem therefore that the complexity of endotoxicity in gram-negative infections is related primarily to the complexicity of the factors altering the susceptibility of the host rather than to be actual mechanisms of the lethal action of endotoxin.

The identification of the factors leading to sensitization and their underlying mechanisms would contribute to a better understanding of the complex phenomenon endotoxicity.

REFERENCES

Galanos C, Lüderitz O, Rietschel ETh, Westphal O (1977) Newer aspects of the chemistry and biology of bacterial lipopolysaccharides, with special reference to their lipid A component. In Intern Rev of Biochem, Biochemistry of Lipids II 14: 239-335. Goodwin TW (ed), Univ Park Press, Baltimore, London, Tokyo

Rietschel ETh, Galanos C, Lüderitz O, Westphal O (1982) The chemistry and biology of lipopolysaccharides and their lipid A component. In Immunopharmacology and the regulation of leukocyte function 183:229, Webb DR (ed), Dekker M, New York.

Galanos C, Freudenberg MA, Reutter W (1979) Galactosamine-induced sensitization to the lethal effects of endotoxin. Proc Natl Acad Sci USA 76: 5939-5943.

Freudenberg M, Keppler D, Galanos C (1986) Requirement for lipopolysaccharide-responsive macrophages in D-galactosamine-induced sensitization to endotoxin. Infect Immun 51: 891-895.

Rosenstreich DL, Glode LM, Wahl LM, Sandberg AL, Mergenhagen SE (1977) Analysis of the cellular defects of endotoxin-unresponsive C3H/HeJ mice. pp. 314-320. In Schlessinger D (ed), Microbiology. Am Soc Microbiol, Washington, DC.

Beutler B, Milsark IW, Cerami AC (1985) Passive immunization against Cachectin/Tumor necrosis factor protects mice from lethal effect of endotoxin. Sci 229.

Carswell E, Old L, Kassel R, Green S, Fiore N, Williamson B (1975) An endotoxin-induced serum factor that causes necrosis of tumors. Proc Nat Acad Sci USA 72: 3666-3670.

Lehman V, Freudenberg MA, Galanos C (1987) Lipopolysaccharide and TNF express similar lethal toxicity in D-galactosamine-treated mice. J Exp Med 165: 657-663.

Bartoleyns J, Freudenberg MA, Galanos C (1987) Growing tumors induce hypersensitivity to endotoxin and tumor necrosis factor. Infect Immun 55: 2230-2233.

Schramek S, Kazar J, Sekeyova Z, Freudenberg MA, Galanos C (1984) Induction of hyperreactivity to endotoxin in mice by Coxiella burnetii. Infect Immun 45: 713-717.

Bacterial Endotoxins: Pathophysiological Effects, Clinical Significance, and Pharmacological Control, pages 309–318

INTERACTIONS OF BACTERIAL LIPOPOLYSACCHARIDES WITH SERUM PROTEINS

Richard J. Ulevitch and Peter S. Tobias

Department of Immunology, Scripps Clinic and Research Foundation, 10666 North Torrey Pines Road, La Jolla, California 92037

During gram-negative sepsis it is hypothesized that the endotoxins (LPS) from the invading organisms are released from outer membranes and initially gain entry into the vascular compartment of the host. Sufficient examples are to be found in the literature and from presentations at this meeting to support this concept, although unambiguous identification of LPS in the blood of patients by specific chemical or immunologic procedures is lacking. Nevertheless, the rationale is good for studying the interaction of purified LPS with cells and plasma proteins in animal models and *in vitro* systems to give insight into the pathogenic mechanisms of LPS-induced injury. During the past several years our understanding of the interactions of LPS *in vivo* has moved from the phenomenological to a level where the molecules involved have been identified.

A central focus of our laboratory has been in the interactions of LPS with serum proteins and particularly with lipoproteins. The use of density gradient ultracentrifugation of LPS/serum mixtures incorporating radiolabeled LPS as a tracer provides a quantitative method to determine serum effects of LPS. This technique is useful since most LPS preparations have buoyant densities in the range of 1.4 g/cm^3 and after binding to HDL the LPS is converted to a form with a density of less than 1.2 g/cm^3 (Munford et al, 1981a, 1981b; Ulevitch et al, 1978, 1979). The use of this technique yielded observations that led to a greater understanding of the mechanism of LPS binding to HDL, and this technique continues to provide data for insights into the mechanisms of LPS interactions with plasma proteins.

In studies performed in our laboratory (Ulevitch et al, 1978, 1979, 1981), we noted that after injection of LPS into rabbits or mixing LPS with serum *in vitro*, the buoyant density of LPS decreased. Our observations *in vivo* were obtained with LPS from *Escherichia coli* 0111:B4 radioiodinated (Ulevitch, 1978) and then injected into rabbits (Ulevitch et al, 1978). LPS/plasma mixture from these experiments was subjected to equilibrium density gradient ultracentrifugation in CsCl. The data derived from analysis of LPS/plasma mixtures indicated that the LPS in plasma rapidly undergoes changes in buoyant density. The *E. coli* 0111:B4 LPS is partitioned into at least two forms with a density (d) of approximately 1.4 g/cm^3 and a density of less than 1.2g/cm^3. The LPS at $d \cong 1.4$ g/cm^3 is either removed from the blood or converted to the $d<1.2$ g/cm^3 form so by 30 minutes more than 90% of the LPS in plasma is at $d<1.2$ g/cm^3. This form of LPS ($d<1.2$ g/cm^3) disappears with a $t_{\frac{1}{2}}$ of 10-12 hours. Similar findings have been reported by Munford et al (1981a, 1982) with *Salmonella typhimurium* LPS in studies performed in rats and squirrel monkeys.

Analysis of the mechanism of the change in density of LPS was aided by the observation that this change occurs *in vitro* when LPS is mixed with plasma or serum. Data obtained with LPS from *E. coli* 0111:B4, *E. coli* 0111:B4:55, *Salmonella minnesota* R595, *Salmonella typhimurium*, and *Salmonella abortus equi*, and with serum from several different species including man, rabbit, mouse and rat reveal qualitatively similar changes in the density of LPS. The addition of serum to LPS produces a time-dependent, small decrease in the density of the parent LPS accompanied by disaggregation of the LPS, followed by the appearance of LPS at $d<1.2$ g/cm^3. The rate of density change to $d<1.2$ g/cm^3 depends on temperature, LPS concentration, and varies with the type of LPS and serum. For some LPS the presence of divalent ions inhibits the reduction in LPS density, although this can be overcome by the addition of EDTA to serum or prior treatment of LPS by electrodialysis (Galanos and Lüderitz, 1976) or chelex resin treatment (Strain et al, 1983; Ulevitch et al, unpublished data). A mechanism to explain the serum-induced decrease in LPS density must include at least two steps as shown below.

Step 1(disaggregation): LPS(aggregated) $\rightleftharpoons$ LPS(disaggregated)
Step 2(binding to HDL): LPS(disaggregated + HDL $\rightarrow$ LPS-HDL
$d<1.2 g/cm^3$

EVIDENCE FOR DISAGGREGATION OF LPS IN SERUM

Several observations provide support for the concept that disaggregation of LPS is produced by components of lipid-free serum. If serum is depleted of lipids by extraction (Cham and Knowles, 1976), or by ultracentrifugation (Munford et al, 1981b), this lipid-depleted serum does not decrease the density of LPS to less than 1.2 g/cm^3. However, exposure of LPS to delipidated serum results in a change in the behavior of LPS in sucrose gradients that is consistent with a decrease in the particle size of LPS aggregates. The identity of the serum proteins responsible for LPS disaggregation is unknown. Johnson et al (1977) described an α-globulin component of LPS disaggregation. The α-globulin is 4.5S, is not a lipoprotein, and is not blocked by inhibitors of serine esterases. The role of this protein in disaggregation of LPS and subsequent binding of LPS to HDL needs to be explored further.

In our studies (Ulevitch et al, 1979) as well as those of Munford et al (1981b), it was demonstrated that addition of LPS (in the absence of delipidated serum) to HDL did not result in binding of LPS to HDL. Munford et al also showed that treatment of LPS with sodium deoxycholate enhanced the association of LPS with HDL (Munford et al, 1981b). Since deoxycholate reduces the size of LPS by disaggregation, this finding adds strength to the contention that a reduction in the particle size of LPS is required for binding to HDL. It is possible that reduction in the particle size of LPS aggregates by removal of cations, which can be accomplished by a number of techniques including electrodialysis (Galanos and Luderitz, 1976), chelex resin treatment (Strain et al, 1983) or a combination of these treatments, might facilitate the binding of LPS to HDL in the absence of other serum components. This question deserves further analysis.

THE ROLE OF HDL IN ALTERATIONS OF LPS BY SERUM

The first evidence that HDL is involved in the reduction of LPS density in serum was obtained in experiments reported by Ulevitch et al (1979). Delipidated serum was mixed with purified lipoproteins: very low density lipoproteins (VLDL), low density lipoproteins (LDL) or HDL, and this "reconstituted" serum was tested for the ability to reduce the density of LPS. Studies using LPS from S. minnesota Re595 or E.coli 0111:B4 showed that HDL and not other lipoproteins resonstituted the delipidated serum. Munford et al (1981b) also showed that purified rat HDL added back to lipoprotein-depleted rat serum reconstituted lipoprotein-deficient serum to lower the density of LPS.

Other evidence that LPS at d<1.2 g/cm^3 is a complex of disaggregated LPS and HDL was obtained by Ulevitch et al (1981) using immunoaffinity chromatography. Antibody to S. minnesota Re595 LPS was immunopurified, linked to Sepharose 4B, and used to isolate LPS at d<1.2 g/cm^3 from LPS/serum mixtures. The properties of the bound fraction are summarized below. This fraction contained LPS with a d<1/2 g/cm^3 when centrifuged in CsCl gradients and contained protein and lipid components in addition to LPS. This SDS-PAGE demonstrated that a polypeptide with an apparent molecular weight of approximately 25.7 kilodaltons is the major protein component. Evidence that this polypeptide is derived from HDL was obtained from amino acid analysis of the 25.7-kilodalton band which showed that the amino acid composition of this 25.7-kilodalton polypeptide of the LPS-HDL complex was identical to that of the apo AI present in rabbit HDL (Ulevitch et al, 1981).

Freudenberg et al (1980) have also noted a role for HDL in interactions of LPS in serum. Using crossed immunoelectrophoresis they showed that the presence of either smooth or rough form LPS changed the electrophoretic mobility of HDL. This alteration in the charge of HDL results from binding of LPS to HDL, with complexes of smooth form LPS and HDL migrating slower, while those of rough form LPS and HDL migrate faster than HDL. These differences may result from the more negative charge of rough form LPS and suggest that the biological properties of LPS-HDL complexes formed with either rough or smooth forms of LPS may differ.

Thus ample evidence exists to support the two-step mechanism shown above and the major role of HDL as a carrier of LPS in serum is well supported by experimental data.

BIOLOGIC PROPERTIES OF THE LPS-HDL COMPLEX

It is known that binding of LPS to HDL results in changes in both the kinetics and specificity of interactions of LPS with blood cells, other plasma proteins and tissues as well as modification of the biologic activities resulting from LPS interactions. Comparisons of the activity of the parent LPS and LPS-HDL may be derived from our data (Mathison and Ulevitch, 1981; Ulevitch et al, 1978, 1979). The summary in Table 1 shows that LPS binding to HDL produces a spectrum of changes in LPS activity ranging from minimal loss to as much as a 1000-fold decrease in activity. The most dramatic changes occur with activities associated with LPS in a highly aggregated state, such as activation of the classical complement pathway. LPS from S. minnesota Re595, a potent activator of the classical complement pathway when bound to HDL, is at least 100-fold less active than the parent LPS. LPS activity known to be dependent on complement activation, such as transient thrombocytopenia observed after an intravenous injection in rabbits, are then inhibited in a similar manner. The ability of LPS to produce a pyrogenic response in rabbits is also markedly diminished as a result of binding of LPS to HDL. In contrast are the lesser effects of the formation of an LPS-HDL complex on a variety of other LPS-dependent activities including mitogenicity in murine B cells, macrophage activation, and the ability to produce hypotension, disseminated intravascular coagulation and death in experimental animals.

TABLE 1: COMPARISON OF THE ENDOTOXIC ACTIVITIES OF PARENT LPS AND LPS-HDL COMPLEXES

	Re595LPS	Re595 LPS-HDL	Relative Activity LPS-HDL vs LPS
Pyrogenicity	50 ng	5000 ng	1/100
Leukopenia	50 ng	50000 ng	1/1000
Immediate, transient thrombocytopenia	10 µg	500 µg	at least 1/50
CH_{50} reduction (50%)	1*	1000*	1/1000
Shock and DIC in rabbits	100-200µg	50-200µg	unchanged
LD_{50} adrenalectomized mice	45 ng	280 ng	1/6

* Relative amount of LPS needed to reduce rabbit CH_{50} level by 50%.

BINDING OF LPS TO HDL IN ACUTE-PHASE SERUM

Because of the effects of complexation of LPS and HDL on the endotoxic properties of LPS it was of interest to study these complexes under conditions where the HDL structure was varied so that LPS-HDL complexes of different compositions could be compared. The serum concentrations of certain acute-phase proteins are known to increase dramatically as a result of infection, after exposure to nonspecific inflammatory substances and following LPS injection. Because of the possibility that acute-phase proteins might regulate the interactions of LPS with serum proteins or cells, we initiated studies to examine the binding of LPS to HDL in acute-phase serum. Consideration was first given to the possibility that one acute-phase serum protein, serum amyloid A (SAA), an apolipoprotein of HDL, might either influence the binding of LPS to SAA-containing HDL particles or the subsequent biologic activity of the LPS-HDL(SAA) complex (Hoffman and Benditt, 1982; Tobias et al, 1982). As noted in recently published studies by Tobias et al (1982), however, evidence to support this hypothesis has not yet been obtained, although these studies are still being pursued.

As part of these studies, Tobias et al (1982) examined the dependence of the formation of the LPS-HDL complex on LPS concentration in acute-phase serum. Careful inspection of these data yielded a striking difference between normal and acute-phase serum. At concentrations of LPS between 1 and 50 ug/ml, the addition of LPS to acute-phase serum resulted in the formation of LPS complexes with a density of 1.3 g/cm^3 (termed complex 1.3) as well as the expected complex of LPS-HDL at d$<$1.2 g/cm^3. In more recent studies, these observations were extended to include a comparison of the kinetics of binding of LPS to HDL in normal and acute-phase serum (Tobias and Ulevitch, 1983).

We have reported that human, murine, and lapine acute phase sera differ functionally from the respective normal sera in at least three ways, insofar as interactions with bacterial LPS are concerned. First, the rate at which LPS binds to HDL is much faster in normal sera than in acute phase sera. For example, in normal sera the half time for binding of _Salmonella minnesota_ Re595 LPS to HDL at 37°C is typically 2-3 min., whereas in acute phase sera, the same reaction can have a half time of up to 100 min. Second, when Re595 LPS is added to serum and spun to equilibrium in a CsCl density gradient, the LPS not bound to HDL is found at a density of 1.33 g/cm^3 in normal sera, whereas in acute phase sera the LPS forms a complex with a density of 1.3g/cm^3. We refer to the form of LPS in acute phase serum as complex 1.3 (C.1.3). Finally, when sera containing LPS not yet bound to HDL are chilled rapidly to 4°C and dialysed against very low ionic strength buffer at 4°C an LPS-protein precipitate is formed from both normal and acute phase sera, but the LPS in the precipitate from normal serum redissolves readily in isotonic saline, while the LPS in the precipitate from acute phase serum does not. These phenomena have been seen after acute phase induction with either subcutaneous silver nitrate, intraperitoneal LPS, or casein (unpublished data). We postulated the existence of a hitherto unrecognized acute phase reactant to explain these findings.

A two-step procedure was developed for the isolation of the acute phase protein that is responsible for the behavior noted above and we have termed this protein lipopolysaccharide binding protein (LBP). This procedure utilizes BioRex-70 and HPLC chromatography on Mono-Q to yield a purified protein that constitutes a 2,300-fold purification with a yield of 10% of the initial LBP activity. While the product is a mixture of two peptides, the peptides appear to be very closely related (Tobias et al, 1986). The two peptides that copurify as LBP have identical partial amino acid sequences, although these are clearly separable by SDS-PAGE. Whether the apparent size difference results from as yet unrevealed differences in amino acid sequence or from differential glycosylation, and whether the apparent size difference reflects any functional differences are all questions that are not now answerable. Given the structural similarity between the two peptides and their recovery in a somewhat variable ratio, it seems unlikely that LBP in serum is a heterodimer of the peptides we purify.

The available amino acid sequence data, when tested for homology against a wide variety of acute phase and normal proteins, do not reveal any clear correlations. Thus, although $<5\%$ of the sequence of LBP is available, LBP does not appear to be homologous to any known normal or acute phase protein. The diagram shown below summarizes our current understanding of the interactions of LPS with HDL and LBP in serum. An initial disaggregation step allows binding of LPS to either LBP or HDL to form C.1.3 or C.1.2, respectively. However, since the final state of LPS in serum is C.1.2, C.1.3 must also dissociate LPS. Whether LPS transfers directly from LBP to HDL or goes via some other intermediate form is unknown.

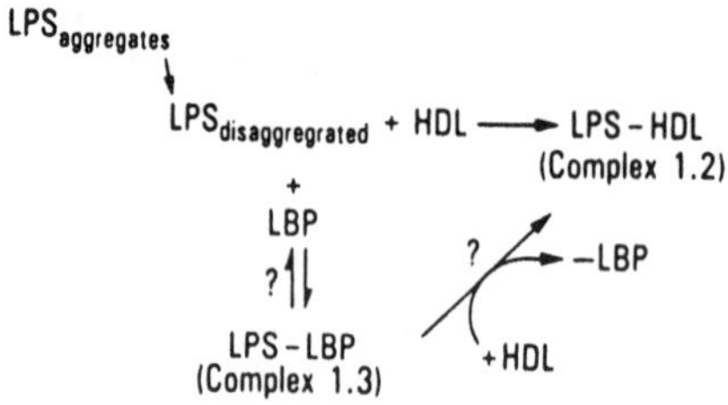

Schematic diagram of LPS interactions wtih HDL and LBP in serum.

Recent evidence suggests that LBP interacts with LPS through Lipid A. This has been obtained from studies using a solid phase binding assay for LBP and ligand inhibition analysis (Tobias and Ulevitch, manuscript in preparation). This LBP appears to be one of the few known proteins with a lipid A binding site. We speculate that LBP probably plays some role in modifying the endotoxic effects of LPS. It is known that binding of LPS to HDL significantly alters the endotoxic activity of LPS in at least three ways. First, the time course of disappearance of a bolus dose of LPS from the circulation *in vivo* is biphasic; uncomplexed LPS and HDL-complexed LPS are removed from plasma with approximate $t_{1/2}$ of 2 min and 12 h, respectively. Second, the endotoxic activities of LPS and LPS-HDL complexes are different; the complexes are less pyrogenic and less able to activate complement, but can still induce lethal hypotensive shock and activate B cells and macrophages. The third consequence of LPS-HDL complexation is the cellular targeting of LPS and LPS-HDL complexes; stated simply, LPS itself is taken up by tissues metabolizing cholesterol, notably the adrenal gland. These observations underscore the point that the biochemistry and endotoxic activity of LPS are subject to modification by plasma proteins in important ways. Complexation of LPS to LBP provides a mechanism to modulate the endotoxic properties of LPS in ways not previously suspected.

Future studies should point out the ways by which LBP modifies the host response to LPS.

REFERENCES

Cham BE, Knowles BR (1976) A solvent system for delipidation of plasma or serum without protein precipitation. J. Lipid Res. 17:176-181.

Freudenberg MA, Bog-Hansen TC, Back U, Galanos C (1980) Interaction of lipopolysaccharides with plasma high density lipoproteins in rats. Infec. Immun. 28:373-380.
Galanos C, Luderitz O (1976) The role of the physical state of lipopolysaccharides in the interaction with complement. Eur. J. Biochem. 65:403-408.
Hoffman JS, Benditt EP (1982) Changes in high density lipoprotein content following endotoxin administration in the mouse. Formation of serum amyloid protein-rich subfractions. J. Biol. Chem. 257:10510-10517.
Johnson KW, Ward PA, Goralnick S, Osborn MJ (1977) Isolation from human serum of an inactivator of bacterial lipopolysaccharides. Am. J. Pathol. 77:559-574
Mathison JC, Ulevitch RJ (1981) In vivo interaction of bacterial lipopolysaccharide (LPS) with rabbit platelets: modulation by C3 and high density lipoproteins. J. Immunol. 126:1575-1580.
Munford RJ, Andersen JM, Dietschy JM (1981a) Sites of tissue binding and uptake in vivo of bacterial lipopolysaccharide-high density lipoprotein complexes. J. Clin. Invest. 68:1503-1513
Munford RJ, Hall CC, Dietschy JM (1981b) Binding of Salmonella typhimurium lipopolysaccharides to rat high density lipoproteins. Infect. Immun. 34:835-843.
Munford RJ, Hall CC, Lipton JM, Dietschy JM (1982) Biological activity, lipoprotein-binding behavior, and in vivo deposition of extracted and native forms of Salmonella typhimurium lipopolysaccharides. J. Clin. Invest. 70:877-888.
Strain SM, Fesik SW, Armitage IM (1983) Characterization of lipopolysaccharide from a heptoseless mutant of Escherichia coli by Carbon 13 nuclear magnetic resonance. J. Biol. Chem. 258:2906-2910.
Tobias PS, Ulevitch RJ (1983) Control of lipopolysaccharide-high density lipoprotein binding by an acute phase protein(s). J. Immunol. 131:1913-1916.
Tobias PS, McAdam KPWJ, Ulevitch RJ (1982) Interactions of bacterial lipopolysaccharide with acute-phase rabbit serum and isolation of two forms of rabbit serum amyloid A. J. Immunol. 128:1420-1427.
Tobias PS, Soldau K, Ulevitch RJ (1986) Isolation of a lipopolysaccharide-binding acute phase reactant from rabbit serum. J. Exp. Med. 164:777-793.
Ulevitch RJ (1978) The preparation and characterization of a radioiodinated bacterial lipopolysaccharide. Immunochemistry 15:157-164.

Ulevitch RJ, Johnston AR (1978) The modification of the biophysical and endotoxic properties of bacterial lipopolysaccharide by serum. J. Clin. Invest. 62:1313-1324.
Ulevitch RJ, Johnston AR, Weinstein DB (1979) New function for high density lipoproteins. Their participation in intravascular reactions of bacterial lipopolysaccharides. J. Clin. Invest. 64:1516-1524.
Ulevitch RJ, Johnston AR, Weinstein DB (1981) New function for high density lipoproteins. Isolation and characterization of a bacterial lipopolysaccharide-high density lipoprotein complex formed in rabbit plasma. J. Clin. Invest. 67:827-837.

Bacterial Endotoxins: Pathophysiological Effects, Clinical Significance, and Pharmacological Control, pages 319–325

IMMUNOLOGICAL INTERVENTION IN ENDOTOXIN SHOCK

Elizabeth J. Ziegler

Division of Infectious Diseases, Department of Medicine, University of California at San Diego, UCSD Medical Center, San Diego, California 92103

INTRODUCTION

Small amounts of endotoxin (LPS) stimulate protective responses in man and animals but large amounts introduced into the circulation - by direct injection, by leaching from the bacterial surface during bacteremia or by release during antibiotic-mediated bacterial lysis - generate harmful reactions which sometimes result in shock and death (Morrison and Ulevitch, 1978). It has been well-established that antibodies to type-specific oligosaccharide LPS side chains protect animals against the toxic effects of LPS and against infection by the same (homologous) strain but not by other (heterologous) strains of gram-negative bacteria. Antibodies raised against oligosaccharide-deficient rough strains of *E. coli*, *S. minnesota* and *S. typhimurium* have afforded heterologous protection against LPS and experimental infections with smooth bacterial strains of Enterobacteriaciae and *Pseudomonas aeruginosa* (Braude and Douglas, 1972; McCabe, 1972; Ziegler et al., 1973; Ziegler et al., 1975; Marks et al., 1982).

The immunologic basis of this protection has been controversial. It has been difficult to remove protective antibody from core LPS sera because the adsorption procedures left considerable amounts of LPS in the sera (Marks et al., 1982); thus adsorbed sera were themselves toxic. Immunization with core LPS is much more effective with whole bacterial vaccines than with purified LPS but it is likely that sera obtained after administration of whole bacterial vaccines contains antibodies to many other bacterial antigens. An additional theoretical problem with the animal protection experiments has been that LPS stimulates many non-antibody responses which could be reflected in the core antisera and which might mediate protection (Morrison and Ulevitch, 1978; Tracey et al, 1986). Monoclonal antibody technology offered an opportunity to demonstrate than antibody directed specifically against LPS determinants was capable of protecting animals against the toxic effects of purified LPS as well as against gram-negative infection (Kirkland and Ziegler, 1984). With respect to the difficulty some investigators have experienced in demonstrating heterologous protection with core LPS antisera (Ng et al., 1976), the nature of the animal model and the amount of antiserum given have been shown to be critical factors (Sakulramrung and Domingue, 1985).

There also have been difficulties in demonstrating cross-reactive antibodies in core LPS antisera in vitro (Siber et al., 1985). However, two recent studies have yielded evidence that there is broadly cross-reactive antibody in these sera. It has been shown that such antisera contain large amounts of type-specific antibody to core determinants which, probably because of its relatively high avidity, obscures less avid cross-reactive antibodies, also present in high titer, unless assays are performed at limiting antibody concentrations. These antibodies are much more efficiently adsorbed from serum by LPS displayed on the whole bacterial surface than by purified LPS in liquid suspension (Baumgartner et al., 1987). Core LPS antisera seem more reactive with LPS from heterologous smooth bacteria extracted in an early phase of growth (3 h) than with LPS from the same strain extracted after more protracted growth (19 h), when O side chain substitution is more nearly complete (McCallus and Norcross, 1987). This observation may explain how core LPS antisera can protect against heterologous smooth strains, namely, by binding to incomplete LPS in rapidly-multiplying organisms.

The foregoing discussion makes it logical to assume that most of the cross-protection observed after passive immunization of animals with core LPS antisera is due to antibody cross-reactive with LPS core determinants, either sugars or lipid A, which are highly conserved among gram-negative bacteria of clinical importance.

CLINICAL STUDIES WITH POLYCLONAL HUMAN CORE-LPS ANTISERUM

The question arises whether immunity to LPS determinants can be applied to clinical gram-negative infections. Passive immunization is more appropriate in the majority of susceptible patients than active vaccination because of time constraints and because patients at risk of severe gram-negative bacteremia often respond poorly to immunization. Therefore, all human studies to date have employed passive transfer of antibody. Most efficacy studies in patients have involved human antiserum to the J5 mutant of E. coli O111, a mutant which lacks UDP-galactose epimerase and also the capability to incorporate exogenous galactose into its LPS. This mutant has been stable and rough for many years.

A randomized controlled clinical trial in bacteremic patients showed that human J5 antiserum lowered the death rate from gram-negative bacteremia in patients treated near the onset of bacteremia compared to pre-immune serum from the same donors (Ziegler et al., 1982). This protection was evident even in patients on pressors for profound endotoxin shock. Attempts to conduct passive prophylaxis trials in neutropenic patients have been hampered by low rates of proven gram-negative infection, probably in large part because of the early use of broad-spectrum empiric antibiotics given when fever begins (McCutchan et al., 1983; Pizzo and Ziegler, unpublished). When J5 plasma was administered prophylactically to abdominal surgery and trauma cases in a randomized controlled fashion, there was no benefit in terms of prevention of gram-negative focal infections or secondary septic episodes from those infections. However, the incidence of gram-negative shock and death was significantly lower in the J5 plasma group than among controls treated with nonimmune (preimmune) plasma (Baumgartner et al., 1985). This study suggests that core LPS antisera are antitoxic rather than antibacterial in their protective activity. This corroborates other laboratory studies which show no participation of J5 antibody in opsonization or complement-mediated bacteriolysis. A recent study of therapeutic J5 immune human IgG to treat established endotoxin shock showed no benefit, however (Calandra, this conference).

Taking the existing information as a whole, one can conclude that if antibody to core LPS is going to be useful clinically, it probably should be applied in treatment of bacteremia or in an anticipatory fashion to hospitalized patients at immediate risk of bacteremia. At least, it is clear that efficacy testing must be performed in these settings. Since there is no antibacterial activity in these sera, antibody probably is not needed until the body's own detoxification mechanisms and antibacterial defenses are near exhaustion and until the patient's own natural and acquired core and type-specific antibacterial antibodies are consumed at the site of local infection.

DEVELOPMENT OF A PROTECTIVE MONOCLONAL ANTIBODY AGAINST LIPID A

The advantages of delivering core LPS antibodies in the form of monoclonal immunoglobulins include a high degree of specificity, ease of production and standardization, and avoidance of the risks of transfusion. Human monoclonals theoretically present less risk of recipient sensitization than monoclonals from other species. With these considerations in mind, our laboratory in conjunction with Henry Kaplan's group at Stanford decided to try to produce human monoclonal antibodies against J5 LPS determinants and to test them for cross-reactivity and protective potency against heterologous smooth LPS and gram-negative infections. Briefly, a patient with Hodgkins disease scheduled for staging splenectomy was immunized with E. coli J5 heat-killed cells by the scheme routinely employed for human vaccinees. At surgery, no tumor was found in the spleen. Splenic B cells were separated from T cells and infected with EB virus. These transformed cells were fused with a heteromyeloma derived from fusions between human myeloma line FU266-El and mouse myeloma line X63-Ag.8653. Hybridomas were screened by immunodot for broad-spectrum cross-reactivity with LPS and lipid A preparations from smooth and rough gram-negative bacteria. The supernatant from one of the hybridomas (A6(H4C5) was broadly reactive and therefore it was chosen for further study. This line, which produces only pentameric IgM (k), has been a stable high-level antibody producer for over four years (Teng et al., 1985). This antibody prevents the dermal Shwartzman reaction from E. coli and K. pneumoniae LPS and protects against lethal bacteremia with E. coli (3 smooth strains), K. pneumoniae, and P. aeruginosa (2 strains) in a murine mucin-enhanced peritonitis model.

(Teng et al., 1985, Ziegler et al., unpublished). When given after the onset of Pseudomonas bacteremia in neutropenic rabbits (Ziegler and Douglas, 1979), 300 ug of this monoclonal antibody (in 1 kg rabbits) is highly protective against bacteremic deaths (0/12 deaths in the antibody group versus 9/12 deaths in controls given nonimmune polyclonal human IgM; p = 0.001, chi-square with Yates correction). In these animals, the monoclonal antibody has a half-life of 17.6 hours. The antibody levels and half-life associated with protection in this rabbit model are, as expected, lower than the levels and half-life in human volunteers given comparable levels of antibody (Ziegler et al., 1987, Khazaeli et al., 1987). In humans there were no adverse reactions to antibody administration and there was no evidence of immunogenicity up to 42 days afterwards. Clinical treatment trials of this and other monoclonal antibodies are about to begin. Such antibodies, if they prove effective, given singly or in combination, will not replace aggressive medical and surgical management of shock and infection. However, they might supplement current treatment by neutralizing the toxicity of LPS before it can activate mediators of endotoxin shock.

REFERENCES

Baumgartner JD, McCutchan JA, Glauser MP, Vogt M, Muehlen E, Ziegler E, van Melle G, Luethy R, Chiolero R, Geroulanos S (1985). Prevention of shock and death in surgical patients by prophylaxis with antibody to core glycolipid. Lancet 2: 59-63.

Baumgartner JD, O'Brien TX, Kirkland TN, Glauser MP, Ziegler EJ (1987). Demonstration of cross-reactive antibodies to smooth gram-negative bacteria in E. coli J5 antiserum. J Infect Dis, in press, July.

Braude AI, Douglas H (1972). Passive immunization against the local Shwartzman reaction. J Immunol 108: 505-512.

Khazaeli M, Kubagawa H, Teng N, Ziegler E, Rogers K, Peterson L, Saleh M, Bogard C, Carrano R, Berger H, and LoBuglio A (1987). Phase I trial and pharmacokinetics of a human IgM monoclonal antibody to endotoxin (LPS). Clin Res 35:615A.

Kirkland TN, Ziegler EJ (1984). An immunoprotective monoclonal antibody to lipopolysaccharide. J Immunol 132: 2590-2592.

Marks MI, Ziegler EJ, Douglas H, Corbeil LB, Braude AI (1982). Induction of immunity against lethal *Haemophilus influenzae* type b infection by *Escherichia coli* core lipopolysaccharide. J Clin Invest 69: 742-749.
McCabe WR (1972). Immunization with R mutants of *S. minnesota*. I. Protection against challenge with heterologous gram-negative bacteria. J Immunol 108: 601-610.
McCallus DE, Norcross NL (1987). Antibody specific for *Escherichia coli* J5 cross-reacts to various degrees with an Escherichia coli clinical isolate grown for different lengths of time. Infect Immun 55: 1042-1046.
McCutchan JA, Wolf JL, Ziegler EJ, Braude AI (1983). Ineffectiveness of single-dose human antiserum to core glycolipid (*E. coli* J5) for prophylaxis of bacteremic gram-negative infections in patients with prolonged neutropenia. Swiss Med J 113(suppl 14): 40-45.
Morrison DC, Ulevitch RJ (1978). The effect of bacterial endotoxins on host mediation systems. Am J Pathol 93: 525-617.
Ng A-K, Chen C-L H, Chang C-M, Nowotny A (1976). Relationship of structure to function in bacterial endotoxins: Serologically cross-reactive components and their effect on protection of mice against some gram-negative infections. J Gen Microbiol 94: 107-116.
Sakulramrung R, Domingue GJ (1985). Cross-reactive immunoprotective antibodies to *Escherichia coli* O111 rough mutant J5. J Infect Dis 151: 995-1004.
Siber GR, Kania SA, Warren HS (1985). Cross-reactivity of rabbit antibodies to lipopolysaccharides of *Esherichia coli* J5 and other gram-negative bacteria. J Infect Dis 152: 954-964.
Teng NNH, Kaplan HS, Hebert JM, Moore C, Douglas H, Wunderlich A, Braude AI (1985). Protection against gram-negative bacteremia and endotoxemia with human monoclonal IgM antibodies. Proc Natl Acad Sci USA 82: 1790-1794.
Tracey KJ, Beutler B, Lowry SF, Merryweather J, Wolpe S, Milsark IW, Hariri RJ, Fahey TJ III, Zentella A, Albert JD, Shires GT, Cerami A (1986). Shock and tissue injury induced by recombinant human cachectin. Science 234: 470-474.
Ziegler EJ, Douglas H, Sherman JE, Davis CE, Braude AI (1973). Treatment of *E. coli* and *Klebsiella* bacteremia in agranulocytic animals with antiserum to a UDP-Gal epimerase-deficient mutant. J Immunol 111: 433-438.
Ziegler EJ, McCutchan JA, Douglas H, Braude AI. Prevention of lethal *Pseudomonas* bacteremia with epimerase-deficient

E. coli antiserum (1975). Trans Assoc Am Physicians 88: 101-108.

Ziegler EJ, Douglas H. Pseudomonas vasculitis and bacteremia following conjunctivitis; A simple model of fatal Pseudomonas infection in neutropenia (1979). J Infect Dis 139: 288-296.

Ziegler EJ, McCutchan JA, Fierer J, Glauser MP, Sadoff JC, Douglas H, Braude AI (1982). Treatment of gram-negative bacteremia and shock with human antiserum to a mutant Escherichia coli. New Engl J Med 307: 1225-1230.

Ziegler EJ, Teng NNH, Douglas H, Wunderlich A, Berger HJ, Bolmer SD (1987). Treatment of Pseudomonas bacteremia in neutropenic rabbits with human monoclonal IgM antibody against E. coli lipid A. Clin Res 35: 619A.

Bacterial Endotoxins: Pathophysiological Effects, Clinical Significance, and Pharmacological Control, pages 327–338

MONOCLONAL ANTIBODIES THAT RECOGNIZE EPITOPES IN THE CORE AND LIPID A REGION OF LIPOPOLYSACCHARIDES

Matthew Pollack, John K.S. Chia, Nancy L. Koles, Maryalice Miller and Gretchen Guelde

Department of Medicine, F. Edward Hebert School of Medicine, Uniformed Services University of the Health Sciences, Bethesda, Maryland 20814

It has been proposed that broadly cross-reactive antibodies exist which recognize conserved epitopes in the core and lipid A region of lipopolysaccharides (LPS) from phylogenetically diverse Gram negative bacteria (Chedid et al., 1968). This hypothesis predicts that because these structures represent the toxic moieties of LPS, antibodies reactive with the core oligosaccharide or lipid A would possess antiendotoxic properties. Moreover, it is argued, assuming LPS (or endotoxin) plays an important pathogenic role in Gram negative sepsis, endotoxin-neutralizing antibodies might be expected to exert an important protective function in the septic host (McCabe, 1972; Ziegler et al., 1973).

Our laboratory has attempted to evaluate the immunochemical basis for the above hypothesis by preparing monoclonal antibodies (mAbs) that recognize distinct epitopes in the core and lipid A regions of enterobacterial LPS, and by defining their specificity, cross-reactivity, and functional properties. Reported here are the preliminary results of such studies.

MATERIALS AND METHODS

<u>LPS, lipid A</u>. Rough and smooth LPS purified by the methods of Galanos (Galanos et al., 1969) and Westphal (Westphal and Jann, 1965), respectively, and further purified chromatographically, were obtained from List Biological Laboratories, Campbell, CA. Lipid A prepared by acid

hydrolysis of Re mutant LPS was obtained from List, or from Ribi Immunochem Research, Inc., Hamilton, MT. Monophosphoryl lipid A derived from *S. minnesota* R595 LPS was obtained from Ribi. The lipid A precursor compound designated lipid X (Raetz, 1984) was purchased from Lipidex, Middleton, WI. Synthetic *E. coli* lipid A (LA-15-PP, or compound 506) (Imoto et al., 1984) was obtained commercially from Daiichi Pure Chemicals, Co., Ltd, Tokyo, Japan.

ELISA. These assays were performed as previously described (Pollack et al., 1983). The coating concentrations of LPS and lipid A were 10 μg/ml and 25 μg/ml, respectively.

Passive hemolysis assay (PHA) (Galanos et al., 1971). Sheep red blood cells (SRBC) were sensitized (coated) with LPS or lipid A in phosphate buffered saline, pH7.3, at 37°C for 30 min, washed and resuspended to a concentration 0.5% v/v in veronal buffer. The optimal concentration of sensitizing antigen, determined in each case, was usually about 250 μg/ml. Mouse ascites fluid containing mAb was diluted, in four-fold increments, over a range of 10^{-1} to 2.6×10^{-6}. To the V-shaped wells of a 96-well polystyrene microtiter plate were added 50 μl of sensitized SRBC's, 50 μl of diluted antibody, and 25 μl of guinea pig complement previously absorbed with SRBC's. The plates were gently agitated, sealed, and incubated at 37°C for 1 hr. Fifty microliters of cell-free incubation mixture was then removed carefully from each well, transferred to U-wells of a second microtiter plate, and free hemoglobin detected by determining optical density at 405 nm in an automated microELISA spectrophotometer. Titration endpoints were taken as the highest dilution yielding an optical density $\geq$ twice background (the latter assessed in the absence of antibody.

Preparation of mAbs. BALB/c mice (National Cancer Institute, Bethesda, MD) were immunized with purified rough mutant LPS (or lipid A), heat-killed rough mutant bacteria, or live wild type parent strains (see Table 1). Immunizations were carried out by the intraperitoneal route and repeated at several weekly intervals; mice were boosted three days prior to fusion in most instances. Spleen cells were obtained in standard fashion and fusions performed as previously described (Gefter et al., 1977). The P3-X63/Ag8 or SP 2/0-Ag14 mouse myeloma fusion partners

Table 1. Fusion Strategies

Immunization
1. *E. coli* J5 (Rc) or *S. minnesota* R595 (Re) LPS
2. *E. coli* J5 or *S. minnesota* R595 heat-killed bacteria
3. Viable *E. coli* 0111:B4 or *S. minnesota* wild type bacteria
Screening antigens
1. *E. coli* J5 and *S. minnesota* R595 LPS
2. *E. coli* J5 and *S. minnesota* R595 lipid A
3. *S. minnesota* R60 (Ra), R345 (Rb), R5 (Rc), and R7 (Rd) LPS
4. *E. coli* 0111:B4 and *S. minnesota* wild type LPS

were employed, and hybridomas selected in hypoxanthine-, aminopterin-, and thymidine- containing (HAT) medium. Hybridomas were screened by ELISA 10-14 days post-fusion, employing the antigens shown in Table 1. Hybridomas of interest were cloned by limiting dilution and adapted as ascites tumors in pristane-primed BALB/c mice.

LPS hydrolysis (Galanos et al., 1984). One milligram of purified smooth LPS was mixed with 3 ml of 1% acetic acid in a 10-ml capacity round-bottom flask. The mixture was refluxed for 2 hrs at 100°C, allowed to cool, and adjusted to pH 7.5 with 2N NaOH. The hydrolyzed LPS was diluted 1:10 in coating buffer and used to coat the wells of microtiter plates for determination of antibody binding in the ELISA.

RESULTS AND DISCUSSION

Twelve fusions yielded hybridoma growth in a total of 3,391 microtiter wells containing selective medium. Of these, 164 (approximately 5%) secreted antibodies reactive with rough mutant LPS and/or lipid A. Fifty-three hybridomas were cloned by limiting dilution, and their mAb products further characterized.

Extensive analyses by means of ELISA revealed a number of distinct antibody specificities against rough and smooth LPS and isolated lipid A. Binding data obtained with eight representative mAbs are shown in Table 2. Many of the mAbs recognized structural elements of the core oligosaccharides of *E. coli* and *Salmonella* LPS corresponding to the hexose (Ra and Rb chemotypes), heptose (Rc chemotype), and 3-deoxy-D-manno-octulosonic acid (KDO) region (Re chemotype).

Table 2. Binding Activity in ELISA of Core- and Lipid A-Reactive Monoclonal Antibodies.

	Monoclonal antibodies (isotype)							
LPS/lipid A	S1-3A2 (IgG3)	J7-3G7 (IgG2a)	J16-1B4 (IgG3)	Y1-2H9 (IgG2a)	J8-5G4 (IgG2b)	R6-3F2 (IgG3)	J8-4C10 (IgG2a)	O4-1H6 (IgM)
Smooth LPS								
E. coli 026:B6	-	-	-	-	-	-	+++	-
E. coli 055:B5	-	-	-	-	-	-	+++	-
E. coli 0111:B4	+	-	-	-	-	-	-	-
E. coli 0127:B8	-	-	-	-	-	-	-	-
E. coli K235	-	+	-	-	-	-	-	-
E. coli K12, BB26-36	+++	-	-	-	-	-	-	-
S. minnesota	+	-	-	-	-	-	-	-
S. typhimurium	++	-	-	-	-	-	-	-
Rough LPS								
E. coli K12, mm294 (Ra)	++	-	-	-	-	-	-	-
S. minnesota R60 (Ra)	++	-	-	-	-	-	-	-
S. minnesota R345 (Rb)	++	-	-	-	-	-	-	-
E. coli J5 (Rc)	-	+++	++	+++	+++	-	-	-
S. minnesota R5 (Rc)	-	-	-	+++	+++	-	+	-
S. minnesota R7 (Rd)	-	-	-	-	-	-	-	-
S. minnesota R595 (Re)	-	-	-	-	-	++	-	++
E. coli K12, D31m4 (Re)	-	-	-	-	-	++	+++	+++
Lipid A								
E. coli K12, D31m4	-	-	-	-	-	-	-	++
E. coli J5	-	-	-	-	+	-	-	++
S. minnesota R595	-	-	-	-	-	-	-	+++
S. typhimurium G30/C21	-	-	-	-	++	-	-	+++
Synthetic Lipid A								
LA-15-PP (506)	-	-	-	-	-	-	-	++
Monophosphoryl Lipid A								
S. minnesota	-	-	-	-	+	+	-	+++

Note. Tissue culture supernatants from hybridoma clones were assayed by ELISA at a 1:25 dilution. Optical density readings were graded as follows: $\leq$0.49, - ; 0.50-0.99, +; 1.0-1.49, ++; $\geq$1.50, +++. Background readings were <0.10.

Other mAbs, like O4-1H6, demonstrated specificity for the lipid A moiety. It is apparent from the data shown in Table 2 that relatively few of the core- or lipid A-reactive mAbs examined were able to recognize their respective epitopes in smooth LPS. This observation, along with the exquisite specificity demonstrated by mAbs for particular rough mutant chemotypes, underscored the importance of epitope accessibility as a major determinant of antibody reactivity. In contrast, certain core-specific mAbs, like S1-3A2 and J8-4C10 (see Table 2), reacted selectively with particular smooth LPS's. Immunoblots of electrophorectically separated samples of such LPS's, developed with these mAbs, localized the epitopes in question to the fast migrating core region. We speculated in the case of S1-3A2, that its ability to react with some smooth E. coli and Salmonella LPS may have been related to the relatively exposed position of its putative epitope in the outer portion of the core oligosaccharide. In contrast, J8-4C10 recognized a more proximal core determinant, which encompassed at least one of two KDO residues. This epitope was surprisingly accessible to antibody attack on E. coli O26:B6 and O55:B5 smooth LPS, as well as more predictably on LPS from the E. coli D31m4 Re chemotype rough mutant strain.

Monoclonal antibody specificities demonstrated by ELISA were confirmed in several other assays. As mentioned above, immunoblotting provided a useful means not only for corroborating the specificity of mAbs for particular antigens, but also for delineating those portions of the LPS macromolecule which contained particular epitopes. Antibody specificity was also confirmed by extensive inhibition experiments, in the ELISA, with both rough and smooth LPS and lipid A.

Results of the PHA were considered of particular significance since the solid phase upon which antigen was displayed represented a biological membrane not unlike the outer membrane of Gram negative bacteria. It was reasonable to assume, therefore, that the hydrophobic long chain fatty acids of LPS (i.e., lipid A) inserted into the LPS-sensitized RBC membrane in a three dimensional configuration similar to that found on the bacterial surface (Rietschel et al., 1984). Moreover, since the biological endpoint measured by PHA, bacteriolysis, depended upon complement fixation via the Fc portion of the immunoglobulin molecule, one predicted that the assay would detect only mAbs "physiologically" oriented on the RBC surface through specific covalent attachment, by means of the Fab domain, to sensitizing (LPS) antigen. This, in turn, provided some assurance that the PHA detected specifically rather than nonspecifically bound antibody.

The representative data shown in Table 3 demonstrate the usually close correspondence between mAb binding detected in the PHA and ELISA systems. Nonetheless, potentially important differences were noted between the two assays in respect to particular antigen-antibody pairings. For example, the *E. coli* J5-reactive mAb, J7-3G7, showed activity in the PHA against *S. minnesota* Ra and Rb LPS, whereas little or no reactivity was demonstrated against these antigens by ELISA. The ability of J7-3G7 to recognize a determinant in more complete forms of the core oligosaccharide, and in a number of smooth *Salmonella* and *E. coli* LPS's, was confirmed in immunoblotting experiments. These findings suggested the existence of a second epitope, similar to but distinct from the epitope recognized by J7-3G7 on *E. coli* J5 LPS. A heptose residue α-1,7 linked to the more distal of two heptoses in the main chain of *S. minnesota* Ra and Rb LPS (Luderitz et al., 1982) may represent the second epitope in question, since an analogous configuration is found on *E. coli* J5 LPS comprised of a heptose residue α-1,7 linked nonstoichiometrically to the more proximal of the two heptose residues in the main chain (Fuller et al., 1973). PHA and ELISA also yielded different results in regard to the cross-reactivity of J7-5C10 (and other mAbs not shown) between *E. coli* J5 and *S. minnesota* R5 LPS. PHA indicated a strong cross-reaction between these two Rc chemotype LPS's, while the reactivity of J7-5C10 in the ELISA was limited to *E. coli* J5 LPS. This was in marked contrast to the lack of cross-reactivity of other J5-reactive mAbs (e.g., J7-4D2) by either assay, suggesting the likelihood that J7-5C10 and J7-4D2 recognized different epitopes. Another, perhaps quantitative, difference between PHA and ELISA was demonstrated in the case of mAbs J8-5G4 (Table 2) and J18-8D2 (Table 3), which cross-reacted in both assays with *E. coli* J5 and *S. minnesota* R5 LPS. Although these mAbs showed some degree of cross-reactivity with lipid A in the ELISA, this reactivity was more prominent in the PHA and was confirmed in immunoblotting experiments. This may have represented another instance in which mAbs demonstrated multispecifi-

Table 3. Comparative Binding Activities in Passive Hemolysis Assay and ELISA of Core- and of Lipid A-Reactive Monoclonal Antibodies.

	Monoclonal antibodies (isotype)							
LPS/lipid A	S1-3A5 (IgG3)	J7-3G7 (IgG2a)	J7-4D2 (IgG2a)	J7-5C10 (IgM)	J18-8D2 (IgG2a)	Y1-4H6 (IgG3)	J8-1C2 (IgM)	J23-4E12 (IgM)
S. minnesota LPS	160 ++	<10 -	<10 -	<10 -	<10 -	<10 -	<10 -	<10 -
S. minnesota R60 (Ra) LPS	10,240 +++	40 -	<10 -	<10 -	<10 -	<10 -	<10 -	<10 -
S. minnesota R345 (Rb) LPS	160 +++	40 -	<10 -	<10 -	<10 -	<10 -	<10 -	<10 -
E. coli J5 (Rc) LPS	<10 -	40,960 +++	40,960 ++	163,840 ++	10,240 +++	160 -	<10 -	<10 -
S. minnesota R5 (Rc) LPS	<10 -	<10 -	<10 -	40,960 -	2,560 +++	160 -	<10 +	<10 -
S. minnesota R7 (Rd) LPS	<10 -	<10 -	<10 -	<10 -	<10 -	160 -	<10 -	<10 -
S. minnesota R595 (Re) LPS	<10 -	<10 -	<10 -	<10 -	<10 -	2,560 +++	<10 -	<10 ++
E. coli D31m4 (Re) LPS	<10 -	<10 -	<10 -	<10 -	<10 -	2,560 +++	<10 ++	<10 +++
S. minnesota R595 lipid A	<10 -	<10 -	<10 -	<10 -	640 -	<10 -	<10 -	40,960 ++
E. coli D31m4 lipid A	<10 -	<10 -	<10 -	<10 -	640 -	<10 -	<10 -	40,960 ++
Synthetic _E. coli_ lipid A	<10 -	<10 -	<10 -	<10 -	640 -	<10 -	<10 -	40,960 +

Note. Shown are reciprocal passive hemolysis titers of ascites fluid and graded optical density readings in ELISA of tissue culture supernatant assayed at a 1:25 dilution.

city by recognizing similar but distinct epitopes on different antigens. It is possible, in this case, that J8-5G4 and J18-8D2 reacted with an epitope on E. coli J5 LPS comprised in part by the L-glycero-D-mannoheptose-4-phosphate moiety (Luderitz et al., 1982) and cross-reacted with the highly analogous 2',3'-diacylglucosamine-4' -phosphate in the glucosamine backbone of lipid A (Rietschel et al., 1984). The ability of Re-reactive mAbs like Y1-4A6 (Table 3) to recognize by PHA, but not ELISA, determinants on Rc and Rd chemotype LPS represented another illustration of assay differences. These cross-reactions, which sometimes included Ra chemotype LPS as well, were reminiscent of those previously attributed to certain Re-reactive mAbs by Appelmelk and his colleagues (Appelmelk et al., 1987). Another apparent KDO-reactive mAb, J8-1C2, reacted strongly with E. coli D31m4 (Re) LPS by ELISA but not PHA, further distinguishing the specificity of this mAb from that of Y1-4A6, which reacted with Re chemotype LPS in both assays. Finally, in the case of lipid A-reactive mAbs like J23-4E12 (Table 3), which reacted with Re chemotype LPS as well as lipid A in the ELISA, reactivity was limited to isolated lipid A in the PHA. Overall, differential binding in the ELISA and PHA provided additional insights concerning mAb specificity by drawing distinctions between mAbs not always apparent in either assay alone. These assay differences also underscored the importance of physical-chemical factors, which obviously varied between the two assays, as determinants of epitope recognition by particular mAbs.

Smooth LPS was subjected to acid hydrolysis under conditions which presumably cleaved lipid A from the core oligosaccharide (Galanos et al., 1984). Evaluation by ELISA of mAb binding to various LPS's before and after hydrolysis shed additional light on antibody cross-reactivity and on epitope recognition as a function of epitope exposure. As shown in Table 4, antisera raised against intact (whole) LPS reacted equally well with homologous hydrolyzed and unhydrolyzed LPS. Core- and lipid A-reactive mAbs, on the other hand, yielded varied and distinct patterns of reactivity, which depended on the identity and location of their respective epitopes. At one extreme were E. coli J5-specific mAbs like J16-1G1, which reacted with neither hydrolyzed nor unhydrolyzed smooth LPS. At the other extreme were lipid A-reactive mAbs, represented by J22-4E4 in Table 4, which showed no reactivity against

Table 4. Effect of Acid Hydrolysis of Smooth LPS on Binding Activity in ELISA of Core- and Lipid A-Reactive Monoclonal Antibodies (mAbs).

LPS	mAb (Specificity Group) J16-1G1 (E)	J8-5G4 (D)	J8-1C2 (G)	J22-4E4 (H)	Homologous Antiserum
E. coli 026:B6	-/-	-/+	++/++	-/+++	++++/++++
E. coli 0111:B4	-/-	-/++	-/-	-/+++	++++/+++
S. minnesota	-/-	-/+	++/+	-/+++	+++/+++
S. typhimurium	-/-	-/++	-/-	-/+++	NA
Y. enterocolitica	-/-	-/-	-/-	-/++++	++++/++++
K. pneumoniae	-/-	-/-	-/-	-/++++	++++/+++
S. marcescens	-/-	-/-	-/-	-/+	NA
V. chlolerae	-/-	-/-	-/-	-/++	++++/++
P. aeruginosa	-/-	-/-	-/-	-/+	++++/++++

Note. Graded ELISA reactivity of mAbs with intact and acid-hydrolyzed LPS is indicated to left and right of virgules, respectively. Hydrolysis conditions are indicated in text. NA, not available.

unhydrolyzed LPS, in contrast to striking reactivity against hydrolyzed LPS's from diverse species and genera. This reactivity was presumably directed toward free lipid A released by acid hydrolysis, although no attempt was made to isolate lipid A from reaction mixtures. An intermediate result was obtained with J8-5G4, a mAb which recognizes an epitope shared by *E. coli* and *Salmonella* Rc mutants, but also cross-reacts with isolated lipid A's. This mAb demonstrated no reactivity with various unhydrolyzed smooth LPS's, but did show modest reactivity restricted to hydrolyzed *E. coli* and *Salmonella* LPS's. Finally, J8-1C2, which in a sense represents a special case, reacted almost equally well with hydrolyzed and unhydrolyzed *E. coli* 026:B6 and *S. minnesota* LPS, emphasizing the ability of this mAb (like J8-4C10) to recognize its core epitope in selected smooth LPS's. Perhaps the most significant outcome of the above analyses was the rather broad cross-reactivity demonstrated by lipid A-reactive mAbs, like J22-4E4, among phylogenetically diverse bacteria. Equally significant, however, was the physical-chemical manipulation of smooth LPS required for the full expression of this cross-reactivity.

Summarized in Table 5 are eight distinct specificity patterns demonstrated by the mAbs described above, and a total of eighteen mAbs representing these various specificities. With one exception (Group C, J7-3G7), each specificity pattern was expressed by more than one mAb in this analysis, and multiple isotypes were often represented

Table 5. Distinguishable Specificities of Monoclonal Antibodies (mAbs) Reactive with the Core/Lipid A Region of LPS.

Group	mAb	Isotype	Specificity
A	S1-3A5	IgG3	*E. coli*, *Salmonella* Ra, Rb, selected
	S1-3A2	IgG3	smooth LPS
B	Y1-2H9	IgG2a	*S. minnesota* Rc, *E. coli* J5 (Rc) LPS
	J7-5C10	IgM	
C	J7-3G7	IgG2a	1. *Salmonella* Ra, Rb, selected smooth *E. coli*, *Salmonella* LPS 2. *E. coli* J5 (Rc) LPS
D	J8-5G4	IgG2b	1. *S. minnesota* Rc, *E. coli* J5 (Rc) LPS
	J18-8D2	IgG2a	2. *E. coli*, *Salmonella* lipid A
E	J16-1B4	IgG3	*E. coli* J5 (Rc) LPS
	J7-4D2	IgG2a	
	J18-1C12	IgG2b	
F	Y1-1F11	IgG3	*Salmonella*, *E. coli* Re LPS
	R6-3F2	IgG3	
	Y1-4A6	IgG3	
G	J8-1C2	IgM	*E. coli* D31m4 (Re) LPS;
	J8-4C10	IgG2a	*E. coli* O26:B6, O55:B5 smooth LPS
H	J22-4E4	IgM	*E. coli*, *Salmonella*, other lipid A;
	J23-4E12	IgM	*E. coli*, *Salmonella* Re LPS;
	O4-1H6	IgM	monophosphoryl lipid A, lipid X

within a given specificity group. An exception was Group H; all lipid A-reactive mAbs in this group were of the IgM class.

The results of analyses conducted to date have emphasized the immunochemical complexity of the core/lipid A region of bacterial LPS, and the exquisite specificity of mAbs which recognize different structural elements in this portion of the LPS macromolecule. Cross-reactive mAbs have been identified, particularly among those directed against the lipid A moiety. However, cross-reactivity has been as much dependent upon adequate epitope exposure as upon antigenic conservation among heterologous species. Most core- and lipid A-reactive mAbs have been restricted in their access to their respective epitopes by more distally placed sugars in the O-side chain and outer core oligosaccharide. Whether these restrictions on mAb accessibility to active sites on isolated LPS are applicable in the septic patient depends, in part, on the physical-chemical forms assumed by

LPS in vivo. Moreover, the therapeutic potential of core- and lipid A-reactive mAbs will also depend upon the functional consequences of the interactions between these mAbs and their respective epitopes, even if in vivo conditions permit such interactions.

REFERENCES

Appelmelk BJ, Verweij-vanVught AMJJ, Maaskant JJ, Schouten WF, Thijs LG, MacLaren DM (1987). Monoclonal antibodies detecting novel structures in the core region of Salmonella minnesota lipopolysaccharide. Infect Immun 55:462-466.

Chedid L, Parant M, Parant F, Boyer F (1968). A proposed mechanism for natural immunity to enterobacterial pathogens. J Immunol 100:292-301.

Fuller NA, Wu M-C, Wilkinson RG, Heath EC (1973). The biosynthesis of cell wall lipopolysaccharide in Escherichia coli VII. Characterization of heterogeneous " core " oligosaccharide structures. J Biol Chem 248:7938-7950.

Galanos C, Freudenberg MA, Jay F, Nerkar D, Veleva K, Brade H, Strittmatter W (1984). Immunogenic properties of lipid A. Rev Infect Dis 6:546-552.

Galanos C, Luderitz O, Westphal O (1969). A new method for the extraction of R lipopolysaccharides. Eur J Biochem 9:245-249.

Galanos C. Luderitz O, Westphal O (1971). Preparation and properties of antisera against the Lipid-A component of bacterial lipopolysaccharides. Eur J Biochem 24:116-122.

Gefter ML, Margulies DH, Scharff MD (1977). A simple method for polyethylene glycol-promoted hybridization of mouse myeloma cells. Somatic Cell Genetics 3:231-236.

Imoto M, Yoshimura H, Kusumoto S, Shiba T (1984). Total synthesis of lipid A, active principle of bacterial endotoxin. Proc Jpn Acad Ser B Phys Biol Sci 60:285-288.

Luderitz O, Freudenberg MA, Galanos C, Lehmann V, Rietschel E Th, Shaw DH (1982). Lipopolysaccharides of gram-negative bacteria. Current Topics in Membranes and Transport 17:79-151.

McCabe WR (1972). I. Protection against challenge with heterologous Gram-negative bacilli. J Immunol 108:601-610.

Pollack M, Huang AI, Prescott RK, Young LS (1983). Enhanced survival in Pseudomonas aeurginosa septicemia associated with high levels of circulating antibody to Escherichia coli endotoxin core. J Clin Invest 72:1874-1881.

Raetz CRH (1984). The enzymatic synthesis of lipid A: molecular structure and biologic function of monosaccharide precursors. Rev Infect Dis 6:463-471.

Rietschel ETh, Wollenweber H-W, Brade H, Zahringer V, Lindner B, Seydel U, Bradaczek H, Barnickel G, Labischinski H, Giesbrecht P (1984). Structure and conformation of the lipid A component of lipopolysaccharides. In Rietschel ETh (ed): "Handbook of Endotoxin, Vol. 1: Chemistry of Endotoxin," Amsterdam: Elsevier, pp 187-220.

Westphal O, Jann K (1965). Bacterial lipopolysaccharides. Extraction with phenol-water and further applications of the procedure. Methods in Carbohydrate Chemistry 5:83-90.

Ziegler EJ, Douglas H, Sherman JE, Davis CE, Braude AI (1973). Treatment of E. coli and Klebsiella bacteremia in agranulocytic animals with antiserum to a UDP-Gal epimerase-deficient mutant. J Immunol 111:433-438.

Bacterial Endotoxins: Pathophysiological Effects, Clinical Significance, and Pharmacological Control, pages 339–349

IMMUNOGLOBULIN THERAPY IN PATIENTS WITH ENDOTOXEMIA AND POSTOPERATIVE SEPSIS - A PROSPECTIVE RANDOMIZED STUDY -

Reinhart Grundmann and Michael Hornung

Chirurgische Universitätsklinik Köln-Lindenthal, Joseph-Stelzmann-Str. 9, D-5000 Köln 41, Germany

INTRODUCTION

In a prospective randomized trial the effect of immunoglobulin therapy was evaluated in patients treated in the intensive care unit (ICU) for endotoxemia and postoperative sepsis. The effectiveness of this therapy has not been established. The assessment of success of immunoglobulin treatment depends on several factors including patient selection (Just et al. 1986). Especially, the influence on the mortality of postoperative septic complications can hardly be assessed, in view of the fact that the mortality rate mainly depends on the possibilities of surgical reinterventions to remove the focus of sepsis in case of anastomosis dehiscence, peritonitis or local infections. Under these conditions the administration of a single drug may be considered as adjuvant only. An other difficulty with studies concerning patients treated for septic complications is related to the fact that the severity of sepsis is difficult to define. In the investigation presented here therefore the method described by Elebute and Stoner (1983) for grading of sepsis was used to monitor the progress of severe surgical infections.

MATERIAL AND METHODS

PATIENTS

Seventy-four patients who were admitted postoperatively to the surgical intensive care unit for severe septic complications were studied during a 18 months period. Fifty-two patients had undergone abdominal surgical interventions, 4 patients vascular surgery, 10 patients other operations and 8 patients were admitted with multiple injuries due to trauma. The mean age of the patients was 53 years (range: 11-86), the male:female ratio was 47:27.

ENDOTOXIN MEASUREMENT

The postoperative course of all patients was documented and endotoxin was measured in the plasma once a day using a limulus lysate assay, as described earlier (Grundmann and Ingenhoff, 1986). For measurements the LAL-reagent (Byk-Mallinckrodt) was applied. After centrifugation of the heparinized plasma a dilution of the plasma of 1:10 to 1:10,000 was performed. 100ul of the diluted plasma samples were each incubated with 100ul LAL-reagent in sterile tubes for 1 hr in a water bath. Definite gel formation was considered positive. Aqua ad inj. served as negative control; as positive control an E. coli-endotoxin standard preparation (endotoxin content 0,25ng/ml) was utilized.

Only patients who had a positive reaction at a plasma dilution of 10^{-4} were called "endotoxin-positive". This definition was developed in order to register severely septic patients exclusively. Thus we found in a former study a high percentage of patients showing lower titers of plasma endotoxin to be non septic, since plasma endotoxin titers of 10^{-2} or 10^{-3} could be observed as well due to infection as due to restricted liver and kidney function or could be caused by intraoperative bowel mobilization or by large retroperitoneal hematomas (Grundmann and Ingenhoff 1986).

SEPSIS SCORE

Immediately after admittance to the intensive care unit and daily thereafter the clinical signs of sepsis were scored using the system of Elebute and Stoner (1983) . The sepsis score is based on the numerical scoring of the local and systemic effects of infection. In principle, 4 parameters are graded: a) the local tissue infection (e.g. of the wound, peritonitis or pneumonia) b) the body temperature, c) the secondary effects of sepsis (e.g. jaundice and renal failure), and finally d) the laboratory data (e.g. positive blood cultures, leucocyte count and plasma albumin level). The clinical sepsis score was performed by one of the authors (M.H.). He was not informed about the endotoxin assay and the administration of the immunoglobulins.

IMMUNOGLOBULIN THERAPY TRIAL

Patients were supposed to have a gram-negative bacterial sepsis when endotoxin was present in the plasma at a dilution of 10^{-4} during two subsequent days and when pyrexia of at least 38.5°C was present. Patients fulfilling these criteria were assigned using a randomized envelop technique to the therapy or control group. The random list was prepared by a computer. Patients of the treatment group (n=24) were given 0.25g/kg b.w. immunoglobulin (Intraglobin F[R]) i.v. the day of entering the study and on the following day. Patients of the control group (n=22) were not treated.

A total of fourty-six patients were included in the study. Twenty-eight patients did not comply with the eligibility criteria showing either no endotoxemia on two following days or pyrexia above 38.5 was absent. Nevertheless, the clinical evaluation and the sepsis score indicated a septic complication. Therefore these patients were also prospectively followed to evaluate clinical significance of sepsis scoring and endotoxin measurements.

RESULTS

INITIAL CONDITIONS

Patients of the therapy and control group did not differ with regard to the initial characteristics at entrance into the study. This was true as well for the number of reinterventions, the duration of intensive care treatment, the duration of mechanical ventilation and the initial sepsis score (tab. 1).

TABLE 1: Clinical parameters at the time of randomization

	therapy group (n=24)		control group (n=22)	
patients age (years)	46,9(45(11-73))		52,8(55(18-68)	
male:female	14:10		17:5	
cancer patients (n)	6	25%	5	23%
surgical interventions (n)				
- abdominal	17	71%	13	59%
- thoracic	2	8%	5	23%
- vascular	1	4%	1	5%
- polytrauma	4	17%	3	14%
patients with reintervention (n)	10	42%	11	50%
patients with antibiotic therapy (n)	23	96%	19	86%
intensive care (days)	6.5 ± 4.1		6.1 ± 5.1	
patients with mechanical ventilation (n)	24	100%	22	100%
mechanical ventilation (days)	5.4 ± 3.4		4.7 ± 3.3.	
sepsis score	17.0 ± 4.0		18.9 ± 4.8	

RESULTS OF IMMUNOGLOBULIN THERAPY

A study patient was considered to have completed the trial, when either discharged from the intensive care unit or died during the stay at the unit.

TABLE 2: Clinical parameters at the end of the study

	therapy group (n=24)	control group (n=22)
mortality (n)	15 (63%)	19 (86%)
patients with re-interventions after randomization (n)	3 (12%)	1 (5%)
intensive care (days)	22.8 ± 16.7	21.3 ± 14.1
mechanical ventilation (days)	19.0 ± 15.4	15.1 ± 11.9
sepsis score (day of death)	22.5 ± 4.9	23.1 ± 5.7

As the results in table 2 indicate no differences were observed between the therapy and control group with regard to the number of reinterventions, the duration of intensive care treatment and of mechanical assistance. The sepsis score in both groups was above 20. However, the mortality rate was lower in the therapy group as compared to the control group although this difference did not reach statistical significance. This might be due to the relatively small number of patients in each group, particularly since an additional positive effect was seen in the therapy group: after immunoglobulin treatment the number of endotoxin-positive patients decreased significantly in the therapy group. Nevertheless the sepsis score was not affected by the therapy (fig. 1).

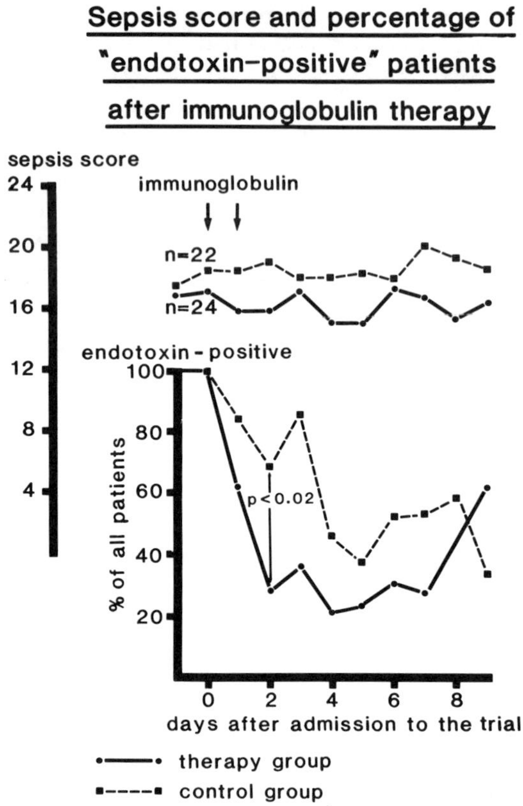

FIGURE 1: Sepsis score and percentage of "endotoxin-positive patients after immunoglobulin therapy

ENDOTOXIN TITER IN SURVIVING AND NON-SURVIVING PATIENTS

The endotoxin titers were analysed in surviving and non surviving patients - independently of whether the patients had been on the therapy protocol or not. Significant differences between survivors and non-survivors could not be detected (fig. 2).

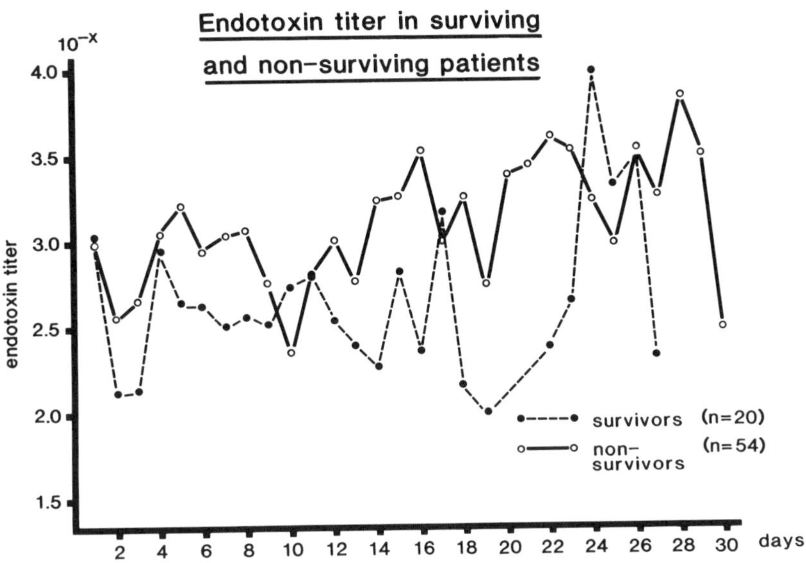

FIGURE 2: Course of endotoxin in surviving and non-surviving patients. The whole period of intensive care treatment was analysed. 12/20 survivors and 34/54 non-survivors had been on the therapy protocol.

DISCUSSION

A positive effect of an adjuvant immunoglobulin therapy in patients with postoperative sepsis was observed in this study. In particular the mortality rate (86% versus 63% in control and treated groups respectively) was noted to be decreased (reduction 23%). However due to small sample size this difference was not statistically significant. Therefore it may be questioned whether the criteria for admission to this study were to rigid. Patients in this study showed high endotoxin titers on two following days only after a relatively long period of treatment in the intensive care unit (mean 6.3 days). At this moment the prognosis had become rather poor, independently of whether the patients received an adjuvant treatment or not, resulting in a negative patient selection and a mortality rate of 74% in the combined study groups.

Furthermore endotoxemia is only seen in patients with gram-negative bacterial sepsis, the gram-positive septicemia cannot be detected by this test. Consequently, 28 of 74 patients prospectively documented here and showing clear-cut clinical signs of sepsis with a sepsis score exceeding 15 could not enter the therapy study because endotoxin could not be detected in the plasma of these patients on two subsequent days in the required concentrations. On the one hand it is conceivable that the majority of these patients had gram-positive infections - according to Rosenthal (1986) 45% of all septicemias are caused by gram-positive cocci - but on the other hand also the criterium of two consecutive days with a positive endotoxin test was to rigid.

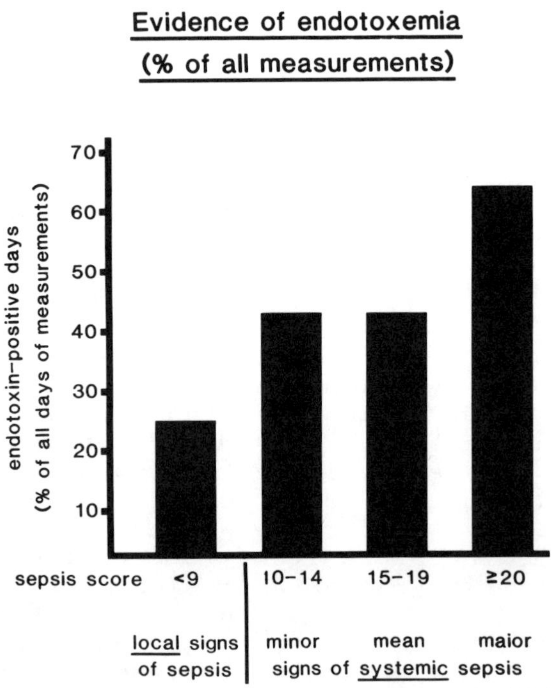

FIGURE 3: Evidence of endotoxemia at different sepsis score levels. All investigated patients were included in this analysis, independently of entering the therapy protocol or not (801 days of measurement in study patients, 229 days of measurement in the remaining patients).

The frequency of endotoxin detection increased gradually with the severity of sepsis, even in seriously ill patients (i.e. a sepsis score of more than 20) endotoxin was found in only 64% of all days of measurement (fig. 3).

These findings demonstrate the usefullness of the sepsis score for defining the septic condition. In agreement with the experiences of Dionigi et al. (1985 and 1986) we were also able to monitor the severity of sepsis more precisely with this grading system than with other measurements.

A definite correlation was found between sepsis score and mortality rate. The mortality rate was 50% in the case the score did not exceed 15 at the time of entrance into the study, but increased to 76% when the initial score was 20 and more (fig. 4). However, in contrast to the findings of Levin et al. (1972), surviving and non-surviving patients did not differ significantly in the endotoxin titers (fig. 2).

Sepsis score and mortality

(all groups, n=74)

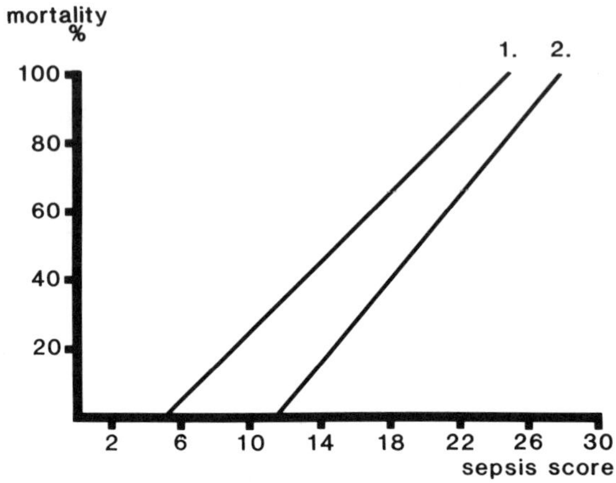

1. initial value(day of admission to the study)
   y=5.08x−25.8   r=0.76

2. Maximum during the course of illness
   y=6.21x−72.3   r=0.85

FIGURE 4: Correlation between sepsis score and mortality rate on two different moments. (All investigated patients)

CONCLUSIONS

1. The data of this investigation suggest that an adjuvant immunoglobulin therapy in patients with postoperative sepsis may be useful. In the prospective study presented here a lower mortality rate was found in treated patients as compared to untreated patients. Furthermore, the endotoxin titers fell significantly after immunoglobulin administration. Further large randomized studies are needed before definitive conclusions can be reached.

2. Endotoxin measurement alone is not

sufficient for defining postoperative sepsis in patients treated in the intensive care unit. This may be due to an inadequate sensitivity of the test applied here, however, it must be also taken into consideration that gram-positive septic illness cannot be detected by this method. Therefore, in future therapy studies further parameters should be introduced for defining septic illness: the study presented here demonstrates that the sepsis score elaborated by Elebute and Stoner (1983) could be such a parameter.

REFERENCES

Dionigi R, Dominioni L, Jemos V, Cremaschi R; Bobbio-Pallavicini F, Ballabio A (1985). Sepsis score and complement factor B for monitoring severely septic surgical patients and for predicting their survival. Europ Surg Res 17: 269-280

Dionigi R, Cremaschi R, Jemos V, Dominioni L, Monico R (1986). Nutritional assessment and severity of illness classification systems: A critical review on their clinical relevance. World J Surg 10: 2-11

Elebute EA, Stoner HB (1983). The grading of sepsis. Br J Surg 70: 29-31

Grundmann R, Ingenhoff E (1986). Postoperative Bestimmung des Endotoxinverlaufes. Prognostischer und diagnostischer Wert auf der Intensivstation. Deutsch Med Wschr 111: 457-462

Just HM, Metzger M, Vogel W, Pelka RB (1986).Einfluß einer adjuvanten Immunglobulintherapie auf Infektionen bei Patienten einer operativen Intensiv-Therapie-Station. Ergebnisse einer randomisierten kontrollierten Studie. Klin Wochenschr 64: 245-256

Levin J, Poore E, Young NS, Margolis S, Zauber NP, Townes AS, Bell WR (1972). Gram-Negative Sepsis: Detection of Endotoxemia with the Limulus Test. Ann Int Med 76: 1-7

Rosenthal EJK (1986). Septikämie-Erreger 1983-1985. Ergebnisse einer mulitzentrischen Studie. Deutsch Med Wschr 111: 1874-1880

Bacterial Endotoxins: Pathophysiological Effects, Clinical Significance, and Pharmacological Control, pages 351–359

OBSERVATIONS ON THE ROLE OF ENDOTOXIN IN GRAFT-VERSUS-HOST DISEASE

I A Lampert*, R H Moore+, R Huby*, J Cohen+

Dept of *Pathology and +Medicine
Hammersmith Hospital, Royal Postgraduate
Medical School, Ducane Road, London, W12 OHS

SUMMARY

We have used a conventional murine model of bone marrow transplantation (BMT) to make a detailed study of the histological features of graft-versus-host disease (GvHD), and to study the effects of oral antibiotic decontamination of the gut flora.

Comparison of the histological appearances of the small and large bowel indicated that the features of GvHD were more severe in the large bowel. Bowel sterilisation caused a striking reduction in the severity of GvHD, both in the gut and at a distant site (the skin of the ear).

The different effects of GvHD in the large as compared to the small bowel suggested that a factor was operative locally, and the effect of sterilisation linked this to the bacteria in the bowel. The fact that sites remote from the bowel were also affected suggested that a diffusible factor from the bowel produced by the bacteria of the gut was responsible.

Studies elsewhere have suggested that the factor involved is likely to be endotoxin. In these experiments however it was evident that systemic administration of endotoxin did not aggravate the disease process and that in some instances the reverse was obtained. If therefore endotoxin is important the dose and its mode of its administration are likely to be critical.

INTRODUCTION

Much experimental and clinical data suggests that infection and graft-versus-host disease (GvHD) are intimately associated. Unlike conventional mice, germfree recipients of allogeneic bone marrow do not develop GvHD (Jones et al, 1971). Similarly gastrointestinal decontamination of recipients with oral non-absorbable antibiotics prevents GvHD (van Bekkum et al, 1974). We have used such a model to study the features of this disease in the bowel and the skin. In addition we have assessed the effects of bowel sterilization and the administration of endotoxin in this model.

METHODS

Mice Female C57BL/B6 ($H-2^b$) and CBA/Ca ($H-2^k$) mice were purchased from Olac, Oxford, UK, and were housed at the Royal Postgraduate Medical School (RPMS), London, UK. Mice were aged 8-16 weeks at the time of transplantation.

Irradiation Whole body X-irradiation was delivered at 240 kv with an HVL of 1.12nm copper. On day 0 after an overnight fast, the mice were positioned in a clear perspex chamber and received 0.05 Gy/min to a total of 8.0 Gy.

The mice were maintained in a protective environment throughout the experiment. Sterile polypropylene cages were fitted with autoclaved metal grills and filter tops. Irradiated food and bedding were changed twice weekly. Mice entered the protective environment on day -14 and received either acidified water (pH 2.8) or antibiotic water (see below) ad libitum.

Transplantation Recipient CBA/Ca mice were irradiated and on day +1 were grafted with $1x10^7$ viable bone marrow cells harvested from donor C57BL/B6 mice. Donor mice were killed by cervical dislocation and femora and tibiae were used as the source of bone marrow cells. A single cell suspension was prepared in RPM1 1640 medium buffered with 20mM HEPES and supplemented with penicillin (50 iu/ml) and streptomycin (50 μg/ml) (Flow Labs Ltd, Rickmansworth, UK). The bone marrow cells were washed three times and resuspended at a concentration of $5x10^7$ viable cells/ml. Viability was determined by trypan blue exclusion and was normally 85-95 percent. Each mouse received $1x10^7$ viable bone marrow cells as an intravenous injection via a tail vein.

Mice were transplanted in groups of five. Different groups (Table 1) received acidified water to prevent opportunistic infections, or oral non-absorbable antibiotics for bowel sterilisation. Some groups also received intra-peritoneal injections of purified E.coli J5 endotoxin either as a single dose of 25μg at 5 days after transplantation, or as multiple doses of 2.5μg on alternate days from 5 to 15 days after transplantation.

TABLE 1

Group	n	Experimental procedure
0	5	Normal animals
1	5	Syngeneic BMT
2	5	Allogeneic BMT
3	10	Allogeneic BMT and antibiotic decontamination
4	5	Allogeneic BMT and ET 25μg i.p. x 1
5	5	Allogeneic BMT and ET 2.5μg i.p. x 5
6	10	Allogeneic BMT and antibiotic decontamination and ET 2.5μg i.p. x 5

TABLE 1: Details of experimental groups.
n = number of animals BMT = bone marrow transplantation
ET = endotoxin i.p = intraperitoneal

Antibiotic decontamination Some groups were given oral non-absorbable antibiotics chosen to selectively eliminate the aerobic gram-negative bacteria from the gastrointestinal tract. The antibiotics were added to the drinking water starting 14 days before transplantation and continuing throughout the observation period. The first week of treatment was: neomycin sulphate BP 5g/l (The Wellcome Foundation Ltd, Crewe, Cheshire, UK), streptomycin sulphate BP 5g/l (Evans Medical Ltd, Greenford, Middlesex, UK) and natamycin 100mg/ml (Brocades (Great Britain) Ltd, West Byfleet, Surrey, UK). On day -7 the amount of neomycin and streptomycin was halved. Mice tolerated the antibiotic water well. The progress of antibiotic decontamination was monitored by assaying faecal endotoxin levels.

Faecal endotoxin Fresh faeces were collected twice weekly from mice and weighed in sterile pyrogen free plastic bijoux. Faecal endotoxin (ET) was assayed using a quantitative microassay described previously (Rogers et al, 1985).

Autopsy and histology Animals were weighed and killed 25 days after bone marrow transplantation. The wet weight of the spleen was measured, and tissue samples from the proximal, mid and distal regions of the large and small bowel, as well as longitudinal sections of ear, were fixed in 10% neutral buffered formalin. These tissues were processed and embedded in paraffin wax, cut at 4μ and stained with haematoxylin and eosin for histological examination. Splenic indices were measured as described previously (Simonsen, 1962).

Quantification of GvHD in the skin The cross section of ear was traced onto paper using a Leitz microscope and drawing tube. Areas of epidermis showing lymphocytic infiltration or extensive cell death were marked as GvHD positive. The total and affected length of epidermis for each ear was measured using a map measurer and recorded as a ratio. The mean ratios of the different groups were compared using a Student's t test or a modified Z test as appropriate.

RESULTS

Faecal endotoxin Mice tolerated the antibiotic water well; the level of endotoxin fell sharply within four days and remained low throughout the observation period. Ten days prior to transplantation the faecal endotoxin in antibiotic treated mice was 5.94 + 1.37μg/wet weight faeces compared to 61.71 + 11.49μg/g in mice receiving acidified water ($p=0.0085$).

Splenic indices There was a significant difference between the splenic index of the mice receiving allogeneic BMT's compared to syngeneic controls ($p<0.005$, Group 1 vs 2 Fig 2a). In animals who had allogeneic BMT and received oral antibiotics the splenic index was reduced to the levels of the syngeneic controls (Group 3 vs 1, $p>0.5$). Neither of the two groups challenged with endotoxin had splenic indices different from the unmodified allogeneic BMT group (Groups 4 and 5 vs 2). The addition of endotoxin to transplanted, antibiotic-treated animals did not affect the splenic index (Group 6 vs Group 3).

Histopathology of GvHD in the bowel Specimens from normal control and syngeneically transplanted animals were indistinguishable (group 0 and 1). In animals which had received allogeneic bone marrow transplants (group 3) there was extensive histological evidence of GvHD. In the small bowel of these animals there was a dense infiltrate of active lymphocytes in the cores of villi, considerable hyperplasia of the neck of the crypts with mucin depletion, with numerous mitoses. A few apoptotic bodies were seen. There was no infiltrate between the bases of the crypts and the muscularis mucosae (Fig 1a). In the large bowel the features were more severe. Mucin depletion, crypt cell hyperplasia and cell death including apoptotic bodies were seen. In places there was complete obilteration of the crypts. Elsewhere a dense infiltration of active lymphocytes resulted in crypts being widely separated. This infiltration extended beyond the base of the crypts, with extensive invasion of the muscularis mucosae, the muscular wall and the subserosal tissue. In these cases, collections of polymorphs were seen in the subserosal fat, suggesting that bowel perforation had occurred. Considerable oedema was also common, resulting in a wide separation of the muscularis mucosae and the muscularis propria (Fig 1b).

Bowel sterilisation with antibiotics greatly reduced but did not prevent these features. As in non-decontaminated animals, symptoms were more severe in the large bowel compared to the small bowel. The effect of a single large dose of endotoxin was to cause the histological features of GvHD to appear less severe. Multiple smaller injections of endotoxin did not appear to influence the disease.

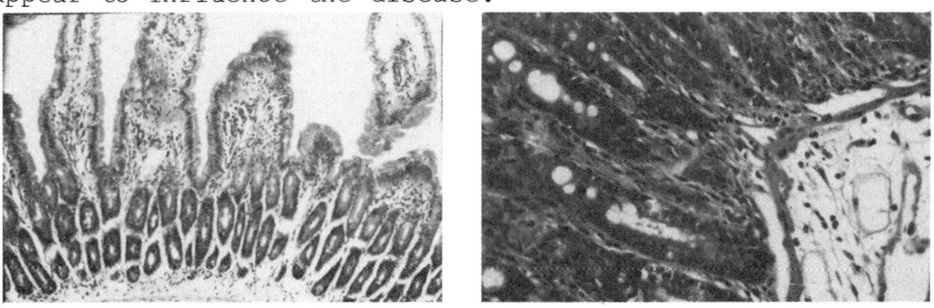

Fig 1. H/E sections of upper small bowel (left panel, x25 obj) and mid large bowel (right panel, x40 obj) from an animal with GvHD (group 2). See text for details.

Histopathology of GvHD in the skin The manifestations of GvHD in the skin were heavy lymphoid infiltration in the dermis and epidermis, causing displacement of keratinocytes. Acanthosis of the keratinocytes, spongiosis and individual cell death in the form of eosinophilic bodies was seen. There was considerable basal vacuolation. Such features were minimal or absent in control animals. Bowel sterilisation of allografted animals dramatically reduced the occurrence of lymphoid infiltrates in both the dermis and epidermis, such that sections were indistinguishable from controls. A single dose of endotoxin produced an intermediate degree of severity. Administration of endotoxin to bowel sterilised animals produced no observable effects on the pathological appearances in the ear.

These observations were quantified using the ear skin index (ESI) described here (Fig 2b). Control groups 0 and 1 were indistinguishable (p>0.1). The ESI in allografted animals (Group 2) was markedly abnormal (p=0.01 compared to group1), but allografted animals receiving antibiotics (Group 3) had virtually normal indices. Transplanted animals that were given a single dose of endotoxin (Group 4) had a low ESI, reflecting the histological changes in the gut but not the splenic index. No effect of endotoxin administration on bowel sterilised allografted animals could be seen (Group 6 vs 4, p>0.1).

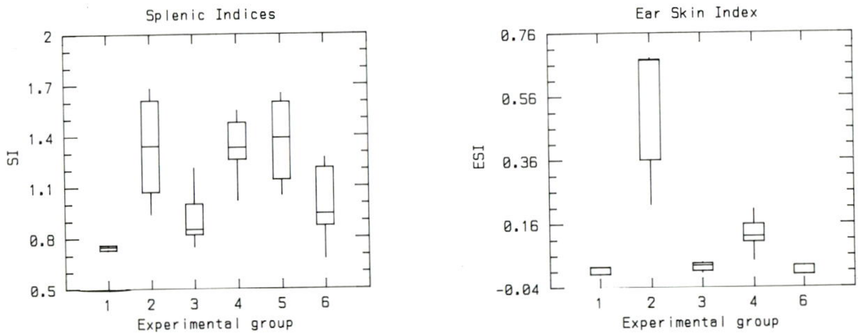

Fig 2. Splenic indices (left panel) and ear skin indices (right panel) in groups of experimental animals. For details see text.

DISCUSSION

The success of bone marrow transplantation depends on adequate engraftment, treatment of opportunistic infection and control or prevention of GvHD. We have shown elsewhere that exposure to endotoxin before bone marrow transplantation exacerbated GvHD produced across major and minor histocompatibility barriers in mice, and that passive immunisation with 'anti-endotoxin' protects against the effects of GvHD in the 'minor HLA-mismatch' model (Moore et al, 1987a; Moore et al, 1987b). In these experiments we have used a well established murine model in which bone marrow transplantation is achieved across major MHC barriers resulting in characteristic clinical and histological features of GvHD in the recipient mice. We have also attempted a hitherto unused measure of GvHD, ie the length of the basal epidermis occupied by lymphocytes, which is in effect a measure of lymphoproliferative activity. These studies have shown that this 'ear skin index' correlates well with other indices of GvHD. A further measure of its value is that the sterilisation of the bowel caused a profound improvement in the effect of the lymphoid infiltration in the epidermis.

Van Bekkum and co-workers were the first to observe that sterilisation of the bowel reduced GvHD (van Bekkum et al, 1974). Their hypothesis was that there was a cross reaction between antigens present in mice tissues and bacterial content and that bowel sterilisation reduced the severity of GvHD by reducing this interaction.

We have shown that GvHD is more severe in the large bowel as compared to the small bowel. Furthermore, the effect of sterilisation of the gut is even more striking at a 'distant site' (the skin of the ear) than is seen in the large bowel itself. If van Bekkum's interpretation was placed on our observations it would be necessary to postulate that this bacterial/mouse antigen was differentially expressed in tissues ranging in the order of density of antigen, large bowel greater than small bowel greater than skin and liver. More probably, however, is not the differential expression of an antigen on these tissues but the distance or accessability of these tissues from some diffusible product originating from the large bowel. The large bowel is heavily contaminated by organisms while the small bowel under normal circumstances is sterile. It therefore follows that if bowel sterilisation reduces the systemic effect of GvHD then it is some bacterial product which must be stimulating this process.

In an accompanying paper (Faulkner et al, 1987), we set out the reasons why we believe endotoxin to be the component of the gut flora responsible for these observations. The findings reported here corroborate this association: we have shown that GvHD is strikingly more severe in the large bowel compared to the small bowel, and that GvHD can be quantitatively represented by both the splenic index and the 'ear skin' (lymphocytic infiltrate) index described here. Gut decontamination is associated with a fall in faecal endotoxin levels and a marked reduction in the severity of GvHD, both locally in the gut, and at a distant site (the ear). Interestingly, attempts to provoke GvHD by intraperitoneal administration of endotoxin were unsuccessful; indeed a single massive dose actually improved the histological appearances, an observation that has been made before (Thompson et al, 1978). Clearly, if endotoxin is indeed important, then the dose and route of administration will be critical in determining its effects.

Acknowledgements: This work is supported in part by the Wellcome Trust. We thank Rabia Amin for typing the manuscript.

REFERENCES

Bekkum DW van, Roodenburg J, Heidt PJ, Waaij D van der (1974). Mitigation of secondary disease of allogeneic mouse radiation chimeras by modification of the intestinal microflora. J Natl Cancer Inst 52: 401-404.

Faulkner L, Rapson N, Moore RH, Cohen J (1987). The influence of the gut flora on graft-versus-host disease (GvHD) following allogeneic bone marrow transplantation - experimental observations and possible mechanisms. Prog Clin Biol Res. In press.

Jones JM, Wilson R, Bealmear PM (1971). Mortality and gross pathology of secondary disease in germfree mouse radiation chimeras. Radiat Res 45: 577-588.

Moore RH, Lampert IA, Chia Y, Aber VR, Cohen J (1987a). Influence of endotoxin on graft-versus-host disease after bone marrow transplantation across major histocompatability barriers in mice. Transplantation. In press.

Moore RH, lampert IA, Chia Y, Aber VR, Cohen J (1987b). Effect of immunisation with Escherichia coli J5 on graft-versus-host disease induced by minor histocompatability antigen in mice. Transplantation. In press.

Rogers MJ, Moore RH, Cohne J (1985). The relationship between faecal endotoxin and faecal microflora of the C57BL mouse. J Hyg (Camb) 95: 397-402.

Simonsen M (1962). Graft-versus-host reactions. Their natural history, and applicability as tools of research. Progr Allergy 6: 349-367.

Thompson PD, Rampy PA, Jutila JW (1978). A mechanism for the suppression of graft-versus-host disease with endotoxin. J Imm 120: 1340-1344.

Bacterial Endotoxins: Pathophysiological Effects, Clinical Significance, and Pharmacological Control, pages 361–371

CLINICAL STUDIES ON ADMINISTRATION OF TAUROLIN IN SEVERE SEPSIS: A PRELIMINARY STUDY

A. Christine McCartney and M. Kennedy Browne.
University Department of Bacteriology, Royal Infirmary, Glasgow G4 OSF and Surgical Unit, Monklands District General Hospital, Airdrie ML6 OJS, Lanarkshire, Scotland.

The morbidity and mortality of severe surgical infection has changed little over the past forty years despite advances in surgery and the advent of the antibiotic era (Bohnen et al, 1983). With the emergence of each new antibiotic, hopes of improvement have been anticipated. However, the antibiotics themselves have led to alterations in normal bacterial flora and the consequent emergence of opportunistic pathogens. Furthermore, widespread antibiotic usage has allowed the development of multiple-resistant strains of bacteria with the capacity to transfer readily the genetic markers for resistance to other bacteria.

In severe sepsis or septic shock, the associated multiple organ failure appears to be related to circulating free endotoxin, which is part of the cell envelope of Gram-negative bacteria, rather than the bacteria *per se* isolated from clinical material (Levin, 1982; McCartney et al, 1987). While Gram-negative bacteria are associated with the majority of cases of septic shock (Ledingham et al, 1983), there is increasing evidence that septic shock may occur in patients with severe sepsis due to Gram-positive bacteria. The endotoxaemia may be derived from a cryptic focus of infection and not necessarily be associated with septicaemia. Indeed, it may be associated with leakage of microorganisms from normal bowel flora into the circulation in the presence of a poorly functioning reticuloendothelial system (Jacob et al, 1977; Harris et al, 1984)

Little is known of the factors either *in vivo* or *in*

vitro which control the release from, or the retention of, endotoxin in the bacterial cell. Antibiotics have been shown *in vitro* to release endotoxin (Goto and Nakamura, 1980) and there is now evidence that bactericidal antibiotics may release endotoxin in an experimental model of Gram-negative sepsis (Shenep and Mogan, 1984; Shenep et al, 1985).

Consequently, there is a need for an anti-endotoxin agent and a chemotherapeutic agent in severe sepsis. Taurolin (Geistlich Sons Ltd., Chester, England) is such an agent.

Taurolin (4,4'-methylene-bis-tetra-hydro 1,2,4 thiadiazine, 1,1-dioxide) is formed by the condensation of two molecules of the amino acid taurine with three molecules of formaldehyde (Figure 1). Taurolin has been shown to be a potent antibacterial agent (Browne et al, 1976), protecting wounds from infections subsequent to peritonitis (Browne et al, 1977; Browne et al, 1978) It also possesses antiendotoxic activity. This has been demonstrated in experimental animals given Taurolin before or after injection of endotoxin (Pfirrmann and Leslie, 1979).

TAUROLIN

4,4'-METHYLENE-BIS-(TETRAHYDRO-

2H-1,2,4-THIADIAZINE-1,1-DIOXIDE)

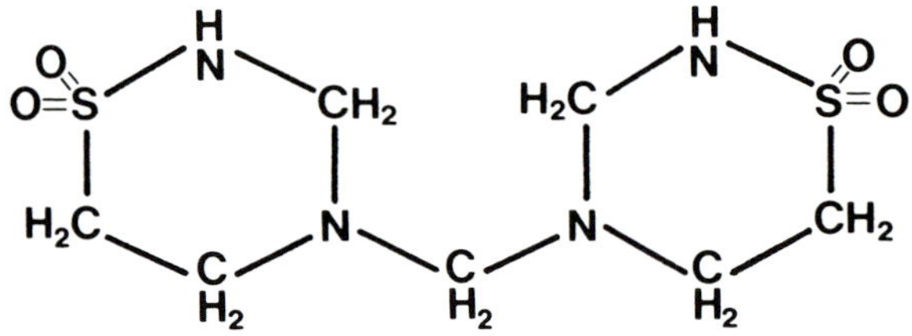

Figure 1 Molecular structure of Taurolin

In contrast to most antibiotics, Taurolin acts by a chemical reaction, releasing methylol groups (CH_2OH) which become available to attach themselves to bacteria or endotoxin as the taurolin is metabolised through taurultam to taurinamide and eventually to taurine (Figure 2).

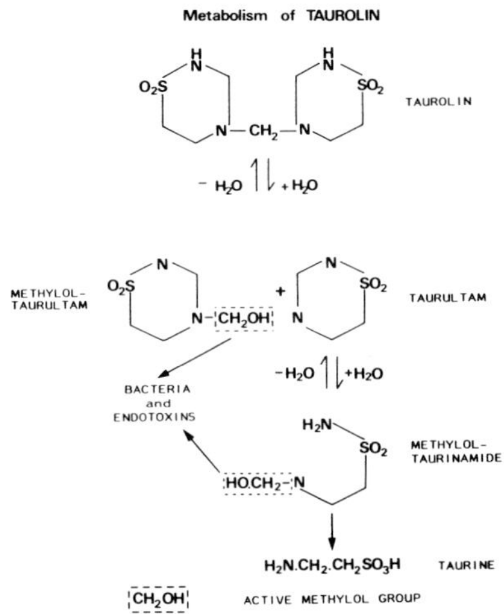

Figure 2 Metabolism of Taurolin

The activity against endotoxin appears to lie in the inter- and intra-molecular cross-linking of the lipopolysaccharide-protein complex and this is enhanced by enzymatic activation (Pfirrmann and Leslie, 1979). Similarly, chemically pure lipopolysaccharide (LPS) (and lipid A) is often found to be considerably less toxic than LPS with a high protein content. In vitro experiments have shown that 7.5mg LPS reacted irreversibly with 94.6 g (0.33 μmol) of taurolidine and 135μg (0.99μmol) of taurultam which agrees well with the theoretical calculations (Browne et al, 1985). Further evidence for the activity lies in the use of radio-marked taurolin synthesized from taurinamide and ^{14}C labelled formaldehyde. When mixed with bacteria or toxins, the bacterial cell wall as well as the toxins and LPS, were cross-linked irreversibly by ^{14}C methylene bridges. Subsequent separation by thin-layer chromatography resulted in radioactive bacteria and toxins and "cold" taurinamide (Pfirrmann, 1985).

Pharmacokinetic studies have demonstrated that it is rapidly absorbed from serous cavities and has a short half-life of $T_{\frac{1}{2}}\beta=5.09\pm0.58$ hours (Knight et al, 1981a; Knight et al, 1981b). Intraperitoneal administration is equivalent to intravenous dosing and the blood levels achieved are similar to those of direct intravenous administration (Erb et al, 1983).

In a prospective study, the clinical efficacy of administration of intravenous Taurolin was evaluated in a group of critically ill patients admitted to the Intensive Therapy Unit, with severe surgical infections or septic shock. Eleven patients (6 males and 5 females: age range 53 - 72 years, mean 65.5 years) were studied. Taurolin was given as a 5 g intravenous infusion over two hours and the infusion repeated at four-hourly intervals for 24 hours. In addition to the usual clinical and laboratory investigations, plasma endotoxin levels were measured using a chromogenic modification of the Limulus amoebocyte lysate (LAL) assay with a sensitivity of 1pg endotoxin per ml in plasma (Piotrowicz et al, 1985). Endotoxin was assayed before, during and after Taurolin infusion. It has been shown that Taurolin does not influence the chromogenic LAL assay (Thomas et al, 1985). During the study, patients were given appropriate surgical management and antibiotic therapy. It was not considered ethical in these seriously ill patients to withold the antibiotics which would constitute the usual management of the presenting condition, although this may have been of value in preventing further release of toxins.

Details of the 11 patients and their primary pathology are given in Table 1. Three of the 11 patients died.

Four case histories are presented to illustrate the effects of Taurolin infusion:

Patient DS was a 71 year old man who was hemiplegic and managed with an indwelling urinary catheter. He presented at the Casualty Department with acute retention of urine. He was shocked, confused and in peripheral failure. A diagnosis of septic shock was made and he was admitted to the Intensive Therapy Unit (ITU). On admission, a central line was inserted and intravenous infusion of Taurolin commenced. Plasma endotoxin initially was 8pg/ml or 0.01EU/ml and during Taurolin infusion fell rapidly to 0 pg/ml (Figure 3). A corresponding good clinical response was observed and by the end of the Taurolin infusion his blood pressure had risen to 170/90mm Hg.

TABLE 1 Clinical information on 11 patients with severe sepsis

PATIENT	SEX	AGE	PRIMARY DIAGNOSIS	OUTCOME
W.R.	M	72	Perforated duodenal ulcer	Recovered
T.D.	M	72	Perforated gastric ulcer	Recovered
D.S.	M	71	Obstructive uropathy	Recovered
M.McA.	F	53	Ascending cholangitis	Recovered
M.O'B.	F	66	Perforated ulcerative colitis	Died
R.J.	M	60	Sigmoid volvulus	Died
C.C.	F	60	Obstructive uropathy	Recovered
G.McC.	F	71	Dehiscence of pyloroplasty	Recovered
R.A.	M	72	Perforation of caecum	Died
A.G.	F	60	Perforated diverticular disease	Recovered
T.McG.	M	64	Leak following anterior resection of rectum	Recovered

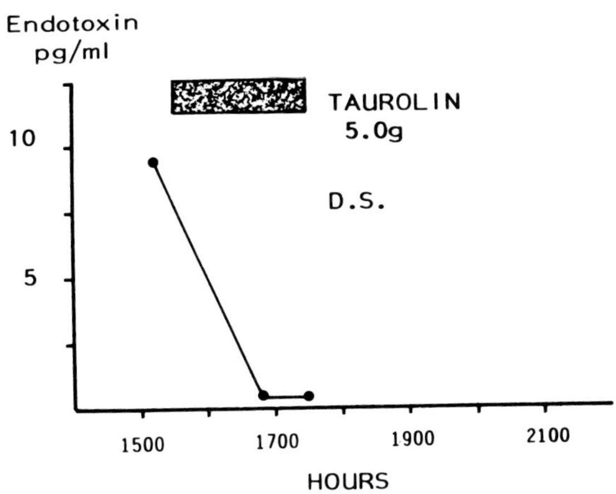

Figure 3 Endotoxin levels correlated with Taurolin infusion in patient DS.

Patient WR was a 72 year old man who was seen in the Casualty Department with severe upper abdominal pain. He was sent home to the care of his general practitioner. However, 24 hours later he was admitted to the Intensive Therapy Unit with peritonitis due to a ruptured viscus. At laparotomy, a perforated duodenal ulcer was repaired and a drainage gastrojejunostomy performed. Gross soiling was present. Antibiotic therapy with cefotaxime and metronidazole was administered concurrently with the Taurolin infusion. The initial high level of endotoxin, 29pg/ml or 0.35EU/ml, fell during the Taurolin infusion to 5.5pg/ml and then to zero (Figure 4). When the subsequent Taurolin infusion was commenced four hours later, the endotoxin level was 12pg/ml. This second infusion and the subsequent infusion, both resulted in lower endotoxin levels at the end of the infusion. Apart from some cardiac problems, the patient made an uneventful recovery.

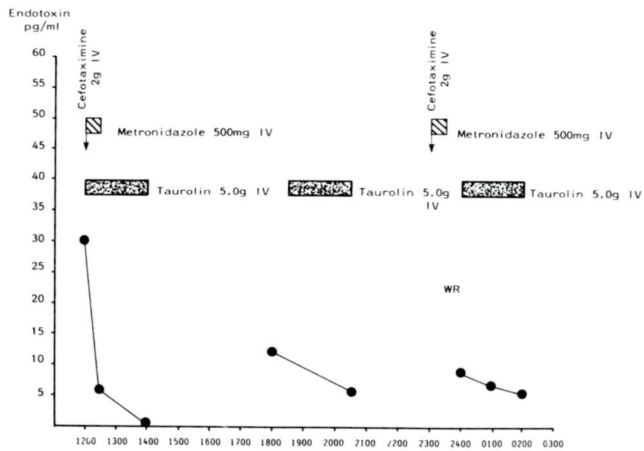

Figure 4 Endotoxin levels correlated with Taurolin and antibiotic therapy in patient WR

TD was a 73 year old man admitted to hospital with a one-week history of abdominal pain, which had become extremely severe several hours prior to admission. He was cold, clammy and collapsed with a rigid silent abdomen. At laparotomy, there was a two-centimetre perforation of the stomach with considerable peritoneal soiling. The peritoneum was washed out with saline, and 2g of Taurolin in 100ml saline were left in the abdomen at closure. An IV infusion of 5g Taurolin in 250ml was set up to run over the next 2 hours. Endotoxin levels, which were very high initially at 159pg/ml, showed a progressive fall to zero after the second dose of IV Taurolin (Figure 5). The patient required ventilation for some days but thereafter made a good recovery.

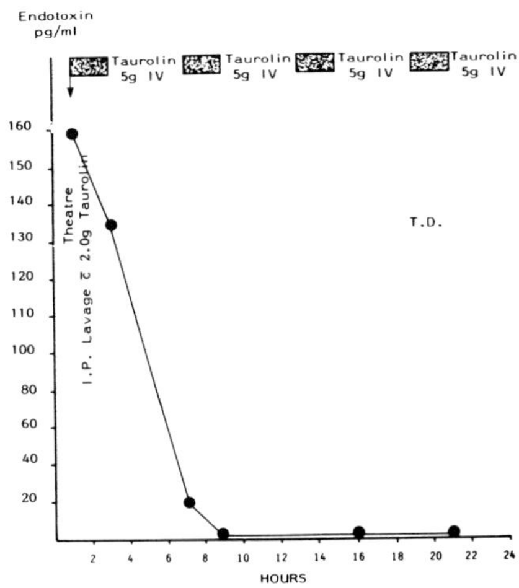

Figure 5 Endotoxin levels correlated with Taurolin administration in patient TD

GMcC was a 71 year old woman admitted to hospital with a haematemesis. Laparotomy was performed and a bleeding prepyloric ulcer oversewn. The operation was completed by pyloroplasty and vagotomy. Post-operatively, she was admitted to ITU for respiratory support. Her condition deteriorated rapidly, and over the next 24 hours she became confused, cyanosed, despite of 100% oxygen, and her pulse rose as her blood pressure became unrecordable. A further laparotomy was performed, and it was found that the pyloroplasty had dehisced allowing drainage of the stomach contents into the peritoneum. Taurolin (2g in 100ml saline) was instilled into the peritoneum and intravenous

Taurolin infusion was commenced. She was also given piperacillin and metronidazole. Before the first intravenous infusion of Taurolin, her endotoxin level was 18.0 pg/ml or 0.216EU/ml (Figure 6). At the end of the infusion it was 17pg/ml or 0.20EU/ml. The level measured in the interval between the first and second infusion was 28pg/ml which fell to 15pg/ml at the start of the second infusion. At the end of this infusion the level decreased to 11pg/ml. A similar pattern was observed with the third infusion. On two occasions, the endotoxin levels rose after the administration of piperacillin. After a prolonged recovery in TTU when she required ventilation and antibiotic therapy, she eventually was discharged well, one month after her original admission.

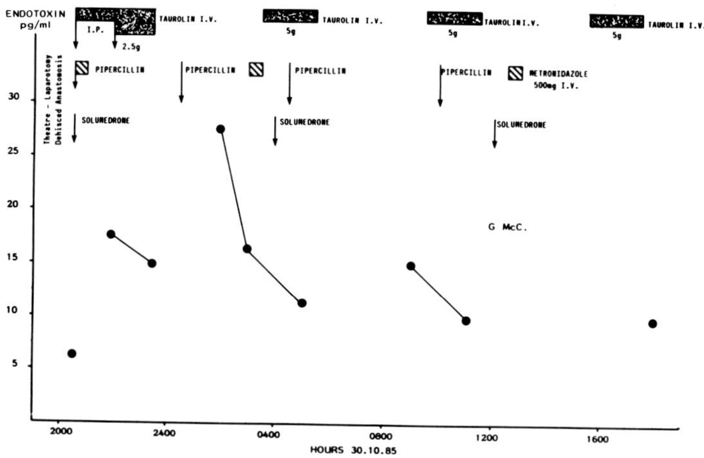

Figure 6 Endotoxin levels correlated with Taurolin and antibiotic therapy in patient GMcC

The above cases illustrate the decline in plasma endotoxin levels which may be achieved by the intravenous administration of Taurolin in patients with multi-system failure due to sepsis.

In severe cases, the level of endotoxin may be high and large doses of Taurolin may well be necessary especially if it is administered concurrently with bactericidal antibiotics. However, Taurolin is non-toxic and doses of more than 40g per day and total doses of 600-700g have been given to patients with pancreatitis (Vankemmel et al, 1985). Furthermore Taurolin differs significantly from the bactericidal antibiotic therapy used in septicaemia, where the bacteria are killed rapidly and may release endotoxin into the circulation. Taurolin not only kills the bacteria slowly but, in addition, is able to block any endotoxin which may be released as a result, thus preventing endotoxaemia (Browne et al, 1976).

In conclusion, our preliminary results indicate that intravenous administration of Taurolin appears to lower the circulating endotoxin levels in patients with endotoxaemia and also to influence these patients' outcome. This supports the claim that Taurolin represents the first "non-toxic" anti-endotoxin agent for use in clinical practice.

ACKNOWLEDGEMENT

The authors are grateful to Miss M McChristie for secretarial assistance.

REFERENCES

Bohnen J, Boulanger M, Meakins JL, McLean PH (1983). Prognosis in generalised peritonitis. Arch Surg 118: 285- 290.

Browne MK, Leslie GB, Pfirrmann RW (1976). Taurolin: a new chemotherapeutic agent. J Appl Bact 41: 363-368.

Browne MK, Leslie GB, Pfirrmann RW, Brodhage H (1977). The in vitro and in vivo activity of taurolin against anaerobic pathogenic organisms. Surg Gyn Obstet 145: 842-846.

Browne MK, MacKenzie M, Doyle PJ (1978). A controlled trial of taurolin in established bacterial peritonitis. Surg Gyn Obstet 145: 721-724.

Browne MK, Leslie G, Pfirrmann RW, McCartney Christine (1985). Studies on the antiendotoxin properties of Taurolin in animals and man. In Ishigami J (ed). "Recent Advances in Chemotherapy". Proceedings of the 14th International Congress of Chemotherapy, Kyoto: University of Tokyo Press, pp2075-2076.

Erb F, Imbenotte M, Huvenne JP, Vankemmel M, Scherpereel P, Pfirrmann RW (1983). Structural investigation of a new organic antiseptic: Taurolidine. Analytical study and application to identification and quantitation in biological fluids. Eur J Drug Metab Pharmokinet 8: 163-174.

Goto H, Nakamura S. (1980). Liberation of endotoxin from Escherichia coli by addition of antibiotics. Japan J Exp Med 50: 35-43.

Harris RI, Stone PCW, Evans GR, Stuart J (1984). Endotoxaemia as a cause of fever in immunosuppressed patients. J Clin Pathol 37: 467-470.

Jacob AI, Goldberg PK, Bloom N, Degenshein GA, Kozim PJ (1977). Endotoxin and bacteria in portal blood. Gastroenterol 72: 1268-1270.

Knight BI, Skellern GG, Browne MK, Pfirrmann RW (1981a) Peritoneal absorption of the antibacterial and antiendotoxin taurolin in peritonitis. Brit J Clin Pharmacol 12: 695-699.

Knight BI, Skellern GG, Browne MK, Pfirrmann RW (1981b). The characterisation and quantitation by HPLC of the metabolites of taurolin. Brit J Clin Pharmacol 12: 439-440.

Ledingham IMcA, Bradley JA, McCartney AC, Redding PJ (1983). Infection and septic shock. In Bain WH, Taylor KM (eds). "Handbook of Intensive Care", London: John Wright, pp196-237.

Levin J (1982). The Limulus test and bacterial endotoxins: some perspectives. In Watson SW, Levin J, Novitsky TJ (eds): "Endotoxins and Their Detection with the Limulus Amebocyte Lysate Test", New York: Alan R Liss, pp7-24

McCartney AC, Piotrowicz BI, Edlin SE, Ledingham IMcA (1987). Evaluation of the chromogenic Limulus lysate assay in septic shock. In Watson SW, Levin J, Novitsky TJ (eds): "Detection of Bacterial Endotoxins with the Limulus Amebocyte Lysate Test", New York: Alan R Liss, pp459-474.

Pfirrmann RW, Leslie GB (1979). The anti-endotoxic activity of taurolin in experimental animals. J Appl Bacteriol 46: 97-102

Pfirrmann RW (1985). Taurolin: Ein neues Konzept zur antimikrobiellen chemotherapie chirurgischer Infektionen Einfuhrung und Ubersicht. In Bruckner WL, Pfirrmann RW (eds): "Taurolin", Munich: Urban and Schwarzenberg, pp3-23.

Piotrowicz BI, Edlin SE, McCartney AC (1985). A sensitive chromogenic Limulus amoebocyte lysate micro-assay for detection of endotoxin in human plasma and water. Zbl Bakt Hyg A 260: 108-112.

Shenep JL, Mogan KA (1984). Kinetics of endotoxin release during antibiotic therapy for experimental Gram-negative bacterial sepsis. J Infect Dis 150: 380-388.

Shenep JL, Barton RP, Mogan KA (1985). Role of antibiotic class in the rate of liberation of endotoxin during therapy for experimental Gram-negative bacterial sepsis. J Infect Dis 151: 1012-1018.

Thomas LLM, Sturk A, Buller HR, Joop K, ten Cate JW (1985). The in vitro inactivation of endotoxins by Taurolin. In Bruckner WL, Pfirrmann RW (eds): "Taurolin", Munich: Urban and Schwarzenberg, pp78-81.

Vankemmel M, Scherpereel PL, Leroy B, Binot D, Kuhle E (1985). Utilisation de le Taurolidine en chirurgie bilio-pancreatique et digestive In Bruckner WL, Pfirrmann RW (eds): "Taurolin", Munich: Urban and Schwarzenberg, pp254-260.

Bacterial Endotoxins: Pathophysiological Effects, Clinical Significance, and Pharmacological Control, pages 373–382

PRODUCTION, CHARACTERIZATION AND BIOLOGICAL EFFECTS OF MONOCLONAL ANTIBODIES TO DIFFERENT PARTS OF THE GRAM-NEGATIVE LIPOPOLYSACCHARIDE CORE REGION.

Ben J. Appelmelk, Marian Verweij-van Vught, Helmut Brade, Janneke Maaskant, Wim Schouten, Bert Thijs and David MacLaren.

Department of Medical Microbiology (B.J.A., M.V.v.V., J.M., W.S., D.McL) and Medical Intensive Care Unit (B.T), Vrije Universiteit, Amsterdam; Forschungsinstitut Borstel (H.B), Borstel, FRG.

INTRODUCTION

The high incidence of, and mortality due to Gram-negative sepsis and shock remain matters of serious concern. One important reason why antibiotic therapy fails frequently in saving a patient's life, is its inability to neutralize endotoxin (lipopolysaccharide, LPS), a substance liberated from the bacterial cell wall and of crucial importance in the etiology of Gram-negative shock. Given the earlier successful use of antisera to neutralize bacterial exotoxins (e.g. tetanus toxin) it was logical to prepare anti-endotoxin antisera for human serotherapy. An antiserum effective against the major pathogens, i.e. *Escherichia coli, Klebsiella pneumoniae* and *Pseudomonas aeruginosa,* would be of great clinical relevance. An obvious target for the preparation of a cross-protective antiserum is the lipopolysaccharide core region, a structure which is to some degree conserved in Gram-negative species. Polyclonal antisera to the core, obtained by the use of rough strains lacking the O-antigen as vaccines have indeed been shown to be cross-protective. However, results have been variable, and moreover the hard proof that anti-core antibodies were the agents responsible for protection could not be provided. In addition it has been difficult, if not impossible to answer questions like: what is the mechanism of protection, i.e. is it mediated by opsonization and killing of bacteria or by neutralization of LPS? Likewise, the question to which part of the core cross-protective antibodies were directed has not been answered satisfactorily. The use of monoclonal instead of polyclonal antisera would make it possible to clarify some of these problems. We will now discuss our experience with the preparation and characterization of murine monoclonal antibodies, directed to different parts of the core, and discuss their biological effects.

MATERIALS AND METHODS

Monoclonal Antibodies

Monoclonal antibodies to J5 LPS or lipid A (Appelmelk *et al.*, 1987b) were obtained by immunization of Balb/c mice with heat-killed (2.5 h) J5 cells ($7x10^8$ colony forming units (c.f.u.), i.v. on days 0, 21, 42 and 45). Monoclonal antibodies to LPS of *Salmonella minnesota* R595 (Appelmelk *et al.*, 1987a) were obtained after immunization with a variety of R595 vaccines (heat-, acetone-, and alkali-killed cells and live bacteria). Fusion of immune B cells with X63-Ag8.653 took place 3 days after the last immunization. Clones were tested for production of antibody by screening in Elisa for reactivity with rough LPS and lipid A (Appelmelk *et al.*, 1985); subcloning was done using conventional limiting dilution methodology. Finally ascites were prepared. The monoclonals were characterized by measuring the reactivity of serially diluted ascites specimens against rough LPS or lipid A, coated on Elisa plates. We used LPS from *S. minnesota* strains R60 (Ra-chemotype), R345 (Rb2), R5 (Rc), R7 (Rd1) and R595 (Re); *E. coli* strain J5 (Rc); lipid A from the *E. coli* strains F515 and D31m4 (both Re), and from *S. minnesota* R595. LPS and lipid A were complexed with bovine serum albumin prior to their use as coating antigens. Likewise, by use of the appropriate antigen and HRP-labelled conjugates, immunoglobulin class and subclass were determined.

Binding of Monoclonal Antibodies to Heat-killed or Live Bacteria.

This was investigated (Appelmelk *et al.*, 1987b) in a competitive Elisa where constant amounts of antibody yielding an OD=1.7 in the absence of competitor were incubated with $5x10^9$ bacteria (*E. coli, K. pneumoniae* or *P. aeruginosa*), followed by centrifugation, and back-titration of antibody; in this competitive assay a drop of Elisa response after incubation with bacteria indicates binding.

The Effect of Monoclonal Antibodies on the Clearing of *E. coli* 0111:B4 from the Mouse Peritoneal Cavity.

Exponentially growing bacterial cells ($5x10^5$ c.f.u./ml) were incubated with an equal volume of antibody (45 min at 37° C), which was followed by injection of the mixture into the peritoneal cavity of Swiss albino mice. At t=0, 30, 60 and 90 min, mice were killed, and the peritoneal cavity was washed with PBS. The number of bacteria in the fluid obtained was determined by viable counting.

Protection Experiments

1. Lethal endotoxemia. The Actinomycin D sensitized mouse model was used. Four groups of 6 mice were used for each LD50 determination. Two variations of the model were employed, i.e. antibody was either injected i.p., 24 h before i.p. challenge with LPS, or antibody was mixed with LPS and injected i.v. The LD50s in this model for rough LPS are about 1 ng.

2. Lethal sepsis. Exponentially growing *E. coli* 0111:B4 cells were serially diluted, mixed with a mucin-hemoglobin suspension and injected i.p. into mice. The LD50s obtained in this model are very low (<1000 c.f.u.). Antibody was injected i.p., 24 h before the bacterial challenge.

For both models, LD50s were calculated according to Spearmann-Kärber, using the number of deaths recorded one week after challenge. Statistical differences between mortality figures were evaluated by means of chi-square analysis.

RESULTS AND DISCUSSION

After immunization with J5 (Appelmelk *et al.*, 1987b), IgG-clones only (IgG1,2a,2b,3) were obtained. In Elisa, two reactivity patterns were seen: the first pattern involved a strong reactivity (ascites titer= $1:10^6$) with lipid A from *E. coli* or *S. minnesota*, and virtually no reactivity with rough LPS. This pattern was seen with two clones. We concluded that these monoclonals are anti-lipid A antibodies, recognizing an epitope exposed in free, but not in covalently linked lipid A as present in intact LPS. The second pattern involved a strong binding (titers up to 1:10^7) with LPS of the Rc-chemotype (strains *E. coli* J5 and *S. minnesota* R5), and virtually no binding with lipid A or with LPS of the other chemotypes. Thus, these monoclonals recognize the terminal part of the Rc-LPS structure, most likely involving terminal glucose. This pattern was seen in 8 out of 10 anti-J5 clones. We conclude that the terminus of Rc-LPS is immunodominant. The latter clones bound very well to heat-killed as well as live cells of 3 *E. coli* smooth strains; no reactivity with a variety of non-*E. coli* strains (*K. pneumoniae, Pseudomonas aeruginosa, Proteus mirabilis, Haemophilus influenzae*) was observed. In contrast, the lipid A clones recognized -not unexpected- a structure common to many Gram-negatives, and a strong binding with heat-killed cells of both *E.coli* and non-*E. coli* strains was observed. An ascitic pool consisting of equal volumes of all anti-J5 clones did not lead to increased clearing of *E. coli* 0111:B4 from the mouse peritoneal cavity, despite the fact that 8 clones bound very well to live bacteria. The control mice received ascites from clone Pen 7, which is an antibody specific for benzylpenicillin; the number of bacteria recovered from these control mice at t= 90 min was not significantly lower than the number recovered at t=0. In contrast, a polyclonal murine antiserum to 0111:B4 caused -within a time span of 90 min- more than 99% clearing (compared to Pen 7 controls at t=90) from the peritoneal cavity. Thus our anti-core monoclonals are not opsonic. Animal sepsis studies showed that neither the ascitic pool, nor individual ascites were able to increase the LD50; the control mice either received PBS or Pen 7 ascites. In contrast, a polyclonal rabbit anti-J5 antiserum caused an increase in LD50 from 826 to 3.6×10^4 c.f.u. (Appelmelk *et al.*, 1986). Tests in the Actinomycin D model were negative too: neither ascitic pool, nor individual clones were protective against J5 LPS; ascites were either injected i.p., 24 h before challenge or i.v., together with LPS. A lipid A specific clone, likewise, was not protective against challenge with lipid A. These findings led us to conclude the following: The observation that clones to the Rc LPS terminus cross-react only with strains of *E. coli* means that more broadly cross-reactive clones probably are directed to epitopes more proximal on the LPS, i.e. nearer to the lipid A part. This assumption is in accordance with the known heterogeneity of the outer core region, with other structural and immunochemical data concerning the LPS core region, and with the finding that our clones to the deepest, lipid A part indeed exhibited such broader reactivity. The fact that clones to the outer core do not opsonize bacterial cells, most likely means that monoclonals to deeper parts of the core will not be opsonic either. In other words, when broadly cross-reactive monoclonals do have a protective effect, this will most likely be mediated not by enhanced opsonization of bacterial cells, but by neutralization of LPS. The additional findings that neither the clones binding to the outer core, not those to free lipid A were capable of neutralizing LPS or lipid A, led us to assume that neutralization of LPS requires binding to an epitope which is neither too far away from lipid A, nor too close to it. We want to emphasize that the logic of the above reasoning is flawed by the absence of clones of the IgM class in the anti-J5 panel. Despite this limitation we decided to make clones possessing the required binding characteristics, i.e. binding to epitopes consisting of parts of lipid A and the core together, or consisting of structures localized in the ketodeoxyoctonate (KDO) region, the latter constituting that part of the core most near to lipid A. LPS from *S. minnesota* R595, which consists of lipid A and two KDO residues was chosen as the antigen to elicit such monoclonals. 16 Fusions were done (Appelmelk *et al.*, 1987a), and out of a total of 10,000 clones 12 were chosen for subcloning, further growth and preparation of ascites. The criteria used to select the clones were high reactivity to R595 LPS, immunoglobulin class and subclass, reactivity to LPS from other rough strains of *S. minnesota*, and reactivity to lipid A of strain R595. Relevant data are presented in table 1, which also contains the antibody titers (Elisa) toward R595 LPS as well as the LPS neutralization results (Appelmelk *et al.*, 1987c). The results are expressed as protection factor, which constitutes the LD50, obtained in animals which had received anti-R595 antibody, divided by the one obtained in Pen 7-immunized mice. As shown, the clones prepared were of the IgG2a, the IgG2b, the IgG3 subclass, or the IgM class. At first sight, the reactivity patterns obtained seem to indicate that all clones except nr. 22 recognize structures present in free, as well as covalently linked lipid A.

TABLE 1. The Effect of Monoclonal Antibodies against *Salmonella minnesota* R595 LPS on Lethal Endotoxemia in Actinomycin D-sensitized Mice

	MAB	Reactivity pattern (ELISA)	Titer (R595 LPS)	Protection Factor	p-Value
14	IgM	R595, R345, LA	27.000	7	NS
15	IgG2b	R595, R345, R7, LA	729.000	69	$p<0.001$
17	IgG2b	R595, LA	243.000	91	$p<0.001$
19	IgG3	R595, R345, LA	9.000	15	$p<0.05$
20	IgM	R595, R345, LA	729.000	46	$p<0.001$
21	IgG3	R595, R5, R7, LA	27.000	21	$p<0.025$
22	IgG3	R595	729.000	5	NS
25	IgG3	R595, R345, LA	729.000	5	NS
26	IgG2a	R595, R345, LA	729.000	22	$p<0.01$
27	IgG3	R595, R345, LA	81.000	1	NS
28	IgM	R595, LA	9.000	2	NS
29	IgG2a	R595, LA	729.000	2	NS

Portions of 0.3 ml ascites were mixed with varying amounts of R595 LPS (0.1, 1, 10 and 100 ng) and 20 μg Actinomycin D and injected intravenously.
Salmonella minnesota R595 (Re), R7 (Rd1), R5 (Rc), R345 (Rb2)
Reactivity pattern= strong reactivity with the LPS indicated
LA= lipid A prepared from strain R595; NS= not significant, $p>0.05$

However, a more detailed study (Brade *et al.*, 1987) by means of passive hemolysis (PH) instead of Elisa showed that clone 20 and 25 recognized the terminal KDO residue and a 2.4 linked KDO-KDO disaccharide respectively, without the involvement of lipid A. This discrepancy results most likely from the fact that hydrophobic structures like lipid A easily interact non-specifically with monoclonal antibodies, which in Elisa shows up as a strong increase in OD. In contrast, PH measures specific binding, which results in complement activation and lysis of the LPS-sensitized red blood cells, such lysis not taking place after non-specific interactions. A second misleading result was obtained by us through using Elisa methodology: in Elisa clones 20 and 25 behaved exactly the same, i.e. a very strong binding to rough LPS of strains R595 (Re) and R345 (Rb2 chemotype), and virtually no reactivity with LPS from the other chemotypes tested. However, in PH it was found that clone 20, unlike clone 25, reacted also with LPS from R7 (Rd1) and R5 (Rc) LPS. Most likely, the latter interactions are of lower affinity compared to the interaction with R595 and R345 LPS, which allows a disruption of antigen-antibody interaction caused by the Elisa plate washing procedure. As shown in the table, 6/12 clones neutralized LPS, causing increases in LD50 from a 15-fold (clone 19) to a 91-fold (17). The other 6 clones did not have significantly protective effects. Then, what determines whether a clone will be protective or not? As shown, antibody class and subclass alone did not determine the outcome of the neutralization experiments. The same conclusion applies to antibody titer to R595 LPS, and to affinity of antibody, when the slope of the Elisa dose-response curve to R595 is taken as a measure for affinity (Fig. 1). An important question is whether or not the neutralizing monoclonals all bind to the same epitope. The findings that clone 20 (protective) binds to the terminal KDO without the involvement of lipd A, and that clone 17 (protective) binds to an epitope requiring parts of lipid A and KDO (Rozalski *et al.*, 1987), indicate that this is not the case. Another possibility could be that the binding sites of protective clones, albeit not the same, have the common property of being very close to each other.

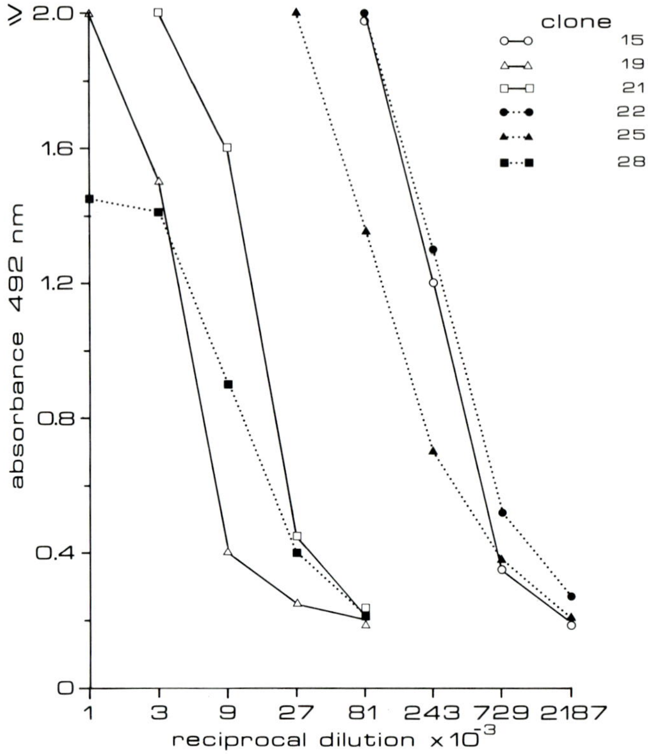

Fig. 1. Representative Elisa dose-response curves of monoclonal antibodies to LPS of *Salmonella minnesota* R595 (Re-chemotype). Protective clones are indicated with open symbols and lines, non-protective ones by closed symbols and dotted lines.

The exact chemical structure of R595 LPS is fully known now, and thus a two-dimensional picture of its structure is available; however, the question whether or not certain binding sites are close to each other or not, cannot yet be answered since sufficient information on the three-dimensional structure of the R595 LPS molecule is lacking. Consequently, whether the binding site of a given monoclonal determines its neutralizing effect, or that protection is determined by yet another, presently unknown parameter, is an open question.

In summary, we have produced monoclonals to different parts of the core, and only those recognizing structures in the KDO-lipid A region showed biological effects, i.e. neutralization of LPS. Despite the fact that a very much simplified model was chosen, i.e. neutralization of homologous LPS, and mixture of antibody and LPS before injection into mice, we have no explanation why some monoclonals were effective and others not. For neutralization of the toxic part of LPS, specific recognition of lipid A -the toxic part- is not required (clone 20). We assume that, by binding to KDO either a shielding or a deformation of lipid A takes place, thus reducing its toxicity. Clone 20, recognizing terminal KDO which is a common structural element of LPS (Tacken *et al.*, 1986) has now been tested for cross-reactivity with heterologous LPS and bacteria. It was found to bind very strongly (Elisa) with LPS and heat-killed bacteria from *K. pneumoniae*, and also with cells of *E. coli* and *P. aeruginosa*. The ability of clone 20 to protect against lethal sepsis is currently being evaluated.

Acknowledgements: We want to thank Dr. E.J. Ziegler for her continuing support of and advice on this project. This work was financed partially by the Praeventiefonds, Den Haag, and by the Deutsche Forschungsgemeinschaft (grant Br 731/4-1).

REFERENCES

Appelmelk BJ, Verweij-van Vught AMJJ, MacLaren DM, Thijs LG (1985). An enzyme-linked immunosorbent assay (ELISA) for measurement of antibodies to different parts of the Gram-negative lipopolysaccharide core region. J Immunol Meth 82:199-207.

Appelmelk BJ, Verweij-van Vught AMJJ, Maaskant JJ, Schouten WF, Thijs LG, MacLaren DM (1986). Use of mucin and hemoglobin in experimental murine Gram-negative bacteremia enhances the immunoprotective action of antibodies reactive with the lipopolysaccharide core region. Antonie van Leeuwenhoek 52:537-542.

Appelmelk BJ, Verweij-van Vught AMJJ, Maaskant JJ, Schouten WF, Thijs LG, MacLaren DM (1987a). Monoclonal antibodies detecting novel structures in the core region of *Salmonella minnesota* lipopolysaccharide. FEMS Microbiol Lett 40:71-74.

Appelmelk BJ, Verweij-van Vught AMJJ, Maaskant JJ, Schouten WF, de Jonge AJR, Thijs LG, MacLaren DM (1987b). Production and characterization of monoclonal antibodies reacting with the lipopolysaccharide core region of Gram-negative bacilli. J Med Microbiol (in press).

Appelmelk BJ, Verweij-van Vught AMJJ, Brade H, Maaskant JJ, Schouten WF, Thijs LG, MacLaren DM (1987c). Prevention of lethal endotoxemia by treatment of mice with monoclonal antibodies to *Salmonella minnesota* R595 lipopolysaccharide. (Submitted for publication).

Brade L, Kosma P, Appelmelk BJ, Paulsen H, Brade H (1987). Use of synthetic antigens to determine the epitope specificities of monoclonal antibodies against the 3-deoxy-D-*manno*-octulosonate region of bacterial lipopolysaccharide. Infect Immun 55:462-466.

Rozalski A, Brade L, Appelmelk BJ, Kosma P, Brade H (1987). Characterization of monoclonal antibodies to the KDO region of bacterial lipopolysaccharide. Poster no. 75, presented at the International Conference on Endotoxins Amsterdam II.

Tacken A, Rietschel E, Brade H (1986). Methylation analysis of the heptose/3-deoxy-D-*manno*-octulosonic acid region (inner core) of lipopolysaccharide from *Salmonella minnesota* rough mutants. Carbohydr Res 149:279-291.

Bacterial Endotoxins: Pathophysiological Effects, Clinical Significance, and Pharmacological Control, pages 383–393

Generation of a Protective Human Monoclonal for the Treatment of Gram-Negative Sepsis

Key Terms: Human MAbs; septic shock

James W. Larrick*, Oanh Gray*, Tracy Deinhart*, Jeff Wang*, Jeffrey Flatgaard†, and K.T. Chong‡

Departments of Immunology*, Process Development†, and Pharmacology-Toxicology‡, Cetus Corporation, Palo Alto*, California 94303 and Emeryville‡, California 94608

ABSTRACT

Using a combination of Epstein-Barr virus transformation and cell fusion to a mouse/human heteromyeloma cell line, we have generated a human monoclonal antibody, D234, that recognizes cross-reactive determinants on the lipopoly-saccharides (LPS) of Gram-negative bacteria. Direct binding to a series of rough mutant LPSs and smooth LPSs demonstrates the broad cross-reactivity of this monoclonal. D234 inhibits Re LPS-induced chemiluminescence. In a murine model of Gram-negative sepsis, D234 given after infection significantly increases survival. This antibody may have therapeutic potential for the treatment of life-threatening Gram-negative infections in humans.

INTRODUCTION

Despite the development of novel antibiotics and aggressive supportive techniques, Gram-negative sepsis continues to be a major cause of mortality and morbidity in modern hospitals. This is due to the inability of current antibiotics to neutralize the major toxic moiety of Gram-negative bacteria, endotoxin, or lipopolysaccharide. LPS is an essential component of the outer membrane of all Gram-negative bacteria. It has a tripartite structure consisting of (a) the variable O-specific carbohydrate side chains, from which the immunological diversity of Gram-negative bacteria arises, (b) the core, a more biochemically conserved region consisting of several sugars, and (c) lipid A, the most biochemically conserved region and the most biologically active, toxic portion of the molecule (14).

Antibodies binding the O-side chains of LPS protect against challenge with the homologous LPS (16,20,26). However, the marked diversity of these antigens makes them poor targets for active or passive immunization.

A number of workers have shown that antisera generated by immunization with rough mutant bacteria such as Salmonella minnesota Re595 or Escherichia coli J5, which display core antigenic determinants on their surfaces, protect animals from challenge by heterologous LPS or bacteria (2,23,15,16,29).

Recently, Ziegler et al. (30) showed that human serum obtained ten days after immunization with heat-killed E. coli J5 bacteria reduced mortality in patients with Gram-negative bacteremia and shock by 50%, compared to a preimmune serum.

Despite these results, some groups have inconsistently shown binding (6) or protection using antisera (7-9, 24) or monoclonals (17) in animals. In fact, the nature of the protective entity in the postimmunization sera has not been firmly established.

The generation of monoclonal antibodies (MAbs) which bind well-defined epitopes in the LPS core region would permit the critical analysis of these conflicting results. We have recently reported the preparation of such human MAbs (13,21) and confirmatory data have come from several other laboratories (10,17,18,19,27).

In this report we demonstrate (a) the binding of a monoclonal antibody to highly conserved antigenic determinants in the core-lipid A region of LPS, (b) the neutralization of LPS-induced activation of the respiratory burst in granulocytes, and (c) protection of mice against lethal Gram-negative sepsis. This human monoclonal antibody may have therapeutic or prophylactic potential for diseases caused by Gram-negative bacteria.

METHODS

Human Monoclonals were prepared as previously described (13,21).

Bacterial Strains. The uridine diphosphate-4-galactose epimerase-deficient Escherichia coli J5 mutant (4) was obtained from A. I. Braude (University of California, San Diego, CA). E. coli 0111:B4 and Salmonella minnesota wild-type strains were obtained from L. Shore (List Biological

Laboratories, Inc., Campbell, CA). E. coli SM18, a type-02 clinical isolate (Saint Mary's Hospital, San Francisco, CA), was cultured overnight in brain-heart infusion broth (BHI) at 37°C, harvested by centrifuging at 6000 rpm for 15 min, washed in phosphate-buffered saline (PBS), resuspended in BHI containing 20% glycerol, and aliquoted and stored at -70°C. The viability of the bacteria was determined by plating ten-fold serial dilutions onto blood agar plates and counting the colony-forming units (cfu) after a 24-hour incubation at 37°C. The titer was expressed as the mean number of cfu/ml of stock bacteria. The bacteria were prepared for injection by diluting a freshly thawed vial in PBS at room temperature and were administered at 1 x minimal LD_{100} as the challenge dose in vivo.

Lipopolysaccharides (LPS), Lipid A, and Part Structures. Rough and smooth LPS, purified by Galanos (5) and modified Westphal procedures (28), respectively, were obtained from List Biological Laboratories or Ribi Immunochem Research, Inc., Hamilton, MT. Most LPS preparations contained less than 1% protein and nucleic acid by weight. Lipid A's, prepared by acetic acid hydrolysis of Re mutant LPS, were obtained from List or from Ribi. Monophosphoryl lipid A (22) derived from Re mutant S. minnesota LPS, was obtained from Ribi. The lipid A precursor molecule designated lipid X (25), was purchased from Lipidex, Middleton, WI.

Inhibition Enzyme-linked Immunosorbent Assays (ELISA). Binding of purified MAbs to LPS on plate. LPS was coated on polystyrene 96-well plate, in 0.05 M bicarbonate buffer pH 9.6, at 10 µg/ml at 4°C overnight. After washing the plates with PBS/0.1% Tween 20, dilutions of MAb in PBS were allowed to bind for 1 hour at room temperature. Plates were washed and the amount of MAb bound to the LPS was determined by detection with a goat antihuman IgM coupled to horseradish peroxidase (Jackson Laboratories, Bar Harbor, ME) in PBS/0.1% Tween 20/1% BSA. Colorimetric reaction was developed with O-phenylenediamine-2HCl. Absorbance was read at 492 nm. Three data points were done per experiment. A total of three experiments were done on three different days.

Blocking of purified MAbs by LPS in solutions. Prior to the actual blocking experiment, a determination of 75% binding for each MAb was done. It was determined by an EIA using LPS bound to polystyrene 96-well plates. E. coli J5 LPS (List) was used for D234 and L118, and E. coli SM18 wild-type LPS (in-house prepared) was used for T88.

Following the blocking experiment, the MAbs at 75% binding concentration were preincubated for 1 hour, room temperature, with blocking LPS in solution at different concentrations. The mixtures were then moved to a plate precoated with E. coli J5 LPS or E. coli SM18 LPS.

After 1 hour at room temperature, the plates were washed with PBS/0.1% Tween 20. The amount of MAb bound to LPS on the plate was determined by detection with a goat antihuman IgM coupled to horseradish peroxidase, in PBS/0.1% Tween 20/1% BSA. Colorimetric reaction was developed with O-phenylenediamine-2HCl. Absorbance was read at 492 nm. Three data points were done per experiment. A total of three experiments were done on three different days.

IgM-Specific ELISA. Immulon II flat-bottom microtiter plates were coated with goat antihuman IgM (1:100) in 50 mM bicarbonate buffer (pH 9.6). After 90 min at 37°C, plates were washed as above and 100 µl of PBS^{++}, 1% BSA, 0.05% Tween 20, and 0.01% thimerosal, followed by antibody containing supernatants was added to each well. Plates were incubated for 30 min at 22°C, washed, and developer and substrate were added as described above.

Measurement of Neutrophil Chemiluminescence was performed as previously described (12).

Mice. Female CD1 mice (Virus-Antibody-Free) were purchased from Charles River Laboratories, Inc., Wilmington, MA.

Determination of LD_{100}. Groups of mice were injected intraperitoneally (i.p.) with 0.5 ml of a serial ten-fold dilution of virulent E. coli SM18 in sterile saline, and mortality was recorded daily. The LD_{100} represented the minimum number of bacteria necessary to kill all mice within a group.

Statistical calculations. The proportion of animals surviving on day seven after bacterial challenge in IL2-treated mice was compared to that of the saline or excipient buffer control-treated animals by use of the one-tailed Fisher's Exact test.

RESULTS

Specificity of Anti-Lipopolysaccharide Human Monoclonal Antibodies. Some binding data for D234 was described previously (21). Figure 1 shows the direct binding of D234 to smooth and rough mutant lipopolysaccharides. Binding to smooth LPS is less uniform than binding to rough LPS. The ability of the MAb to recognize shared determinants on various glycolipids was studied using a competitive inhibition ELISA. The maximum reactivity of the MAb for E. coli J5 core glycolipid was determined (12.5 µg/ml). The concentration of MAb giving 75% of this signal was used for inhibition studies. This concentration of MAb was premixed with various concentrations of lipid X, lipid A, S. minnesota R595, E. coli J5 and smooth E. coli (SM18) LPS. The binding of MAb was

Figure 1: Binding of D234 to various purified endotoxins. Three experiments (each the mean of triplicates) are shown for each endotoxin. The ordinate gives the amount of D234 bound. The abscissa refers to the endotoxins listed in the key.

Key:

Left: Smooth LPS

1. E. coli 026:B6
2. E. coli 055:B5
3. E. coli 0111:B4
4. E. coli 0127:B8
5. E. coli K235
6. S. minnesota
7. S. typhimurium
8. K. pneumoniae
9. S. marsescens
10. V. cholerae
11. Y. enterocolitica
12. P. aeruginosa IT-1
13. P. aeruginosa IT-2
14. P. aeruginosa IT-4
15. E. coli SM18

Right: Rough LPS

1. S. minnesota Ra
2. S. minnesota Rb
3. S. minnesota Rc
4. E. coli J5
5. S. minnesota Rd
6. S. minnesota Re
7. E. coli K12, D21m4

Lipid A

8. E. coli K12
9. S. minnesota
10. S. typhimurium

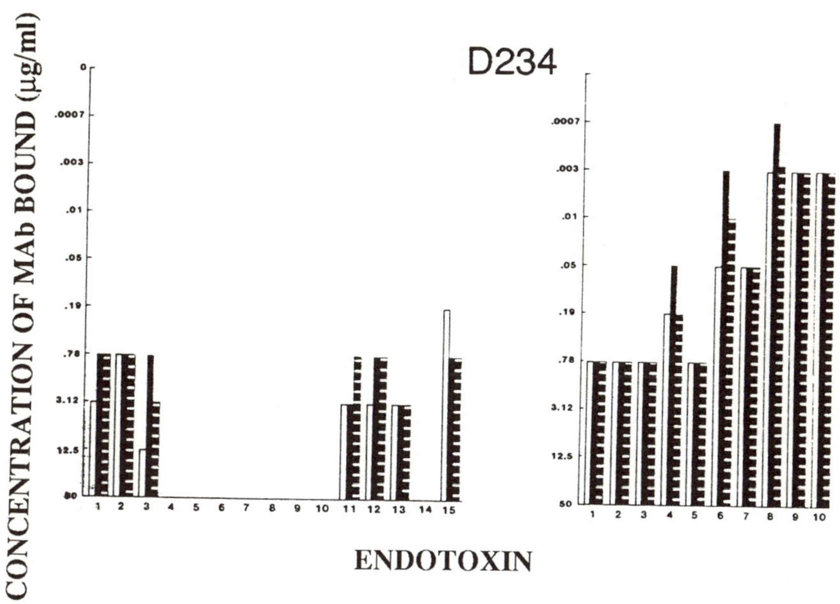

Figure 2: LPS Blocking of Purified D234 on J5 LPS

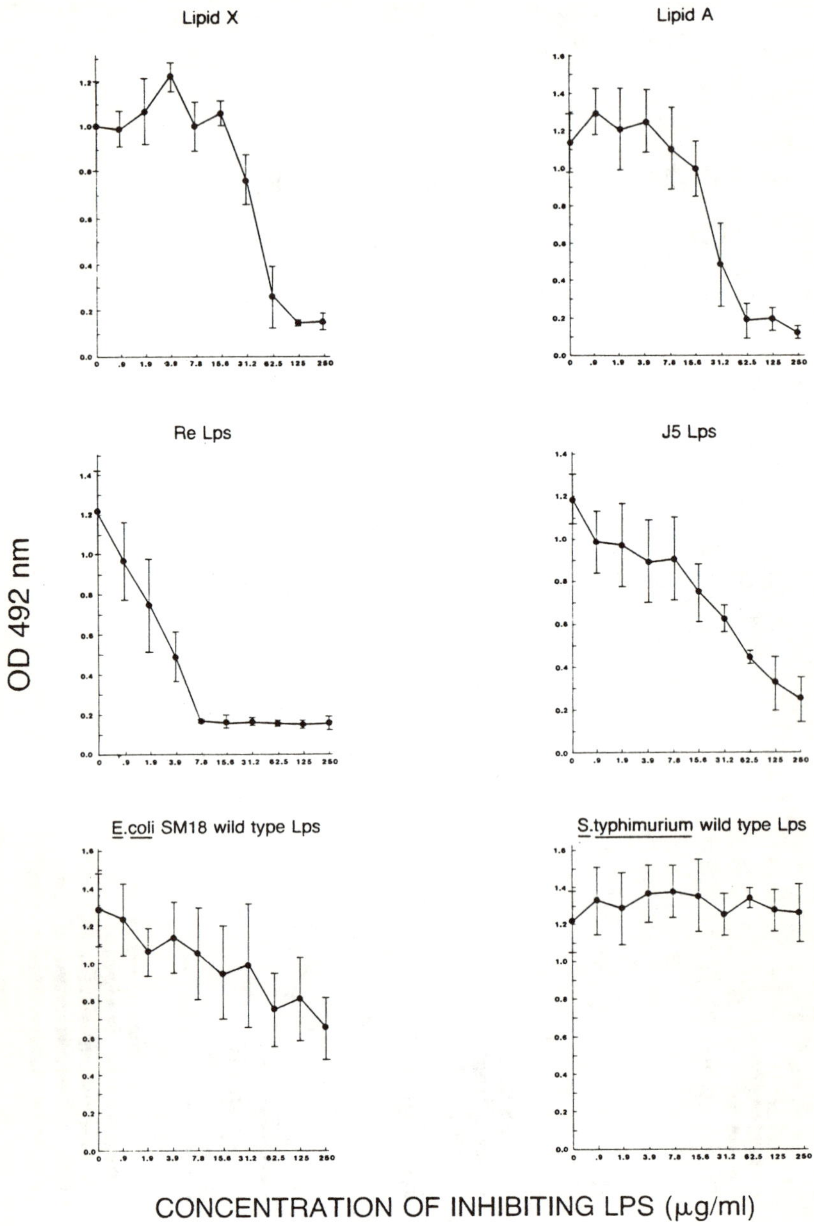

then measured by ELISA on the appropriate LPS. Figure 2 shows that the binding of D234 is best inhibited by the rough mutant lipopolysaccharides. A higher quantity of smooth LPS is required to give 50% inhibition. Direct binding, as well as inhibition studies, demonstrate that antibody D234 binds to a conserved glycolipid region of lipopolysaccharide. This epitope is exposed in some, but not all, LPSs from a panel of Gram-negative bacteria. Fifty percent inhibition were as follows: E. coli lipid X, 31.2 μg/ml; S. minnesota R595 lipid A, 32 μg/ml; S. minnesota R595 Re LPS, 2.6 μg/ml and SM18 LPS, 23.4 μg/ml.

Neutralization of LPS-Induced Chemiluminescence. Lipopolysaccharide causes a rapid transient increase in neutrophil chemiluminescence. Chemiluminescence of PMNs stimulated by S. minnesota Re-LPS (10 ng) preincubated with D234 is shown in Figure 3. D234 blocked the ability of LPS to stimulate neutrophil chemiluminescence. A nonbinding IgM myeloma protein did not diminish the LPS-induced signal (data not shown).

Figure 3: Anti-LPS Human Monoclonal Antibody D234 Blocks LPS-induced Chemiluminescence. Control: No MAb, no LPS. LPS - Re LPS 10 ng.

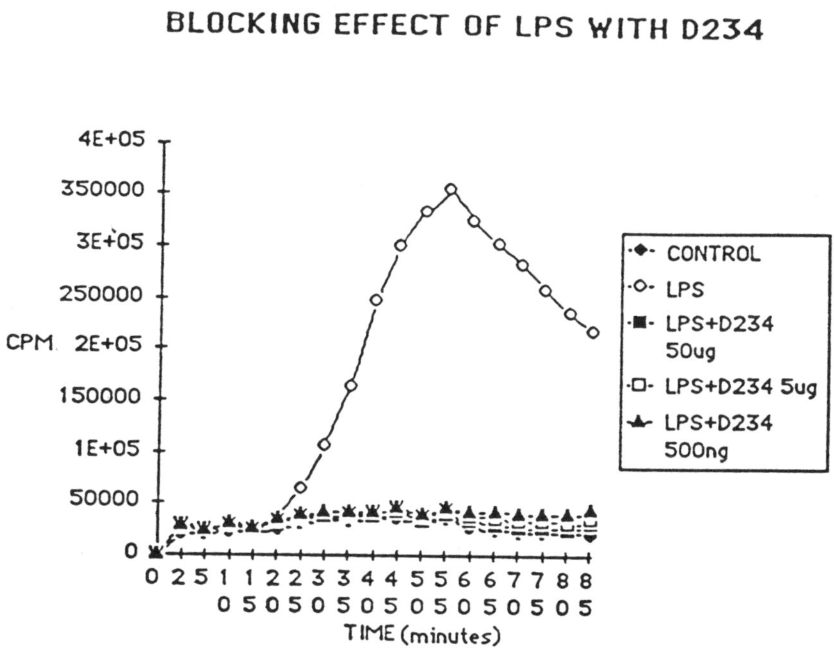

Therapeutic Efficacy of Cross-Reactive Human Monoclonals. Most workers reporting protective antibacterial antisera or monoclonals, use animal models in which the antibodies are given prior to challenge. Because of our previous work demonstrating the capacity of low quantities of endotoxin administered prior to challenge to protect lethally infected mice (3), we designed a treatment model. In this model, mice are given a peritoneal infection with an LD_{100} (5×10^7 CFU) of a clinical bacteremic E. coli strain followed at one and three hours by a dose of gentamicin (3.2 mg/kg).

D234 was administered intravenously one hour after infection. Table 1 shows the results of this experiment using antibody D234. Compared to groups of mice receiving only gentamicin, significant protection is provided by 2.5 mg of antibody mouse and higher doses of the monoclonal. Insignificant protection is given by the pool of nonspecific, nonbinding control myeloma proteins (Rockland IgM).

Table 1

Therapeutic Effect of D234 in E. coli SM18 Infection Model

Test	% Survival Day 3	% Survival Day 7
Bacteria control	0	0
Gentamicin + D234 (10 mg)*	65	60
Gentamicin + Rockland IgM (10 mg)	30	30
Gentamicin alone	25	20

* $P \geq 0.01$ for D234 vs. gentamicin alone

DISCUSSION

These studies demonstrate three points. First, it is possible to generate human monoclonal antibodies that bind to cross-reactive determinants on the lipopolysaccharides of Gram-negative bacteria. Second, this binding can neutralize the ability of Re core mutant LPS to stimulate the respiratory burst in granulocytes. Finally, at high doses, these antibodies provide protection from lethal Gram-negative sepsis in mice.

It is apparent from the binding studies that a cocktail of monoclonals will probably be required to cover all Gram-negative species.

The large doses of monoclonals given to protect mice deserves comment. It is unclear how much more effective the MAbs would be when used in their species of origin. Mice are known to be less sensitive to LPS than humans and might require more MAb/kg. Large doses of core-specific antisera (vs. type-specific antisera) are often required to show protection in infection models (2,16,23,29). Other workers have also reported that large amounts of the core-reactive MAb were required to demonstrate protection in mice (D. Dunn, personal communication). If similar quantities of endotoxin are released in mice and in humans, then the human dose of antibody would not be a direct mass equivalent to the dose required to protect mice.

These studies do not address the problem of how the MAbs have their effect and how truly cross-reactive they are in vivo. If the outer carbohydrate is shaved off of LPS when it is processed in vivo, then a single antibody might bind to all varieties of LPS. Our previous work with D234 (21) showed that this antibody bound much better to hydrolyzed vs. intact LPS. Studies of this antibody to demonstrate protection in other animal models of sepsis are in progress.

REFERENCES

1. Baumgartner, J.D., M.P. Glauser, J.A. McCutchan, E.J. Ziegler, G. van Melle, M.R. Klauber, M. Vogt, E. Muehlen, R. Luethy, R. Chiolero, and S. Geroulanos. 1985. Prevention of Gram-negative shock and death in surgical patients by antibody to endotoxin core glycolipid. Lancet ii:59-63.
2. Braude, A.I., H. Douglas, and C.E. Davis. 1973. Treatment and prevention of intravascular coagulation with antiserum to endotoxin. J. Infect. Dis. 128:S157-164.
3. Chong, K-T. and M. Huston. 1987. The complications of endotoxin contamination in the evaluation of anti-LPS antibodies against Gram-negative sepsis. Infect. and Immunity. (in press).
4. Elbein, A.D., and E.C. Heath. 1965. The biosynthesis of cell wall lipopolysaccharide in Escherichia coli. I. The biochemical properties of a uridine diphosphate galactose-4-epimeraseless mutant. J. Biol. Chem. 240:1919-1925.
5. Galanos, C., O. Luderitz, and O. Westphal. A new method for the extraction of R lipopolysaccharides. Eur. J. Biochem. 9:245-249.

6. Gigliotti, F., and J.L. Shenep. 1985. Failure of monoclonal antibodies to core glycolipid to bind intact smooth stains of Escherichia coli. J. Infect. Dis. 151:1005-1011
7. Greisman, S.E., E.J. Young, and J.B. DuBuy. 1973. Mechanisms of endotoxin tolerance. VII. Specificity of serum transfer. J. Immunol. 111:1349-1360.
8. Greisman, S.E., J.B. DuBuy, and C.L. Woodward. 1978. Experimental Gram-negative bacterial sepsis; reevaluation of the ability of rough mutant antisera to protect mice. Proc. Soc. Exp. Biol. Med. 158:482-490.
9. Greisman, S.E., J.B. DuBuy, and C.L. Woodward. 1979. Experimental Gram-negative bacterial sepsis: prevention of mortality not preventable by antibiotics alone. Infect. Immun. 25:538-557.
10. Kirkland, T.N., D.E. Colwell, S.M. Michalek, J.R. McGhee, and E.J. Ziegler. 1986. Analysis of the fine specificity and cross-reactivity of monoclonal anti-lipid A antibodies. J. Immunol. 137:3614-3619.
11. Larrick, J.W., B.J. Dyer, G. Senyk, S.M. Hart, R. Moss, D. Lippman, M.C. Jahnsen, J. Wang, H. Weintraub, and A.A. Raubitschek. 1985. In vitro expansion of human B cells for the production of human monoclonal antibodies. In: Human Hybridomas and Monoclonal Antibodies. E.G. Engleman, S.K.H. Foung, J.W. Larrick, and A.A. Raubitschek, eds. Plenum Press, New York. pp. 149-165.
12. Larrick, J.W., D. Graham, K. Toy, L.S. Lin, G. Senyk, and B.M. Fendly. 1987. Recombinant tumor necrosis factor causes activation of human granulocytes. Blood 69:640-644.
13. Larrick, J.W., M. Jahnsen, G. Senyk, S. Weiss, and K. Watson. 1986. Production of human monoclonal antibodies recognizing cross-reactive determinants on lipopolysaccharides. In: Szentivangi, A., Friedman, H., Nowotny, A. (eds.) The Immunobiology and Immunopharmacology of Bacterial Endotoxins. Plenum Press, New York. pp. 75-81.
14. Luderitz, O., A.M. Staub, and O. Westphal. 1966. Immunochemistry of O and R antigens of Salmonella and related Enterobacteriaceae. Bacteriol. Rev. 30:192-255.
15. Marks, M.I., E.J. Ziegler, H. Douglas, I.B. Corbeil, and A.I. Braude. 1982. Induction of immunity against lethal Haemophilus influenzae type B infection by Escherichia coli core lipopolysaccharide. J. Clin. Invest. 69:742-749.
16. McCabe, W.R. 1972. Immunization with R mutants of S. minnesota. I. Protection against challenge with heterologous Gram-negative bacilli. J. Immunol. 108:601-610.
17. Miner, K.M., C.L. Manyak, E. Williams, J. Jackson, M. Jewell, M.T. Gammon, C. Ehrenfreund, E. Hayes, L.T. Callahan III, H. Zweerink, and N.H. Sigal. 1986. Characterization of murine monoclonal antibodies to Escherichia coli J5. Infect. Immun. 52:56-62.

18. Mutharia, L.M., G. Crockford, W.C. Bogard, Jr., and R.E.W. Hancock. 1984. Monoclonal antibodies specific for Escherichia coli J5 lipopolysaccharide: cross-reaction with other Gram-negativebacterial species. Infect. Immun. 45:631-636.
19. Nelles, M.J., and C.A. Niswander. 1984. Mouse monoclonal antibodies reactive with J5 lipopolysaccharide exhibit extensive serological cross-reactivity with a variety of Gram-negative bacteria. Infect. Immun. 46:677-681.
20. Pennington, J.E., and E. Menkes. 1981. Type-specific vs. cross-protective vaccination for Gram-negative bacterial pneumonia. J. Infect. Dis. 144:599-603.
21. Pollack, M., A. Raubitschek, and J. Larrick. 1987. Human monoclonal antibodies that recognize conserved epitopes in the core-lipid A region of lipopolysaccharides. J. Clin. Invest. 79:1421-1430.
22. Qureshi, N., K. Takayama, and E. Ribi. 1982. Purification and structural determination of nontoxic lipid A obtained from the lipopolysaccharide of Salmonella typhimurium. J. Biol. Chem. 257:11801-11815.
23. Rietschel, E.Th., and C. Galanos. 1977. Lipid A antiserum-mediated protection against lipopolysaccharide and lipid A-induced fever and skin necrosis. Infect. Immun. 15:34-49.
24. Siber, G.R., S.A. Kania, and H.S. Warren. 1985. Cross-reactivity of rabbit antibodies to lipopolysaccharides of Escherichia coli J5 and other Gram-negative bacteria. J. Infect. Dis. 152:954-964.
25. Takayama, K., N. Qureshi, P. Mascagni, M.A. Nashed, L. Anderson, and C.R.H. Raetz. 1983. Fatty acyl derivates of glucosamine-l-phosphate in Escherichia coli and their relation to lipid A. J. Biol. Chem. 258:7379-7385.
26. Tate, III, W.J., H. Douglas, and A.I. Braude. 1966. Protection against lethality of E. coli endotoxins with "O" antiserum. Ann. NY Acad. Sci. 133:746-762.
27. Teng, N.N.H., H.S. Kaplan, J.M. Hebert, C. Moore, H. Douglas, A. Wunderlich, and A.I. Braude. 1985. Protection against Gram-negative bacteremia and endotoxemia with human monoclonal IgM antibodies. Proc. Natl. Acad. Sci. (USA) 82:1790-1794.
28. Westphal, O., and K. Jann. 1965. Bacterial lipopolysaccharides: extraction with pehnol-water and further applications of the procedure. Methods in Carbohydrate Chemistry. 5:83-90.
29. Ziegler, E.J., H. Douglas, J.E. Sherman, C.E. David, and A.I. Braude. 1973. Treatment of E. coli and Klebsiella bacteremia in agranulocytic animals with antiserum to UDP-GAL epimerase-deficient mutant. J. Immunol. 111:433-438.
30. Ziegler, E.J., J.A. McCutchan, J. Fierer, M.P. Glauser, J.C. Sadoff, H. Douglas and A.I. Braude. 1982. Treatment of Gram-negative bacteremia and shock with human antiserum to a mutant Escherichia coli. N. Eng. J. Med. 307:1225-1230.

Bacterial Endotoxins: Pathophysiological Effects, Clinical Significance, and Pharmacological Control, pages 395–406

CLINICAL STUDIES OF MONOCLONAL ANTI-LIPID A ANTIBODY XMMEN-OE5

Scott Harkonen, M.D., Donald Kennedy, M.D., Richard Greenberg, M.D.

XOMA Corporation, Berkeley, CA and St. Louis University, St. Louis MO

INTRODUCTION

A promising new approach to the treatment of Gram-negative infections is passive immunotherapy with antibodies directed against cross reactive core determinents on gram-negative lipopolysaccharide (LPS) (Young, 1975; Kirkland and Ziegler, 1984; Telzak and Wolff, 1985; Proctor, 1986). Much of the morbidity associated with Gram-negative infections, including vascular collapse and death, is caused by lipid A, the biologically active component of bacterial LPS. Antibiotics are unable to prevent the toxic effects of lipid A, and may even promote its release from bacteria (Shenep and Morgan, 1984). Antibodies directed against LPS, by neutralizing the toxicity of bacterial endotoxin, may be useful as adjuncts to standard antimicrobial therapy.

Endotoxin (LPS) is a complex macromolecule composed of a highly conserved lipid A moiety, a partially conserved and covalently attached core polysaccharide region, and a distally linked polysaccharide O-specific side chain (Galanos, et al., 1977). Cross-reactive antibodies directed against the conserved core glycolipid region of LPS can be prepared by active immunization with rough mutant strains of bacteria like *Escherichia coli* 0111:4 (J5). Antibodies against *E. coli* J5 (anti-J5) have been shown to cross-react with a wide range of clinically important Gram-negative organisms (Braude and Douglas, 1972; Marks et al., 1982; McCabe, 1972).

Clinical studies have demonstrated that anti-J5 antiserum may be effective in the treatment of Gram-negative bacteremia (Ziegler et al., 1982; Baumgartner, 1985). Over the past five years, a number of monoclonal antibodies (MAbs) have been developed with the hope of providing safe and reproducible sources of anti-LPS antibodies. Several of these MAbs are effective in reducing mortality in experimental models of sepsis (Dunn and Ferguson, 1982; Appelmelk et al., 1986; Young, 1984), some of which are undergoing preparations for clinical trials. One of these monoclonal antibodies, designated XMMEN-OE5 (Xoma Corporation, Berkeley, CA 94710) was studied to determine the safety and pharmacokinetics of this agent in patients with Gram-negative bacteremia.

MATERIALS AND METHODS

Monoclonal Antibody

Monoclonal antibody XMMEN-OE5 was developed using boiled J5 (Rc) mutant *E. coli* 0111 cells as an immunogen and standard hybridization techniques (Young, 1984). Immunized BALB/c splenocytes were fused with the parental murine myeloma line P3-X63 Ag8.653, and a hybridoma cell line was selected by appropriate cloning procedures. Specific binding activity to solid-phase J5 LPS (List Biologics, Campbell, CA) was demonstrated by enzyme-linked immunosorbant assay (EIA). This binding is inhibited by free lipid A, indicating that the binding activity is specific for lipid A. XMMEN-OE5 is of the IgM class. Purified antibody was shown to be free of contaminants and was provided in injectable form at a concentration of 1 mg/ml in 5 mM sodium phosphate, 0.9% sodium chloride with 5% dextrose at pH 7.3.

Patient Selection Criteria

Patients enrolled in this study were admitted to St. Louis University Medical Center, moderately to severely ill with clinical evidence of a serious infection usually caused by gram-negative bacteria. All patients continued to receive antibiotic therapy as prescribed by their attending physicians. Patients under eighteen years of age or with pre-existing renal insufficiency (serum creatinine concentration greater than 2.5 mg/dl) were excluded. A skin test was performed on each patient by intradermal injection of 100 ug of XMMEN-0E5 antibody (0.1 ml) and observation of the test site for 30 minutes for signs of immediate hypersensitivity. All patients, or responsible guardians, gave informed consent prior to entry into the study.

Study Design:

In a study designed to determine safety and pharmacokinetics of XMMEN-0E5, patients were scheduled to receive single intravenous infusions of antibody in increasing doses ranging from 0.1 mg/kg to 15 mg/kg. The first patient received 0.1 mg/kg of antibody, and subsequent sets of 2 patients received 0.5 mg/kg, 2.0 mg/kg, 7.5 mg/kg, and 15 mg/kg respectively. Each infusion was of 1-4 hours in duration. Two additional patients each received two doses of 7.5 mg/kg administered 24 hours apart.

A physical examination was performed on all patients upon entry to the study. A chest radiogram, at least two sets of blood cultures, and, if indicated urine and sputum cultures were obtained prior to antibody treatment. Laboratory studies were performed before treatment and 24 hours, 48 hours, 1 week and 3 weeks after antibody administration, and included hematologic, coagulation, blood chemistry, serum complement, and urinalysis assays. Sera were obtained for analysis of the immune response to XMMEN-0E5 before treatment and 1, 7, and 21 days after treatment. Serum samples for pharmacokinetic analysis were obtained at baseline, mid-infusion, and end of infusion, and at 5 minutes, 30 minutes, one hour, 4 hours, 6 hours, 8 hours, 24 hours, and 48 hours after each infusion.

Immunoassays:

Enzyme-linked immunoassays (EIA) were used to determine the serum concentration of XMMEN-0E5 for pharmacokinetic studies and to determine the human immune response to murine antibody. In the first EIA, microtiter plates were coated with 50 ul of u chain specific rabbit antimouse IgM antibody (Zymed, South San Francisco, CA) and aliquots of patient sera were added to the plates and incubated overnight. XMMEN-OE5 bound to solid phase antigen was detected by means of an alkaline phosphatase linked goat antimouse IgM (u chain specific) probe (Zymed, South San Francisco, CA) followed by p-nitrophenylphosphate substrate. Results were read on a Microelisa Autoreader at 405 nm and quantified by comparing the individual patient's data point with the standard curve.

In the second EIA, microtiter plates were coated with XMMEN-OE5 antibody and dilutions of patient sera were added to wells and incubated overnight. Human IgG anti-XMMEN-OE5 antibodies bound to solid-phase XMMEN-OE5 was detected by means of an alkaline phosphatase linked goat anti-human IgG antibody (gamma chain specific) probe (Zymed, South San Francisco, CA) followed by p-nitrophenylphosphate substrate. Results were read on a Microelisa Autoreader at 405 nm and reported as optical density values. Binding activity at the 1:400 dilution was selected to provide a standard of comparison between samples.

Statistical Techniques:

Linear regression analysis was performed using RS/1 statistical software operating on a DEC VAX 780 minicomputer.

RESULTS

Clinical Results

Eleven patients were evaluated in this study (Table 1). Six of these patients had Gram-negative bacteremia confirmed by blood cultures; one patient had Torulopsis septicemia; one patient had Gram-negative meningitis; and three patients were blood culture negative. XMMEN-OE5 monoclonal antibody was administered as scheduled in doses ranging from 0.1 mg/kg to 15 mg/kg by single intravenous infusion or in two doses of 7.5 mg/kg spaced 24 hours apart. All patients received conventional therapy appropriate for their clinical condition at the time of entry into the study.

Despite numerous clinical and laboratory abnormalites

associated with the patients' underlying diseases, there was no evidence of antibody-mediated toxicity. Two patients did not survive; one patient died from progression of metastatic carcinoma of the lung, and the other died of cardiogenic shock. Neither death was associated with a gram-negative infection or with the administration of the antibody. All of the 6 bacteremic patients in this study survived, and several had objective improvements in their clinical status as manifested by stabilization of vital signs, defervescence, and improved mental status within twenty four hours of antibody administration.

TABLE 1: Patient Profile

Pt.#	Age (Years)	Sex	Primary Diagnosis	Blood Culture	Clinical Outcome
1	67	M	Abscess	Torulopsis	Stable
2	66	F	Pyelo-nephritis	E. coli	Improved
3	66	F	Surgical infection	Acineto-bacter	Stable
4	60	F	Bacteremia	Klebsiella	Improved
5	65	F	Small Cell carcinoma	No Growth	Died
6	25	M	Gram-negative meningitis	No Growth	Stable
7	67	F	Surgical infection	Klebsiella	Improved
8	61	M	Cardiogenic shock	No Growth	Died
9	75	F	Surgical infection	Enterobacter aerogenes	Stable
10	75	F	Urosepsis	E. Coli	Improved
11	67	M	CNS Tumor	No Growth	Stable

Pharmacokinetics

Clearance of antibody from the circulation appeared to follow first-order kinetics (Figure 1). Peak serum concentrations of antibody ranged from 1.7 ug/ml after the 0.1 mg/kg dose to 188 ug/ml after the 15 mg/kg dose. Peak concentrations were 77 ug/ml after the first and 100 ug/ml after the second 7.5 mg/kg dose. Total body clearance at each dose level was calculated by the expression: CL = Dose/AUC.

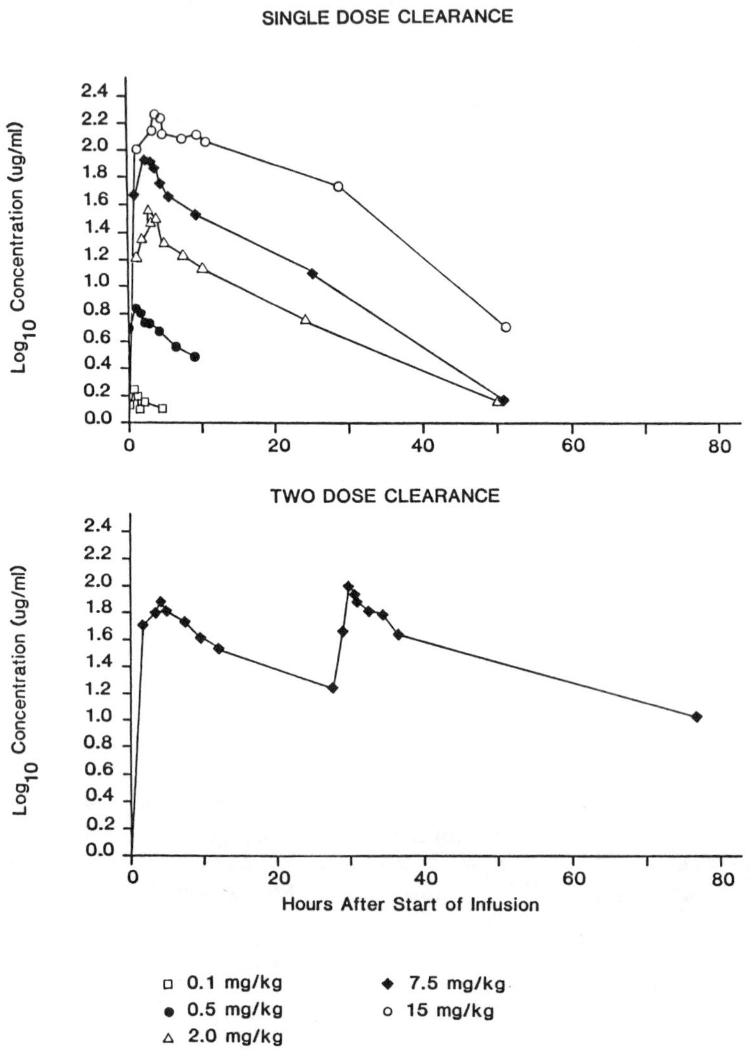

Figure 1: Clearance of XMMEN-OE5 antibody from the serum in patients with single dose (upper) or two-dose regimens (lower).

For doses of 0.1, 0.5, 2.0, and 7.5 mg/kg, serum half-life values ranged from 7.3 hours to 14.1 hours with a mean half life (N=7) of 10.24 hours. For doses of 15 mg/kg, the mean half life (N=2) was 18.8 hours, indicating that at least one non-linear clearance mechanism was present. Individual patients values are listed in Table 2. There were no significant differences in half-life, volumes of distribution, or clearance rates noted in bacteremic patients when compared to non-bacteremic patients.

TABLE 2: XMMEN-0E5 Kinetic Data

Dose (mg/kg)	Clearance (ml/kg-min)	t 1/2 (hours)	Dist. (liters)	Peak Serum Conc. (ug/ml)	Serum Conc. at at 24 Hr. (ug/ml)
0.1	0.02	8.9	5.1	2	0
0.5	0.13	10.7	6.8	7	0
2.0	.10	11.5	6.0	37	5
7.5	0.13	9.2	7.6	77	16
15.0	0.08	18.8	7.8	188	58
7.5 (1st)	0.14	11.8	14.0	77	18
7.5 (2nd)	0.08	14.0	11.6	100	(24)

Immune Response

No immune response to XMMEN-0E5 was detected in patients that received single 0.1 mg/kg or 0.5 mg/kg doses of XMMEN-0E5. Immune responses were detected in several of the patients who received higher doses of antibody (Figure 2). At the 2 mg/kg dose level, one of two patients developed a measurable titer of anti-XMMEN-0E5 antibody at 7 days. The other patient at this dose level had persistant myelosuppression following chemotherapy, and did not have a detectable immune response at 7 days and died from her underlying disease prior to the 21 day time point.

At the 7.5 mg/kg dose level, patients appeared to develop immune responses to XMMEN-0E5. One patient had significant titers of anti-XMMEN-0E5 antibody detected on both day 7 and day 21 after treatment, and the other had an increasing titer on day 7, but was lost to follow up at 21 days. At the 15 mg/kg dose level, one patient developed a significant immune response by day 21. The other patient died shortly after treatment from his underlying disease.

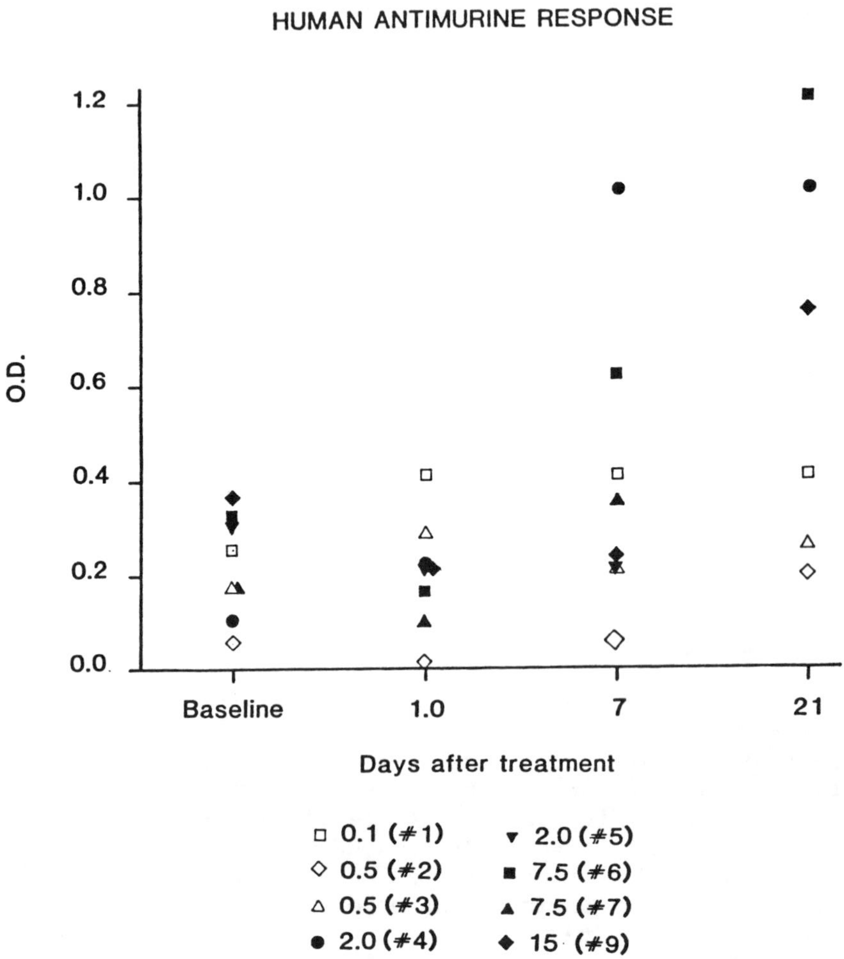

Figure 2: Human IgG anti-XMMEN-OE5 binding activity in patients treated with single infusions of XMMEN-OE5.

DISCUSSION

The results of this study demonstrate that passive administration of XMMEN-OE5, a murine monoclonal antibody directed against the lipid A component of bacterial endotoxin, is well tolerated in patients with clinical evidence of sepsis. These findings, in conjunction with results of numerous other clinical studies of murine monoclonal antibodies, indicate that these agents are generally well tolerated and associated with little toxicity (Sobol, et al., 1982; Sears et al., 1984; Ritz and Schlossman, 1982; Abrams and Oldham, 1985). Despite early concerns of potential allergic reactions, clinical experience has shown that the overall incidence of these reactions is low, even following repeated doses of antibody (Schroff and Stevenson, 1985).

The reticuloendothelial system is the primary clearance mechanism of xenogeneic proteins (Abrams and Oldham, 1985). XMMEN-OE5 antibody has a half-life in humans of approximately 10 hours after doses up to 7.5 mg/kg. A decrease in the clearance of XMMEN-OE5 occurred following the largest dose studied, suggesting a dose-dependent effect on clearance. High doses of XMMEN-OE5 may produce a saturation effect in the RES resulting in decreased clearance of antibody (Fehr et al., 1982). Multiple-dose regimens will compensate for the shorter half life of murine antibodies, and it appears that a two-dose regimen of XMMEN-OE5 provides a measurable level of circulating antibody for at least 48 hours. Since most morbid events associated with Gram-negative sepsis occur within 72 hours of the onset of sepsis, a two-dose regimen of XMMEN-OE5 was selected for ongoing controlled clinical trials.

The development of a human immune response to murine immunoglobulin is well recognized (Kirkman et al., 1983; Schroff and Stevenson, 1985). In the current study, XMMEN-OE5 was immunogenic following single doses of 2.0 mg/kg, 7.5 mg/kg, and 15 mg/kg. No immune responses were detected in three patients treated with less than 2.0 mg/kg of XMMEN-OE5. The appearance of a human antimurine antibody response has generally not been associated with an increase in toxicity or serum sickness reactions, even with continued administration of monoclonal antibody (Jaffers et al., 1983). The reason for this lack of toxicity is not known, but may be related to a relative inability of certain murine monoclonal antibodies to activate the human complement cascade.

Despite the lack of clinically significant toxicity, the human antimurine antibody response may have an impact on the pharmacokinetics, distribution, and efficacy of exogeneously administered antibodies (Jaffers et al, 1983; Foon et al., 1984). These alterations in kinetics can be caused by anti-idiotypic as well as anti-framework host antibodies (Jaffers et al., 1985), implying that treatment with human and chimeric antibodies may also be affected by the development of a host immune response to the exogenous immunoglobulin. In any event, the development of antimurine antibodies should have little, if any, effect on the short-term treatment of Gram-negative sepsis. Controlled clinical trials are ongoing to determine the efficacy of this antibody in reducing mortality and morbidity in patients with Gram-negative sepsis.

Acknowledgement: We wish to thank Carroll Hess for assistance in the preparation of this manuscript.

REFERENCES

Abrams PC, RK Oldham. Monoclonal antibody therapy of solid tumors. In: Foon KA and Morgan AC (eds.) Monoclonal antibody therapy of human cancer. Martinus Nijhoff Publishing, Boston 1985.

Appelmelk BJ, Rapson NT, Verweij-van-Vught AMJJ, Maaskant JJ, Hekker TAM, Peerbooms PGH, MacLaren DM, Thijs LG (1986). Heterogeneity of Escherichia coli J5 vaccines. Lancet 1273-1274

Baumgartner JD, Glauser MP, McCutchan JA, Ziegler EJ, Van Melle G, Klauber MR, Vogel M, Muehler E, Luethy R, Chiolero R, Geroulanos S (1985). Prevention of gram-negative shock and death in surgical patients by antibody to endotoxin core glycolipid. Lancet 2:59-63.

Braude AI, Douglas H (1972). Passive immunization against local Schwartzman reaction. J Immunol 108:505-512.

Dunn DL, Ferguson RM (1982). Immunotherapy of gram-negative bacterial sepsis: Enhance survival in a guinea pig model by use of a rabbit antiserum to Escherichia coli J5. Surgery 92:212-219.

Galanos C, Freudenberg M, Hase S, Jay F, and Ruschmann E. Biological activities and immunological properties of lipid A. p 269-276. In: D. Schlessinger (ed) Microbiology-1977. American Society for Microbiology, Washington, D.C.

Jaffer GJ, Colvin RB, Cosimi JV, Giorgi G, Goldstein TC, Fuller JT, Kurnick C, Lillehi C, Russell PS (1983). The human immune response to murine OKT3 monoclonal antibody. Transplantation proceedings XV:1 646 648.

Kirkland TN, Ziegler EJ (1984). An immunoprotective monoclonal antibody to lipopolysaccharide. J of Immunol 132: 2590-2592.

Marks MI, Ziegler EJ, Douglas H, Corbeil LB, Braude AI (1982). Introduction of immunity against Hemophilus influenzae type b infection by Escherichia coli lipopolysaccharide. J Clin Invest 69:742-749.

McCabe WR (1972). Immunization with R mutants of S. minnesota. I. Protection against challenge with heterologous gram-negative bacilli. J Immunol 108:601-610.

Proctor RA (1986). Role of antibody in the prevention and pathogenesis of endotoxin and Gram-negative shock. In Proctor RA (ed): "Handbook of endotoxin: Vol IV,"Amsterdam: Elsevier, pp 161-84.

Ritz J and Schlossman SF (1982). Utilization of monoclonal antibodies in the treatment of leukemia and lymphoma. Blood 59:1-11.

Schroff RW and Stevenson HC (1985). Human immune responses to murine monoclonal antibodies. In Foon KA and Morgan AC (eds): "Monoclonal antibody therapy of human cancer," Boston: Martinus Nijhoff Publishing.

Sears HF, Herlyn D, Steplewski Z, Koprowski H (1984). Effects of monoclonal antibody immunotherapy on patients with gastrointestinal adenocarcinoma. J Biol Resp Modif 3:138-150.

Shenep JL, and Morgan KA (1984). Kinetics of endotoxin release during antibiotic therapy for experimental gram-negative bacterial sepsis. J Infect Dis 150:380-388.

Sobol RE, Dillman RO, Smith JD, et al. (1982). Phase I evaluation of murine monoclonal anti-melanoma antibody in man: preliminary observations. In Mitchell and Oettgen (eds): "Hybridomas in Cancer Diagnosis and Treatment," New York: Raven Press.

Telzak EE, Wolff SM (1985). Immunotherapy and immunoprophylaxis of Gram-negative rod bacteremia. In Root RK, Sande ME (eds): "Septic Shock," New York: Churchill Livingstone, pp 257-76.

Young LS, Ingram J, Stevens P (1975). Functional role of antibody against core glycolipid of Enterobacteriaceae. J Clin Invest 56:850.

Young L (1984). Functional activity of monoclonal antibodies against lipopolysaccharide (LPS) antigens of gram-negative bacteria. Clin Res 32(2):518A

Ziegler EJ, McCutchan A, Fierer J, Glauser MP, Sadoff JC, Douglas H, Braude AI (1982). Treatment of gram-negative bacteremia and shock with human antiserum to a mutant Escherichea coli. N Engl J Med 307:1225-1230.

Index